ECOLOGICAL STUDIES
NEW HORIZONS

ECOLOGICAL STUDIES
NEW HORIZONS

Editor
Professor (Dr.) Arvind Kumar
Environmental Science Research Unit
Post Graduate Department of Zoology,
S.K.M. University, Dumka – 814 101 (Jharkhand)

2005
DAYA PUBLISHING HOUSE
Delhi - 110 035

ISBN 81-7035-384-X

Published by : **Daya Publishing House**
1123/74, Deva Ram Park
Tri Nagar, Delhi - 110 035
Phone: 27383999
Fax: (011) 23244987
e-mail : dayabooks@vsnl.com
website : www.dayabooks.com

Showroom : 4762-63/23, Ansari Road, Darya Ganj,
New Delhi - 110 002
Phone: 23245578, 23244987

Laser Typesetting : **Classic Computer Services**
Delhi - 110 035

Printed at : **Chawla Offset Printers**
Delhi - 110 052

PRINTED IN INDIA

Preface

Early Indian philosophers and thinkers had some knowledge of ecology as has been revealed by the Indian classical writings, in the Vedic, Upanishadic, Pauranic and Epic literature. Charaka considered the important factors of Vayu (air and gases), Jala (water), Desha (topography) and time in regulating the life of plants. The great classical Indian poet Kalidas has displayed his ecological faculty in his Meghadoot and Ritu Shringar. In the 4th century B. C., Aristotle made references about the habits of animals and environmental conditions prevailing in certain areas. Theophrastus who is regarded as the first true ecologist, wrote about the the association of plant and the relation of plants to each other and the non-living environment.

But now-a-days, the face of ecology has been greatly changed and it attains a separate interdisciplinary faculty in form Environmental Science and has developed multipronged branches such as habitat ecology, ecosystem ecology, conservation ecology, production ecology, radiation ecology, taxonomic ecology, human ecology, space ecology, system ecology, resource ecology and gene ecology. In view of recent advancement in ecological studies, the present book has been undertaken as a treasure of new horizons of ecology which will ceratainly helpful in deciphering the new crux of ecology.

My special thanks and appreciation go to the scientists whose contributions have enriched this volume. I wish to express my sincere gratitude to Dr. P. C. Hembram, Hon'ble Vice Chancellor, S. K. M. University, Dumka who has been a source of constant inspiration. I am especially thankful to Professor (Dr.) B. N. Jha, Hon'ble Pro Vice-Chancellor, S. K. M. University, Dumka for his encouragement. I owe my special thanks to Professor M. C. Dash, Hon'ble Vice Chancellor of Sambalpur University, Professor N. C. Datta of Calcutta University, Professor S. K. Konar of Kalyani University, Professor P. S. Murthy of Bangalore University, Professor P. Natarajan of Trivandrum University, Professor S. P. Roy of Bhagalpur University, Professor A. K. Quereshi of Bhopal University, Professor Tanmay Bhattacharya of Midnapore University, Professor P. C. Mishra of Sambalpur University, Professor B. D. Joshi of Hardwar University, Professor G. C. Pandey of Faizabad University, Professor K. C. Sharma of Ajmer University, Professor M. Raziuddin of Hazaribag University, Professor U. S. Bagde of Mumbai University, Dr. M. P. Sinha of Ranchi University, Professor Gurdeep Singh of I. S. M., Dhanbad, Dr. P. K. Goel of Karad, Dr. M. C. Varma and Shri T. Poddar of Bhagalpur University. I also acknowledge the incentives provided by one of my research scholars, Dr. Chandan Bohra, Head of Zoology, B. S. K. College, Barharwa. for helping me actively in bringing out this book.

I also express my deep sense of gratitude to my parents whose blessings have always prompted me to pursue academic activities deeply. I am also thankful to my sweet wife, Professor Kumari Bimla and my two lovely sons, Kumar Pallav Shivshankaran and Kumar Prasun Ramakrishnan whose natural smiles extended to me relief all through this tiresome endeavour.

Last but not the least, I am also thankful to Mr. Anil Mittal, Proprietor, Daya Publishing House, Delhi for taking keen interest in bringing out of this book. Finally, I will always remain a debtor to all my well-wishers for their blessings, without which this book would not have come into existence.

Dumka **Professor Arvind Kumar**

Contents

1

Waning Wetlands: A Need for its Conservation

Arvind Kumar and C. Bohra***

**Environmental Science Research Unit, Post Graduate Department of Zoology, S.K.M., University, Dumka – 814 101, India*

***Department of Zoology, B.S.K. College, Barharwa (Sahibganj), Jharkhand*

Introduction

Once considered useless and waterlogged unproductive areas and sometimes even as deleterious ecosystems, wetlands are now being looked upon as ecosystems with specific ecological characteristics, functions and values. They are one of the most productive ecosystems of the world and essential life supporting systems, providing a wide array of benefits to human kind. Yet, wetlands are today fast declining and rapidly deteriorating ecosystems in various parts of the world and India is no exception.

The wetland is an ecosystem of lowland area where water is commonly accumulated to induce characteristic water-loving and water-tolerant vegetation and anaerobic humic soils. These inundated or flooded basins are fertile, rich and productive systems that rival the most productive natural and even intensively managed agro-ecosystems. This unique resource has induced the evolution of well adapted taxonomic groups and species that are found in no other place, including water birds, semi-aquatic mammals, amphibians, reptiles and invertebrates. The flora is even more unique and diverse and is the energy-fixing base for complex food chains that result in such diversified wildlife resources (Waller, 1981 and Singh, 2002).

Wetlands include a variety of habitats such as marshes, swamps, lakes, river flood plains and so on. The term "wetland" has grown out or a need to understand and describe the characteristics and values of all waterlogged areas to wisely and effectively manage them. There is no single ecologically sound and universally acceptable definition of wetland ecosystems, primary because of their diversity and for the lack of demarcation between dry and wet environments along a continuum. Hence, a common definition covering all the areas is rather difficult to give. The International Union for Conservation of Nature (IUCN, 1971) defines wetlands as "areas of marsh, fen, peat land or water, whether natural or artificial, that is static or flowing, fresh, brackish including areas of marine water, the depth of which at low tide does not exceed 6 metres. I.B.P. (1972) considered wetland as "part of the surroundings ecological structure and as several stages in the succession from open water to dry land or vice-versa, occurring at site situation as a rule between the highest and lowest water levels, as long as the flooding or water logging of the soil is of substantial ecological significance." Cowardin *et al.* (1979) have defined wetlands as "lands transitional between terrestrial and aquatic system where the water level is usually at or near the surface or the land covered by shallow water." Piecysmaska (1972) is of the opinion that wetland is a store of accumulated matter of terrestrial and aquatic origin. According to them wetlands must have one or more of the following attributes:

1. The substrate is predominantly undrained hydric soil,
2. At least periodically the land supports predominantly hydrophytes,
3. The substrate is non-soil and saturated with water or covered by shallow water at sometime during the growing years.

According to Gopal (1973), wetlands lie at the interface between the land and open waters and include a wide spectrum of habitats. According to a comprehensive definition by the U.S. Fish and Wildlife Service, wetlands are lands transitional between terrestrial and aquatic systems and this definition is similar to Cowardin *et al.* (1979).

Wetlands are distinct from other ecosystems primarily by their hydrological environment characterized by water-logging of the soils often associated with submergence to different depths for varying periods of times, and subjected to water level changes of such large magnitude and frequency that are rarely experienced by the biota in terrestrial or true aquatic ecosystems. Some of the important characters of a wetland are:

1. Water should be present for at least seven successive days in the season.
2. It should support aquatic macrophytes in water and soil at least in some part of the year.
3. Soil should be flooded for a long time and become anaerobic in the upper layers.

Types of Wetlands

The total wetland area of the world is estimated to be around 85,58,000 Sq. km which is about 6/4 per cent of the total area of the earth. India has 2167-recorded natural wetlands, covering an area of 1.5 million hectares. Further, there are 65,254 artificial wetlands, spreading over an area of 0.25 million hectares (Kumar, 1999). Yet, a comprehensive data on the wetlands in India is still not available. There are a great variety of inland and coastal wetland ecosystems in India. The inland wetlands include marshes, swamps, flood plains of rivers and littoral area (marginal zone) of ponds and lakes. Along the coast there are salt marshes, lagoons, backwaters, estuaries, mangrove forests and coral reefs.

Following Shaw and Fredrine (1956), Smith (1974) has divided wetlands into the following 20 categories, which is described here with their characteristics:.

Inland Fresh Areas

Seasonally Flooded Basins and Flats

Soil covered with water or waterlogged during variable periods but well drained during much of the growing season.

Fresh Meadows

Without standing water during growing season; waterlogged within a few centimeters of surface.

Shallow Fresh Marshes

Soil waterlogged during growing season; often covered with 15 cm. or more of water.

Deep Fresh Marshes

Soil covered with 15–90 cm of water.

Open Fresh Water

Water less than 3 m deep. Boardered by emergent vegetation.

Shrub Swamps

Soil waterlogged, often covered with 15 cm or more of water.

Wooded Swamps

Soil waterlogged, often covered with 30 cm of water. Along sluggish streams, flat uplands shallow lake basins.

Bogs

Soil waterlogged; spongy covering by mosses.

Coastal Fresh Area

Shallow Fresh Marsh

Soil waterlogged during growing season, at high tides as much as 15 cm of water. On lowland side deep marshes along tidal rivers, sound, deltas.

Deep Fresh Marsh

At high tide covered with 15–90 cm of water. Along tidal rivers and bays.

Open Fresh Water

Shallow portions of open water along fresh tidal rivers and sounds..

Inland Saline Areas

Saline Flats

Flooded after, periods of heavy precipitation; waterlogged within few cms of surface during the growing season.

Saline Marshes

Soil waterlogged during growing season often covered with 60–90 cm of water; shallow lake basin.

Permanent Areas of Shallow Saline Water

Depth variable.

Coastal Saline Areas

Salt Flats

Soil waterlogged during growing season; sites occasionally to fairly regularly covered by high tide. Landward sides or islands within salt meadows and marshes.

Salt Meadows

Soil waterlogged during growing season, rarely covered with tidewater; landward side of salt marshes.

Irregularly Flooded Salt Marshes

Covered by wind tides at irregular intervals during the growing season. Along shores of nearly enclosed bays, sounds etc.

Regularly Flooded Salt Marshes

Covered at average high tide with 15 cm or 1 more of water; along open ocean and along sounds.

Sounds or Bays

Portion of salt water sounds and bays shallow enough to be dived and tilled. All water landward from average low tide line.

Mangrove Swamps

Soil covered at average high tide with 15–90 cm of water.

Gopal (1988) has grouped Indian freshwater wetlands into two major types depending upon the duration of water logging (*i*) *The perennial wetlands* with water logging throughout the year which include habitats such as river banks and margins of large lakes or reservoir and (*ii*) *The seasonal wetlands* which dry up completely for varying periods of time depending upon the vagaries of monsoon. Examples are numerous–village ponds, fishponds, paddy fields, river flood plains etc.

Ecologically wetlands may be viewed as complex hydrological and bio-geochemical system, endowed with specific structural and functional attributes and performing major ecological role in the biosphere. The studies under I.B.P. and elsewhere during sixties and seventies demonstrated that the wetlands are among the most productive ecosystems (Lieth and Whltaker, 1975). The main interest in wetland studies was earlier emphasized on migratory bird conservation and other aquatic lives. The Ramsar Conference of 1971 adopted a convention on wetlands and made recommendations for the conservation of wetlands as waterfowl habitats. In order to create awareness about the importance of wetlands, particularly as breading grounds of waterfowls. The International Biological Programme (IBP), International Union for Conservation of Nature (IUCN) and International Waterfowl and Wetland Research Bureau (IWWRB) organized a series of international conference in the 1960' s. All these paved the way for the convention on wetlands of International importance of Ramsar, Iran in 1971. This treaty aims at international cooperation for the conservation of wetlands. It also envisages formulation of effective plans and their implementation to protect the dwindling wetlands. A list of wetlands of international importance (Ramsar sites) has been prepared. Under this convention 101 countries have designated a total of 881 wetlands of international importance, occupying an area of 62.77 million hectares. India became a signatory of this convention in 1981. As a signatory a country has to meet obligations such as: (*a*) designate wetlands of international importance in the list of so called Ramsar sites, (*b*) maintain ecological characters of the listed wetlands sites, (*c*) organize planning in such away as to achieve sustainable use of all the wetlands in their territory and (*d*) designate wetlands as natural reserve. Indian wetlands designated as Ramsar sites are Chilka, Keoladeo National Park, Wular lake in T and K, Sambhar in Rajasthan and Loktakin Manipur.

Wetlands occur in areas where precipitation exceeds the potential evapotranspiration leaving an accumulated surplus. Climatically about one-third of Indian subcontinent falls within this category and wetlands are thus common throughout especially in Assam, Bihar, Eastern M.P., East and West coastal regions, Jammu and Kashmir, Jharkhand, North-Eastern Andhra Pradesh, Orissa, U.P., Uttaranchal and West Bengal, Wetlands also occur in areas where local physiography enables certain areas to accumulate water. The variability in climatic conditions and the changing topography is responsible for significant ecological diversity encountered within the wetlands. Biswas (1984) has reported 1193 wetlands covering a total area of 3.9 million hectares in 274 districts of India.

Marshes are shallow water areas with abundant emergent vegetation, such as reeds, rushes, grasses and sedges. These are one of the most productive ecosystems of the world. The terai located along the Himalayan foothills, the Rann of Kutch salt marsh of the Gujarat coast and the Chushul and Henle river marshes of Ladakh are a few good examples. The famous bird sanctuary of India, Keoladeo Ghana National Park in Rajasthan is man-made marshy wetlands. Swamps differ from marshes in possessing woody, shrubs and trees, which are adapted for life in saline waterlogged areas. India has a large number of swamping forests or mangroves along both the east and \vest coasts, representing about 7 per cent of the world's mangroves. The Vedaranyam salt swamping in Tamil Nadu and the Riparian swamps of Dudhwa National Park in U.P. are some of the major swamps.

The major rivers in India have extensive flood plains. The flat land close to the large rivers remains covered with floodwaters due to natural floods during some seasons of the year. These areas remain completely inundated even after the floodwater recede. The flood plains are more diverse along the lower reaches of the rivers. They are the ideal habitats for fish and wildlife. There are also a large number of estuaries, the meeting place between the rivers and the sea. Lakes are yet another form of wetlands. Besides submerged emergent and true plants, plenty of fauna also inhabit the lakes.

Coral reefs are shallow water tropical marine ecosystems exhibiting very high productivity and biological diversity. They are formed by the deposition of calcareous exoskeleton of marine coelenterates called coral polyps over thousand of years. The major reef formation in India are located in the Gulf of Manner, Palk Bay, Gulf of Kutch, Lakhadweep and Andaman and Nicobar Islands.

Reservoirs, tanks. fish ponds. canals and paddy fields could be considered as man-made or artificial wetlands. Although they cannot be compared with natural wetlands in terms of biodiversity and other functions, they may still be ecologically as important as the former.

The main emphasis in wetlands study has so far been towards the realization of their value as habitats for wildlife. Because of this type of approach, other importance aspects have been virtually neglected. Ecological studies on wetland ecosystems in India have been fragmentary and of preliminary nature. Little attention seems to have been given to ecosystem approach in these studies. Very little is known about the nature, extent, status, structure, function and socio-economic importance of wetland ecosystem (Gopal and Sharma, 1982). Thus, the approach for conservation appears non-existent. Our knowledge is totally lacking or too meager to make proper assessment of wetlands utilization and management. We have also not enough data to show the value of wetlands in treatment of wastewaters and regulation of water quality.

On account of inadequate or shallow knowledge of the agro-economical, agro-climatics and other ecological significance of wetlands, we have been neglecting these ecosystems as wastelands. As already stated, wetlands are highly complex water-land interactive systems and are supposed to be amongst the most fertile land productive ecosystems in the world. Their ecological and socio-economic values are only recently being understood. Unfortunately many such areas have been reclaimed for agriculture, industry or for settlements. Alternatively, chemical effluents, household wastes and sedimentation due to ecological degradation in catchment areas have polluted them.

Importance of Wetlands

Man has been associated with wetlands for ages. Man ancient human civilization flourished along the flood plains of major river systems. Even today wetlands continue to help human beings in many ways.

There are millions in India who depend upon wetlands for sustenance activities such as farming, fishing, hunting and for collecting reeds and other economically important plants, including the edible forms. The flood plains are excellent grazing ground for cattle. Mangroves arid corals after a variety of economically important products for the local coastal population. More recently, it has been recognized that in addition to providing wildlife habitat, wetlands playa great role in flood control, recharge of aquifers, regulating water quality treatment of wastewaters, reducing sediment load, production of organic material at rates equaled by few other eco-systems, ground water recharge, dependence of agriculture and animal husbandry in drought prone areas, pollution abatement, bio-fertilizers etc. They provide a source of drinking water for humans as well as animals.

Wetland flora function as natural sewage and wastewater treatment plants. Many wetland plants are known to reduce phosphate and nitrate level by up to 90 per cent and their role in the removal of heavy metals has been widely acknowledged. Sea grasses in many parts of the world playa vital role in cleaning up coastal waters. They reduce the quantity of suspended particles, thus increasing the clarify of water as it runs through the wetlands.

Wetlands are valuable as sinks and transformers of a multitude of chemical, biological and genetic materials. On a global scale, wetlands are considered as CO_2 sinks and climate stabilizers. Considering the function they perform in hydrological and chemical cycles and their function as downstream reservoirs of wastes, wetlands are considered as "the kidneys of the landscape".

Wetlands are also known as "biological supermarkets because of the extensive food chain and immensely rich biodiversity. The diverse type of wetlands harbours many rare, threatened and endangered animals. Major wildlife sanctuaries in India are located close to wetlands.

In coastal areas, the binding action of mangrove vegetation helps to protect the shoreline by preventing soil erosion. The mangrove areas also act as breeding grounds of many aquatic organisms, particularly fish and prawns.

Wetland plants and animals (dragon flies, molluscs etc.) are sensitive to even slightest changes in environmental quality. Their population decreases considerably in polluted waters. Hence they could be used as potential biological indicators to predict pollution of water bodies.

Socio-economic Importance

Historically, wetlands attracted human settlement because of the accessibility of abundant water, fertile soils and other resources such as fish. It is not a coincidence that there is a high concentration of people around important wetlands in Asia. The use of wetlands and their related resources has increased significantly during the fast decades due to factors including increase in human population, people's aspiration and trade with the outside world. Local people have used wetlands wisely for centuries. In the process they evolved community codes systems and beliefs in using and conserving the resources, which internalized within social, economic and religious practices. During the last few decades the pressure of development and the related social, political and legal changes have directly or indirectly undermined the role of communities in natural resources management in general, and in wetlands in particular. Until recently the wetlands were "scientifically" declared to be "wastelands" and relevant technologies were developed to use them productively. Degradation of wetlands is a major concern for local communities, as many people directly depend on wetlands for their livelihood.

Wetlands are often considered low-value land since in their normal condition they cannot be used for most agricultural activities or urban development. However, these ecosystems are of great socio-economic, ecological, academic and recreational importance. Wetlands can be extremely productive not only in the ecologically, but also in the economical sense. They are the breeding ground for many economically important species of fish and shrimp. Besides it may produce other important commodities such as fuel, timber, latex, tannins, alcohol and agricultural products. They can also have excellent prospects for recreation. Often they are of extreme importance for the socio- economic situation of local communities.

Biological diversity in wetland ecosystem play major direct and indirect roles in socio-economics of local inhabitants. Fishery resources and plant resources such as lotus wild varieties of paddy, water, chestnut. *Ipomea aquatica, Furale, ferox, Typha* etc. are the examples of direct roles. Impounded water used for life saving irrigations of crops, improved drainage in the surroundings, religious importance inducing periodic fairs and marketing and deposited clay used for pottery etc. are some of the indirect role of wetlands. Wetlands also help in nutrient dynamics and act as nursery for several kinds of fish and other animals like shrimps, tortoise and turtles, molluscs, crab, water fowls etc. so useful to man. Alcoholic extract of Brahmi plant exhibited anticancer activity (Flagovant *et al.*, 1995). The bitter leaves contain chemical compounds called bacosides, which strengthen, or the ability of neurons in the brain and support learning and recall, important alertness and relieve mental fatigue. *Eichhornia crassipes* (water hyacinth) abundantly found in most of the water bodies which are highly polluted, considered great menace to wetland ecosystem is shown to be useful as a poultry and cattle feed, as a raw material for paper industry, useful in production of biogas and as a water purifiers in polluted ponds by absorbing heavy metals (Kumar and Bohra, 2002). The leaves are rich in protein and may be used as supplement in human diets. Tropical and subtropical wetlands have been exploited and created to produce rice, which feeds more than half the human population. Next to rice cropping, the most widespread-use of wetlands is irrigation, fishing and Trapa cultivation. Some other wetland plants found in use by local people in different ways are food, medicine, religious activities. Fuel, thatching, fodder, fertilizer, ornamental purposes etc. Trapa cultivation is very common. In the month of December and January, local people on large scale man-made water bodies are very much common in suburban and rural areas consume the market price of Trapa ranges between 10 to 20 per kg.

Plants and Plant Products

For Food

The wetland macrophytes which are used as food in different ways by people are: *Amaranthas virdis, Chenopodium album, Ipomoea aquatica, Nelumho nucifera, Nymphaea nouchali, Oryza rufipogon* and *Trapa natans.*

For Medicine

Different wetland macrophytes used as medicine are: *Bacopa monnierii, Coix lochryma-johi, Crinum defixum, Cyperus rotundus, Euphorhia hypercifolia, Ranunculus sceleratus* and *Spilanthes paniculata.*

In Religious Ceremonies

Desmostachya bipinnata, Nelumbo nucifera and *Nymphaea nouachali.*

For Thatching and Hut Making

Arutndo donex, Ipomoea carnea, Phragmiles karka and *Typha angustata.*

For Fuel

Tpomea carnea.

For Fodder

Eichhornia crassipes, Paspalum distichum and *P. scrobiculatum.*

Animal Products

Animals other than fishes used are crab (*Eriocheir* sp.) Pila (*Pila globosa, P. virens,* and *P. nevililana*) and Tortoise (*Trionyx amyda*).

For Pottery

The use of deposited clay from wetlands for pottery is a common practices. There are so many families directly involved in this business and they are known as Kumhars.

Religious Importance of Wetlands

The water bodies are well known for their religious importance. Some festivals and rituals of Hindus are possible only near water bodies, which indicate the importance of water and water reservoirs. Chhath, Jeutia and Karva Chauth festivals of Hindus are celebrated near water reservoirs and along the bank of rivers. From ancient time periodic fair is the common feature near water bodies and it is indirectly linked with many social and economical activities.

So, it is correct to say that the wetlands contain a wealth of biological resources that are interdependent and can be exploited to improve the economic condition of local people.

Threat to Wetlands

Despite the worldwide-heightened interest in wetlands, the wetland cover has been declining consistently over the last few decades. Increase in human population-and rapid urbanization have put considerable pressure on wetlands. The pressure of denuding, polluting, draining, filling and building continue unabatedly at the wetland sites around the globe.

In India, the wetlands are drained for agricultural activities and hence they no longer play their usual ecological functions. Large area of Kolleru lake in Andhra Pradesh, Deepanbil in Assam, Hokarsal in J&K and Vellayani lake in Kerala have already been lost due to agricultural operations. Many wetlands areas in India are now established human settlements, the encroachment continues even now. For instance rapid urban development has shrunken the area of Cochin backwaters in Kerala to 8,000 hectares from the original area of 70,000 hectares. This has also reduced the fishery potential of the backwaters and led to the disappearance of the estuarine crocodiles.

The sewage effluent of Kolkata city was once released into the swamps and marsbes on the eastern edge of the city. These wetlands were functioning as biological filters and flood control centers. This area was later converted into a residential complex by dumping silt from the river Hooghly. Now the city is under threat of floods to manage the colossal amounts of sewage formed every day.

Most of the wetlands in India suffer trom heavy infestation with exotic water hyacinth, pondweed, *Pista, Salvinia* and *Ipomoea*. One-third area of Chilka, the biggest inland lake in India, spreading over 1,100 sq. km and located in the Puri and Ganjam districts of Orissa now remains blanketed with weeds. The process of sedimentation of lakes has been accelerated by human activities such as deforestation and dumping of wastes. This may reduce their water holding capacity and thus pave the way for flooding. Construction of dams and barricades across the rivers has resulted in the loss of flood plains and devastated life in many wetlands.

Ecological degradation of wetlands together with pollution has resulted in the loss of flora and fauna. Hunting and poaching of wild animals including waterfowls continue unabated in many parts of the country. Though millions of water birds visit India every year, only those reaching the protected area enjoy some kind of protection. Many natural wetlands in India have been converted into fishponds, resulting in a host of ecological disturbances. The high amount of fertilizers and other inputs required in aquaculture for increasing the productivity has led to the degradation of the system as a whole.

Thus, as a whole, the rapid industrial and rural development and population explosion of the last century have had an immense impact on wetlands. The destruction and degradation of wetlands worldwide has been massive during the 20th century, and the rate of wetland loss has been highest in recent decades. There are many causes of wetland degradation. At the local level, land tunare and rights of access–which are so often denied the rural poor and it forces farmers to maximize short-term benefits at the expanse of wetland resources. At the National level, wetland degradation often stems from decisions based on short-term political and economic concern and at the global level, the over riding demand for new agricultural land and urban and industrial development is also based on the short-term economic strategies of financial markets rather than on responsible planning of resource use.

Preserving the Wetlands

The country's water resources should be conserved, as they are the most important sources of life and the home of unique flora and fauna. The wetland areas in our country need proper protection as these are under constant threat from ever increasing population. The wetlands received greater public attention as excellent breeding grounds of waterfowls mainly after the Ramsar convention. The need for conserving the precious wetlands is better realized these days and several initiatives have been taken in this regard. The Ministry of Environment and Forests, Government of India set up a National Committee on wetlands, mangroves and coral reefs in 1987 for framing policy guidelines, identifying and monitoring wetlands for intensive conservation and management and for seeking inter-nation co-operation. Based on the recommendations of this committee 20 wetlands have been identified for conservation and management. In 1993, a National Lake conservation plan was framed to protect the endangered lakes; 11 urban wetlands were identified under this scheme for conservation and management. Bhoj wetland of M.P. is now getting assistance from Overseas Economic Co-operation Fund of Japan for Ecological restoration. The Ministry of Environment and Forests has also adopted the National River Conservation Plan.

The present status of wetlands may however be improved by the following measures:

1. Afforestation in catchment area of those wetlands which are in rural areas. This will play an important role in checking the erosion arid sedimentation.
2. Weed control by manual eradication adversely affect biodiversity, hence some other techniques, preferably biological are needed. Further, the proper economic utilization of weeds is also necessary.
3. Mapping of wetlands and their demarcation for checking encroachment for irrigation and agriculture will improve the present status of wetlands. Use of wetlands for irrigation should be forcibly stopped and the feed back mechanism may be used to maintain the water level.
4. Control of discharging sewage and household wastes by administrative and legal measures is necessary. However, awareness among the people regarding the economic value and functions of the wetland ecosystems will also considerably improve the present status of wetlands.
5. Biological treatment of wetlands suffering from eutrophication can be more economic way for the control of nutrient enrichment. The species like *Myriophyllum inducum, Potamogeton crispus,* which have capacity of absorbing 60 per cent nitrogen and quite a high percentage of phosphorus, may introduce. On the bank of wetlands the species like *Typha angustata* may be grown which have capacity of trapping about 50 per cent phosphorus and 25 per cent nitrogen (Singh and Pandey, 1998 and Solbe, 1986).

There is a need to enhance research awareness campaign on economic values of wetlands and their biodiversity. Last but not least, nothing can be achieved without people's participation. In developing countries like India, voluntary organizations, institutions or pressure groups can play an important role in conservation of biological diversity of wetlands and assessment of their socio-economic importance.

However, these efforts are not much more. Many wetlands in India still remain unprotected. A holistic approach towards conservation of wetlands is the urgent need of the hour. It has now become imperative to prevent the ecological degradation of wetlands and to implement appropriate conservation strategies to preserve these natural heritages for the generations to come.

References

Biswas, A. (1984). Wetlands of India. *Water. Qt. Bull. J.*, 13: 15–35.

Cowardin, L.M., Carter, V., Golet, F.C. and La Roe, E.T. (1979). Classification of wetlands and deep-water habitats of United States. U.S. Fish and Wildlife Service, Washington, pp. 103.

Elangovan, V., Ramamurthy, N., Balasubramanyan, K. and Govindaswamy, S. (1995). *In vitro* studies on the anttcancer activity of *Bacopa monnieri. Fitotherapia*, 66: 211–215.

Gopal, B. (1973). A survey of the Indian studies on ecology and production of wetland and Shallow water communities. *Pol. Arch. Hydrobiol.*, 20: 21–29.

Gopal, B. (1988). Wetlands: Management and Conservation in India. *Water Qt. Bull. J.*, 3: 3–6.

Gopal, B. and Sharma, K.P. (1982). Studies on wetlands in India with emphasis on structure, primary production and management. *Aquatic Botany*, 12: 81–91.

IUCN (1971). The Ramsar conference: Final act of the international conference on the conservation of wetlands and water fowl. *Suppl. IUCN Bull.*, 2: 1–4.

Kumar, A.B. (1999). Our Vanishing Wetlands. In: *Science Reporter*, December, 1999, p. 9–15.

Kumar, A. and Bohra, C. (2002). Impact of environmental stress on the growth behaviour of water; hyacinth *Eichhornia crassipes* (Marts) with special reference to removal of pollutants. In: *Ecology and Ethology of Aquatic Biota*, (Ed.) A. Kumar. Daya Publ. House, Delhi, p. 345–353.

Lieth, H. and Whittaker, R.H. (1975). *Primary Productivity of the Biosphere*. Springer-Verlag, Berlin pp. 339.

Piecymaska, E. (1972). Ecology of littoral zone of lakes. *Ekologia Polska*, 22: 637–732.

Shaw, S.P. and Fredrine, C.G. (1956). *Wetlands of United States*. U.S. Fish and Wildlife Serv. Circ., pp. 39.

Singh, A.K. (2002). A floristic and socio-economic study of wetlands of Varanasi (U. P.). In: *Ecology and Ethology of Aquatic Biota*, Voi. II, (Ed.) A. Kumar. Daya Publ. House, Delhi, p. 274–318.

Singh, V. and Pandey, R.P. (1998). A study of some wetland ecosystems of Udaipur and environs, Rajasthan. *Ibid*, 22: 217–224.

Smith, R.L. (1974). *Ecology and Field Biology*, 2nd ed. Harper and Row Publishers, New York, pp. 850.

Solbe, J.F. (1986). *Effect of Land Use on Freshwaters: Agriculture, Forestry Mineral Exploitation, Urbanisation*. Ellis Herwood Ltd. Chinchester.

Waller, M.W. (1982). *Freshwater Marshes*. Univ. Minnesota Press, Minneapolis. U S.A.

Chapter 2

Soil Pollution Due to Municipal Solid Waste Disposal: A Case Study

C. Bala Murali Krishna, R.K. Yaji** and S. Shrihari***

**Department of Civil Engineering, NMAMIT, Nitte, Udipi, Karnataka, India*
***Department of Civil Engineering, NITK, Surathkal, D.K., Karnataka, India*

ABSTRACT

Unscientific methods of disposal of solid waste as fills on low lying areas causes serious environmental geotechnical problems. The leachate generated from the decomposition of solid waste causes the pollution of soil layers. The soil chemical characteristics near municipal solid waste disposal area at various depths were studied. The tests included the measurement of pH, Electrical conductivity, Hardness, Chlorides, Sulphates, Nitrates, Sodium, Potassium, Ammonia Nitrogen and Phosphate contents of soil. The results of this study are presented in this article.

Introduction

All types of pollutants have direct or indirect effect on soil properties. Unless a suitable strategy is planned beforehand, the waste products in the form of liquid, semi-solid or solid can leads to serious geo-environmental problems. Solid waste pollution is often called as third pollution after water and air pollution. Today more than 45 million tonne of solid waste is generated from the urban centers of India which are collected partially, transported improperly and disposed unscientifically. Solid waste management was never taken up seriously by considering a low priority area either by public or by the concerned agencies. Heaps of solid waste can be seen today on the outskirts of many cities. Uncontrolled dumping of solid waste on the outskirts of towns and cities has created overflowing landfills which are not only impossible to reclaim because of the haphazard manner of dumping, but also have serious environmental implications in terms of ground water and soil pollution.

Mangalore city is Headquarter of Dakshina Kannada district and is one of the fast growing cities in Karnataka state. The present area of the Mangalore city is about 115.16 sq. lms and having population of about 3.5 lakh. The Mangalore city is situated between 12° 52′ N′ latitude and 74° 49′ E′ longitude. It has an elevation of 10 to 60 meters above M.S.L. The ambient temperature ranges from 17°C to 37°C. The annual average rainfall is about 3900 mm. The soil of Mangalore city is predominately of lateritic soil. The total quantity of Municipal solid waste (MSW) generated from Mangalore city is estimated to be 225 tonne everyday (0.64 kg/p/day approx). Out of this 215 tonne of MSW is being transported far away from the Mangalore city to an area at Vamanjoor village (nearly 12 kms from the city) and being dumped in the open area. Due to infiltration of water through the solid waste, the organic matter present in waste gets decomposed and releases the water enclosed in their cell wall. This water comes out in the form of leachate, carry numerous contaminants to the soil surface and to the adjacent areas. Contaminant transport through the soil, alters the various properties of the soil. An investigation was undertaken to study the effect of solid waste disposal on chemical properties of soil and the results of the investigation are presented.

Materials and Methods

The field investigations were carried out at solid waste disposal area at Vamanjoor village, Dakshina Kannada district, Karnataka, India, where the Mangalore city municipal solid waste is being dumped from past three decades. The field studies were conducted in two phases. The first phase of field investigations were conducted during February and May 2001. Soil samples were collected approximately 100 meters away from the MSW disposal area. Four soil samples were collected from different locations from lower end of MSW dump area where MSW leachate flow is expected. Correspondingly the soil samples were denoted as S_{11}, S_{12}, S_{13}, S_{14}.

The soil samples were collected from three layers up to a depth of one meter. The top two layers of 30 cm depth each and the bottom layer of depth 40 cm. The second phase of field investigations were conducted during February and May 2002 over a gap of one year after first phase of field investigations. In the second phase again four soil samples were collected with similar procedure adopted in the first phase of field investigations. The soil sampling points in the second phase were very near to the first phase soil sampling points. The soil samples were denoted as S_{21}, S_{22}, S_{23}, S_{24}.

The analysis was conducted as per Standard Methods APHA (1998). The various chemical characteristics analyzed for the soil samples collected from the field are pH, Electrical conductivity, Hardness (as $CaCO_3$), Chlorides, Sulphates (as SO_4), Nitrates (as NO_3), Sodium (Na), Potassium (K), Ammonia Nitrogen (as NH_3–N) and Phosphates (PO_4). The collected field soil samples were oven dried at 100 to 110° C, crushed with a wooden mallet and passed through the 2 mm IS sieve (425 micron IS sieve for measurement of pH). From this dry powdered soil sample a weighed soil sample was taken and maintained a soil-water ratio of 1 : 5 (1 : 2.5 for the measurement of soil pH). The contents were mixed thoroughly once in two to three hours. After 24 hours the supernant was filtered and used for soil chemical analysis.

Results and Discussion

No much variation was observed in pH of soil at various depths in soil samples collected during the year 2002 as compared to the pH of soil samples collected during the year 2001. Except Chlorides concentration in the soil the Electrical Conductivity of the soil, Hardness of the soil, Sulphates, Nitrates, Sodium, Potassium, Ammonia Nitrogen and Phosphates concentration in the soil decreases with increase in depth of the soil in all the soil samples collected during the year 2001 as well as during the year 2002. Except Sodium, Potassium and Ammonia Nitrogen concentration in the soil, the observed Electrical Conductivity of the soil, Hardness of the soil, Chlorides, Sulphates, Nitrates and Phosphates concentration in the soil for different soil samples at various depths was lower during the year 2002 as compared to the year 2001.

Figure 2.1 and Figure 2.2 presents the observed pH of soil at various depths collected near MSW disposal area during the year 2001 and the year 2002 respectively. The minimum and maximum pH of soil observed was 4.7 (30–60 cm depth in soil sample S_{14}) and 5.54 (30–60 cm depth in soil sample S_{12}) respectively during the year 2001. During the year 2002 the observed minimum and maximum pH of soil was 5.11 (60–100 cm depth in soil sample S_{24}) and 5.72 (0–30 cm depth in soil sample S_{23}) respectively. The observed lower pH of soil is possibly due to formation of organic humic acids due to decay of organic matter present in the soil and may also be due to inorganic acids formed by hydrogen ions with Sulphates and Nitrates (sulphates may be reduced to hydrogen sulphides which leads to formation of sulphuric acid which causes the lesser pH to the soil).

The observed EC of the soil for different soil samples at various depths is lower during the year 2002 as compared to the year 2001. The decrease in EC of the soil at various depth possibly due to decrease in ions concentration such as calcium, magnesium etc. Figure 2.3 and Figure 2.4 presents the observed EC of soil at various depths collected near MSW disposal area during the year 2001 and the year 2002 respectively. The observed minimum and maximum EC of soil was 220 μs/ppm (60–100 cm

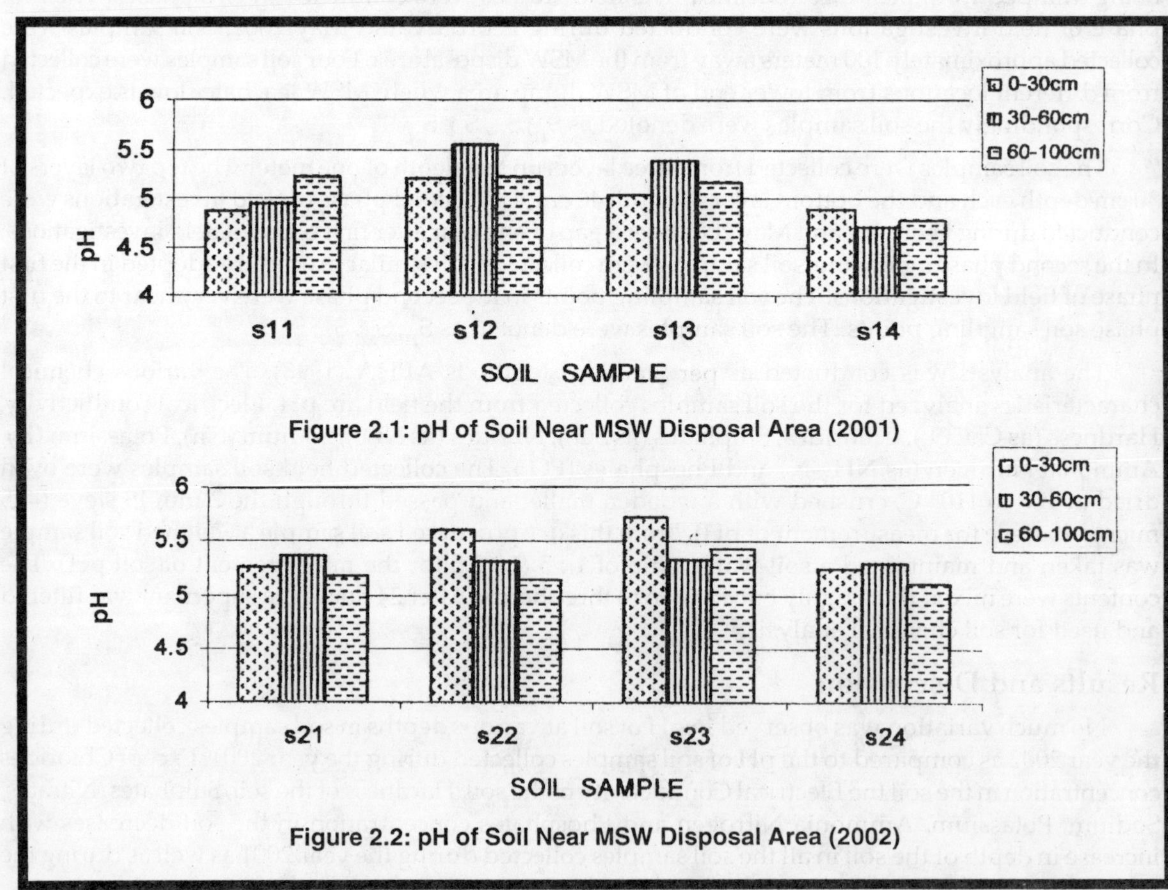

Figure 2.1: pH of Soil Near MSW Disposal Area (2001)

Figure 2.2: pH of Soil Near MSW Disposal Area (2002)

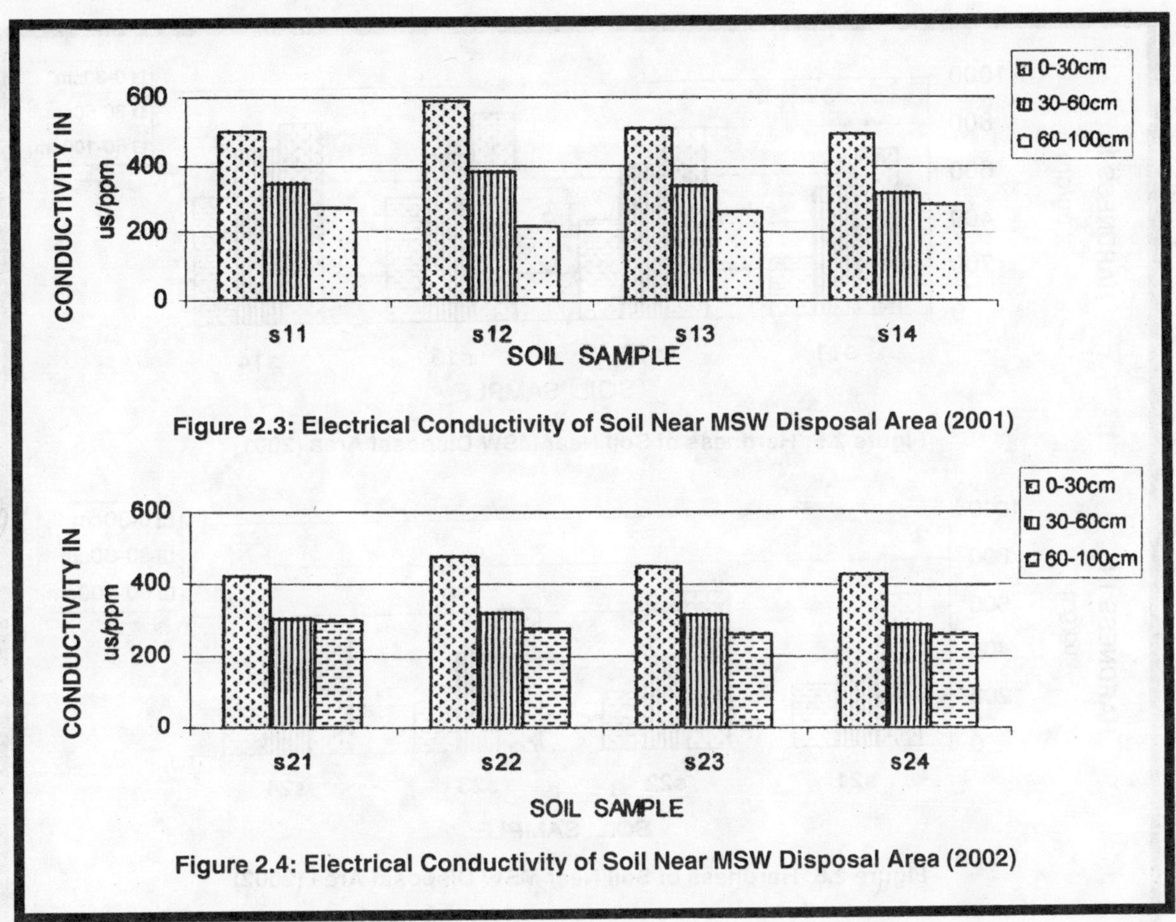

Figure 2.3: Electrical Conductivity of Soil Near MSW Disposal Area (2001)

Figure 2.4: Electrical Conductivity of Soil Near MSW Disposal Area (2002)

depth in soil sample S_{12}) and 590 µs/ppm (0–30 cm depth in soil sample S_{12}) respectively during the year 2001. During the year 2002 the observed minimum and maximum EC of soil was 262 µs/ppm (60–100 cm depth in soil sample S_{23}) and 475 µs/ppm (0–30 cm depth in soil sample S_{22}) respectively.

Figure 2.5 and Figure 2.6 presents the observed Hardness of soil at various depths collected near MSW disposal area during the year 2001 and the year 2002 respectively. The decreasing in Hardness of the soil is possibly due to the replacement of ions (Ca and Mg) which are responsible for the Hardness of soil by Sodium and Potassium ions in the soil. The minimum and maximum Hardness of soil observed was 250 µg/gm (60–100 cm depth in soil sample S_{11}) and 880 µg/gm (0–30 cm depth in soil sample S_{13}) during the year 2001. The observed minimum and maximum Hardness of soil during the year 2002 was 200 µg/gm (60–100 cm depth in soil sample S_{24}) and 675 µg/gm (0–30 cm depth in soil sample S_{22}) respectively. Figure 2.7 and Figure 2.8 presents the observed Chloride concentration in the soil at various depths collected near MSW disposal area during the year 2001 and the year 2002 respectively. The observed minimum and maximum Chloride concentration in the soil was 375 µg/gm (0–30 cm depth in soil sample S_{12}) and 525 µg/gm (60–100 cm depth in soil sample S_{11}) respectively during the year 2001. During the year 2002 the observed minimum and maximum Chloride concentration in the soil was 230 µg/gm (0–30 cm depth in soil sample S_{23}) and 356 µg/gm (60–100 cm depth in soil sample S_{23}) respectively.

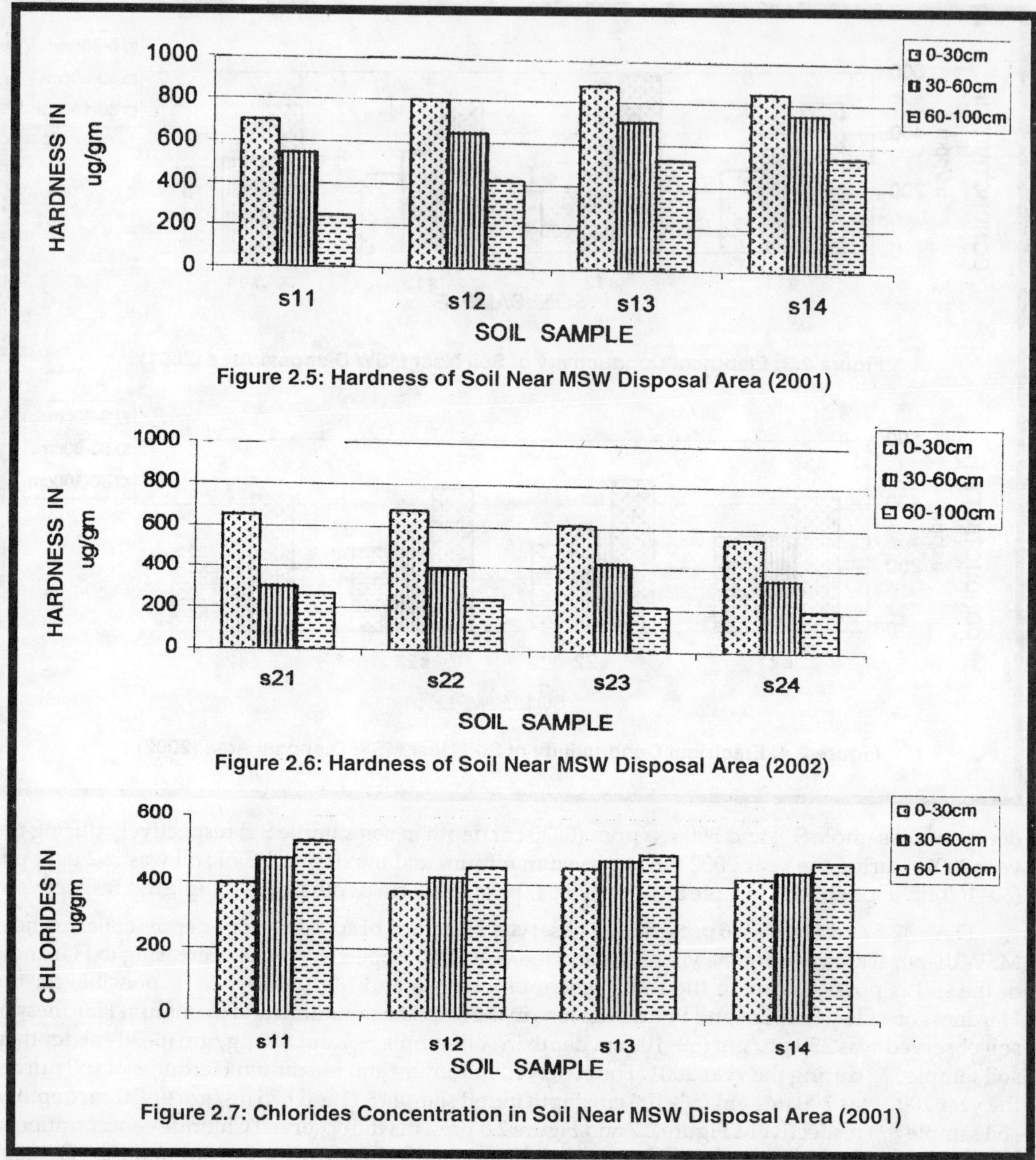

Figure 2.5: Hardness of Soil Near MSW Disposal Area (2001)

Figure 2.6: Hardness of Soil Near MSW Disposal Area (2002)

Figure 2.7: Chlorides Concentration in Soil Near MSW Disposal Area (2001)

Figure 2.9 and Figure 2.10 presents the observed Sulphates concentration in the soil at various depths collected near MSW disposal area during the year 2001 and the year 2002 respectively. The observed minimum and maximum Sulphates concentration in the soil was 38 μg/gm (60-100 cm depth in soil sample S_{11}) and 175 μg/gm (0-30 cm depth in soil sample S_{12}) respectively during the year

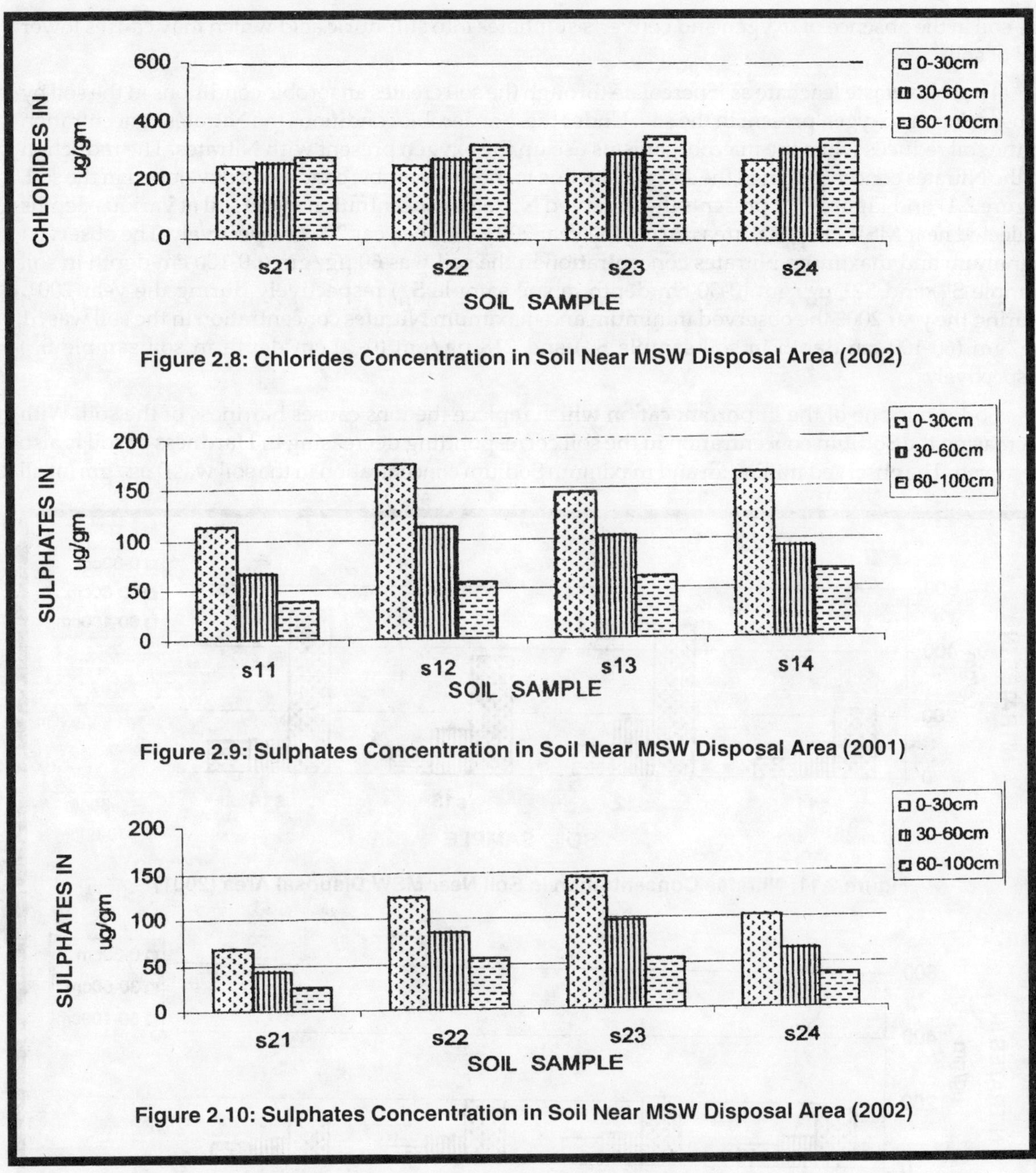

Figure 2.8: Chlorides Concentration in Soil Near MSW Disposal Area (2002)

Figure 2.9: Sulphates Concentration in Soil Near MSW Disposal Area (2001)

Figure 2.10: Sulphates Concentration in Soil Near MSW Disposal Area (2002)

2001. During the year 2002 the observed minimum and maximum Sulphates concentration in the soil was 25 μg/gm (60–100 cm depth in soil sample S_{21}) and 144 μg/gm (0–30 cm depth in soil sample S_{23}) respectively during the year 2002. The decrease in Sulphates concentration in the soil may be due to anaerobic reaction which takes place in the soil during the percolation of solid waste leachate through the soil in the absence of oxygen and converts sulphates into sulphuric acid which may causes lower pH of soil.

The solid waste leachate as it percolate through the soil creates anaerobic conditions in the soil by consuming the oxygen present in the soil. Under these anaerobic conditions the Nitrates concentration in the soil reduces when the microorganisms use up the oxygen present with Nitrates. The reduction in the Nitrates concentration in the soil may be due to these anaerobic conditions developed in the soil. Figure 2.11 and Figure 2.12 presents the observed Nitrates concentration in the soil at various depths collected near MSW disposal area during the year 2001 and the year 2002 respectively. The observed minimum and maximum Nitrates concentration in the soil was 60 μg/gm (60-100 cm depth in soil sample S_{11}) and 521 μg/gm (0-30 cm depth in soil sample S_{14}) respectively during the year 2001. During the year 2002 the observed minimum and maximum Nitrates concentration in the soil was 51 μg/gm (60-100 cm depth in soil sample S_{23}) and 218 μg/gm (0-30 cm depth in soil sample S_{22}) respectively.

Sodium is one of the important cation which replace the ions causes hardness of the soil. With increasing the Sodium concentration in the soil corresponding decreasing in Hardness of soil is also observed. The observed minimum and maximum Sodium concentration in the soil was 0 μg/gm (at all

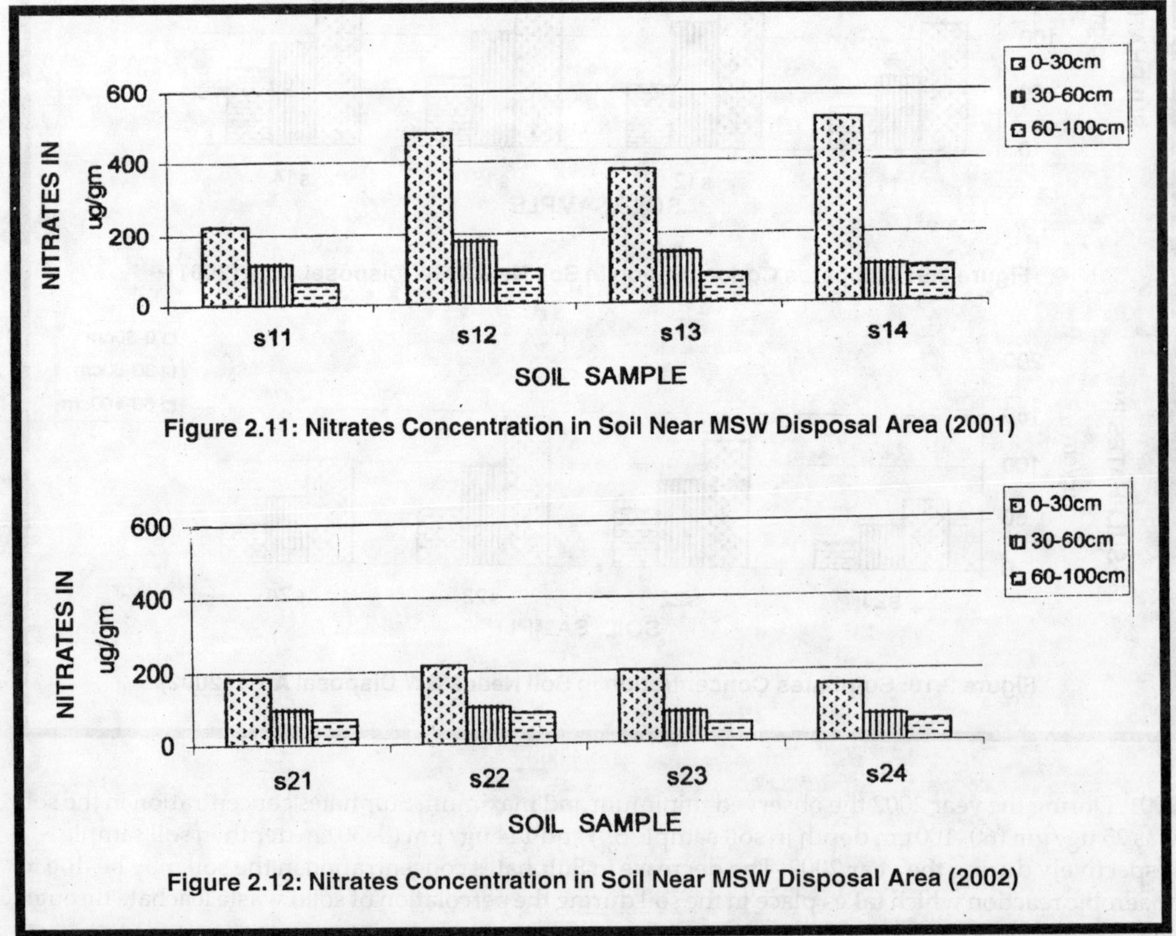

Figure 2.11: Nitrates Concentration in Soil Near MSW Disposal Area (2001)

Figure 2.12: Nitrates Concentration in Soil Near MSW Disposal Area (2002)

depths in soil sample S_{14}) and 27 µg/gm (0-30 cm depth in soil sample S_{13}) respectively during the year 2001. During the year 2002 the observed minimum and maximum Sodium concentration in the soil was 12 µg/gm (60-100 cm depth in soil sample S_{24}) and 67 µg/gm (0-30 cm depth in soil sample S_{22}) respectively. Figure 2.13 and Figure 2.14 presents the observed Sodium concentration in the soil at various depths collected near MSW disposal area during the year 2001 and the year 2002 respectively.

The observed minimum and maximum Potassium concentration in the soil was 14 µg/gm (60–100 cm depth in soil sample S_{14}) and 29 µg/gm (0–30 cm depth in soil sample S_{12}) respectively during the year 2001. During the year 2002 the observed minimum and maximum Potassium concentration in the soil was 27 µg/gm (60–100 cm depth in soil sample S_{21}) and 92 µg/gm (0–30 cm depth in soil sample S_{22}) respectively. Figure 2.15 and Figure 2.16 presents the observed Potassium concentration in the soil at various depths collected near MSW disposal area during the year 2001 and the year 2002 respectively. The observed higher Potassium concentration in the soil compare with sodium concentration in the soil at similar depth may be due to lower geochemical mobility of potassium compared to sodium in the soil.

Figure 2.17 and Figure 2.18 presents the observed Ammonia Nitrogen concentration in the soil at various depths collected near MSW disposal area during the year 2001 and the year 2002 respectively.

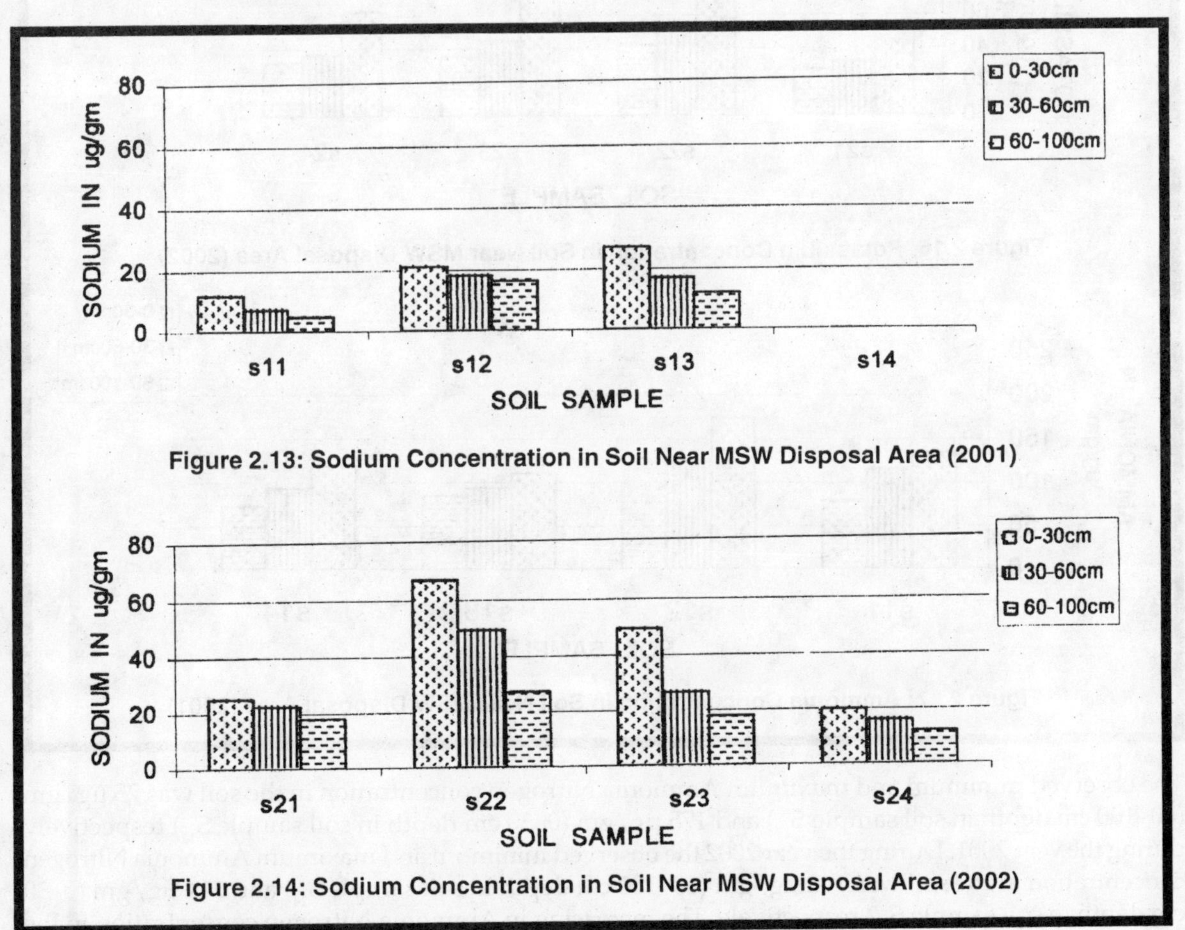

Figure 2.13: Sodium Concentration in Soil Near MSW Disposal Area (2001)

Figure 2.14: Sodium Concentration in Soil Near MSW Disposal Area (2002)

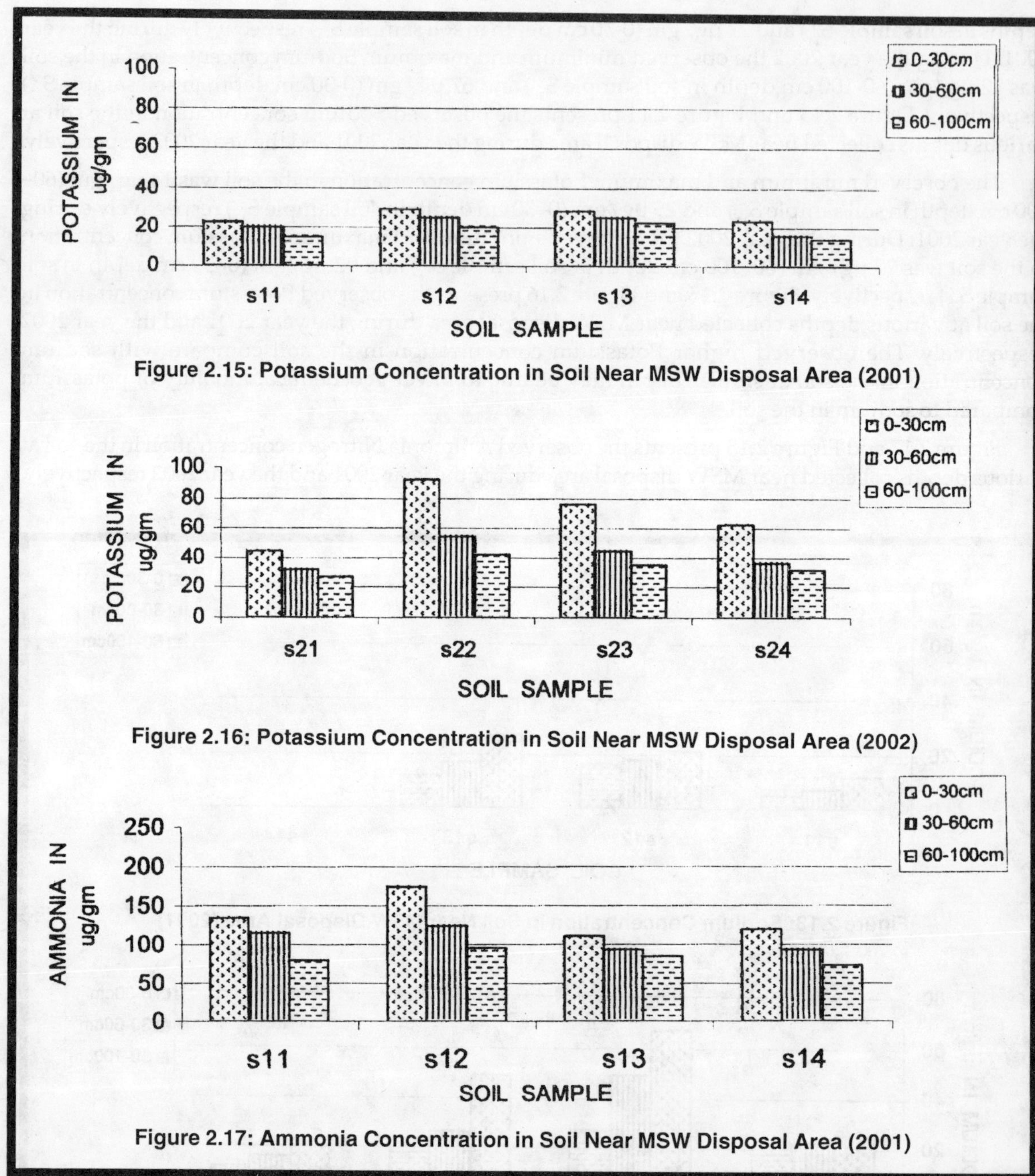

Figure 2.15: Potassium Concentration in Soil Near MSW Disposal Area (2001)

Figure 2.16: Potassium Concentration in Soil Near MSW Disposal Area (2002)

Figure 2.17: Ammonia Concentration in Soil Near MSW Disposal Area (2001)

The observed minimum and maximum Ammonia Nitrogen concentration in the soil was 75 µg/gm (60–100 cm depth in soil sample S_{14}) and 175 µg/gm (0–30 cm depth in soil sample S_{12}) respectively during the year 2001. During the year 2002 the observed minimum and maximum Ammonia Nitrogen concentration in the soil was 102 µg/gm (60–100 cm depth in soil sample S_{24}) and 214 µg/gm (0–30 cm depth in soil sample S_{22}) respectively. The increasing in Ammonia Nitrogen concentration in the soil possibly due to denitrification process which takes place in the soil in the absence of oxygen during the percolation of solid waste leachate through the soil.

Figure 2.19 and Figure 2.20 presents the observed Phosphates concentration in the soil at various depths collected near MSW disposal area during the year 2001 and the year 2002 respectively. The observed minimum and maximum Phosphates concentration in the soil was 23 μg/gm (60–100 cm

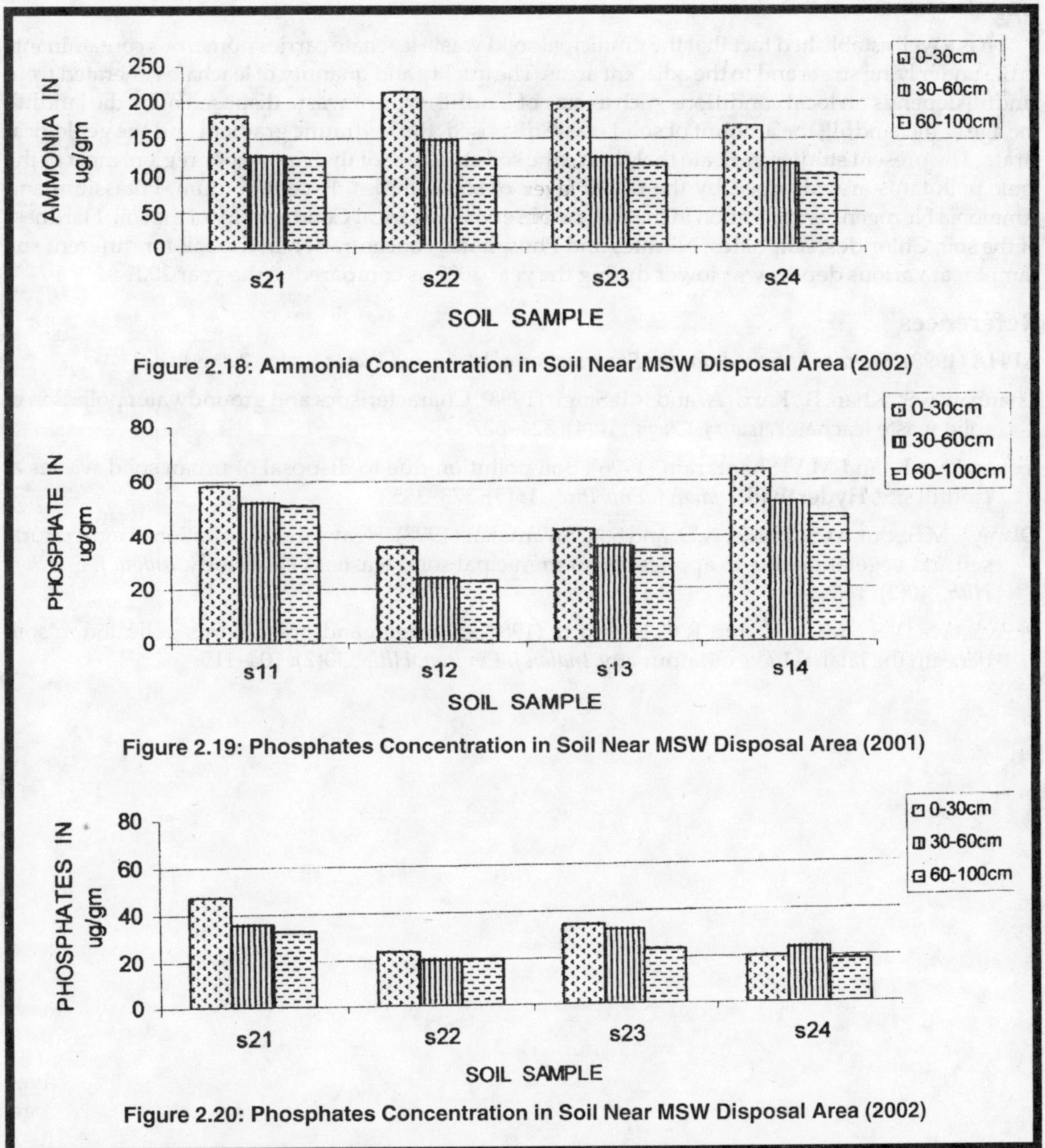

Figure 2.18: Ammonia Concentration in Soil Near MSW Disposal Area (2002)

Figure 2.19: Phosphates Concentration in Soil Near MSW Disposal Area (2001)

Figure 2.20: Phosphates Concentration in Soil Near MSW Disposal Area (2002)

depth in soil sample S_{12}) and 64 μg/gm (0–30 cm depth in soil sample S_{14}) respectively during the year 2001. During the year 2002 the observed minimum and maximum Phosphates concentration in the soil was 18 μg/gm (60–100 cm depth in soil sample S_{24}) and 47 μg/gm (0–30 cm depth in soil sample S_{21}) respectively. The decreasing in Phosphates concentration in the soil possibly due to utilization of Phosphates by the organisms present in the soil as nutrients.

Conclusions

It is a well established fact that the municipal solid waste leachate carries numerous contaminants to the under lying strata and to the adjacent areas. The quality and quantity of leachate generated from landfill depends on local conditions such as size of landfill, type of waste disposed of by the landfill, the age of the landfill, the amount of solid waste disposed, the hydraulic gradient and the geological strata. The present studies indicate that due to the soil structure of the Vamanjoor region, most of the ionic pollutants are adsorbed by the upper layer of soil samples. Except Sodium, Potassium and Ammonia Nitrogen concentration in the soil, the observed Electrical Conductivity of the soil, Hardness of the soil, Chlorides, Sulphates, Nitrates and Phosphates concentration in the soil for different soil samples at various depths was lower during the year 2002 as compared to the year 2001.

References

APHA (1998). *Standard Methods for the Examination of Water and Waste Water*, 20th edn.

Chauhan, B.S., Khan, B., Karri, A. and R.P. Singh (1998). Characteristics and groundwater pollution of solid waste leachate. *Asian J. Chem.*, 10(4): 824–827.

Jeevan Rao, K. and M.V. Shantaram (1996). Soil pollution due to disposal of urban solid wastes at landfill site, Hyderabad. *Indian J. Env. Prot.*, 16(5): 373–385.

Olaniya, M.S., Sur, M.S., Bhide, A.D. and S.N. Swarnakar (1998). Heavy metal pollution of agricultural soil and vegetation due to application of municipal solid waste: A case study. *Indian J. Environ. Hlth.*, 40(2): 160–168.

Shrivastava, V.S., Rai, A.K. and R.C. Mehrotra (1988). Nitrogen and phosphorus pollution in soils beneath the Jalahal Lake of Jaipur city. *Indian J. Environ. Hlth.*, 30(2): 104–115.

Chapter 3
Phytoremediation of Soil Fluoride by Crop Rotation

S.R. Ambika and S. Sumalatha

Plant Physiology Laboratory, Department of Botany, Bangalore University, Bangalore – 560 056
E-mail: pcsraaa@bgl.vsnl.net.in

ABSTRACT

Fluoride when accumulated to appreciable quantities in the soil exerts a definite toxic effect on seed germination and plant growth. Methi var. Bangalore Local and Guar var. PNB was found to be sensitive while Radish var. Pusa Chetki was found to be tolerant to soil fluoride. Crop rotation with fluoride tolerant species could reduce the fluoride toxicity of sensitive crops. In this study Guar when grown in crop rotation with radish showed increased growth and yield.

Keywords: *Crop rotation, Fluoride, Growth, Guar, Methi, Radish, Yield.*

Introduction

Fluoride enters soil through weathering of rocks, precipitation and from contaminated water. Fluoride accumulation and binding to the soil depends on the content of minerals, humus and the pH of the soil (Hansen *et al.*, 1958; Lavado and Reinandi, 1979). Loamy soils are known to retain a high amount of fluoride as it holds more clay and retains more of cations (Mac Intire *et al.*, 1955). Fluoride is known to affect the growth of plants right from germination, seedling growth to vegetative growth until maturity, affecting the yield (Sarkar *et al.*, 1982; Miller, 1993; Miller, 1997). Significant correlation between the concentration of fluoride and reduction in vegetative growth and metabolic content were also reported (Rathore, 1992). The present study was to explore the efficacy of crop rotation methods to phytoremediate the fluoride contaminated soils.

Crop rotation is a definite sequence of crops grown in successive years or growing seasons, on the same land. The statistical analysis of the crop productivity in such rotation experiments, when separately done for each crop when results are combined, there emerged certain meaningful economical comparisons (Preece, 1986).

Materials and Methods

The seeds of guar (*Cyamopsis tetragonoloba* (L.) Taub. var. PNB; Radish (*Raphanus sativus* L. var. Pusa Chetki); Methi (*Trigonella foenum graecum* L. var. PEB1) were obtained from the Farm House of Gandhi Krishi Vignana Kendra, University of Agricultural Sciences, Bangalore. The seeds were stored in clean dry plastic bags at room temperature until use.

Assessment of Phytotoxicity of Soil Fluoride by Crop Rotation of Two Fluoride Sensitive Crops

This experiment was conducted at the field station of Bangalore in the Quadrangle of Department of Botany Department, Bangalore University, and Bangalore during the summer months of May/June. The site had red loamy soil with a thin cover of alluvium. The NPK was maintained at the basal level. The pH was 7.5. The plot was ploughed and manually mixed with 5 kg/m^2 of farmyard manure. The plot was sub-divided into small plots of 1 m^2, separated by a distance of 0.5 m. The plots were bounded on all four sides and 0.1, 0.2 and 0.4 g/m^2 NaF was incorporated into the soil. The control plot had no fluoride mixed into the soil. Methi seeds were sown in 10 rows at a distance of 0.1 m in each plot. A randomized block design with 5 replicates per treatment was adopted and the plots were watered daily. When the plants attained a height of 2 to 3 cm, thinning was done to one quarter so as to maintain a final density of 30 plants/m^2.

Growth analysis was done on the 4th week after emergence prior to the onset of flower buds. Ten plants were harvested and determined the height and dry matter accumulation in the root and shoot. The pods were harvested after drying on the plant for 12 to 14 days. Yield parameters were analysed for the pod and seed. After the Methi plants were harvested Guar seeds [*Cyamopsis tetragonoloba* (L.) Taub. var. PNB] were sown in the plots. No fresh fluoride was incorporated into the soil. The level of NPK was adjusted according to the package of practice. The randomized block pattern was maintained with 5 replications per treatment with final density of 30 plants/m^2. Plants were harvested after the 5th week.

Growth data was recorded on the 5th week after emergence prior to the onset of flower buds. Height and Dry weight were recorded for 10 plants. The pods were harvested at the 7th week. Yield parameters were analysed for the pods and seeds.

The field data from both the trials were analysed by two-way analysis of variance. Significance of the mean was determined by Duncan's hypothesis.

Assessment of Phytotoxicity of Soil Fluoride by Crop Rotation of a Fluoride Resistant Crop with a Fluoride Sensitive Crop

Crop rotation trials were conducted to assess the ability of fluoride resistant crop to alleviate the toxic effect of soil fluoride on a fluoride sensitive crop. The experimental conditions and plot design were similar to that in Expt. 1. NPK were added according to the package of practice. The soil was also mixed with 0.1, 0.2, and 0.4 g/m^2 NaF. Radish seeds (*Raphanus sativus* L var. Pusa Chetki) were sown in 10 rows at a distance of 0.1 m in a randomized block design with 5 replications per treatment and were watered daily. These plants were harvested after 2 weeks. Plant height and dry matter accumulation were determined for two-week old plants. Seeds of Guar *Cyamopsis tetragonoloba* (L.) Taub. var. PEB was sown in the plots where radish was grown and harvested. The experimental design was similar to that executed in previous experiment.

Ten plants from each plot were analysed for the linear and dry matter accumulation on the 5th week after the emergence. Pods were harvested after maturing on the plant at the 7th week stage. Fluoride level of the soil at post harvest condition was determined. Comparison was drawn between the crop rotation trials of the fluoride sensitive species and the resistant species. The entire field trial data were analysed by two-way analysis of variance.

Results and Discussion

Presence of fluoride in the soil besides water plays a vital role in plant nutrition and development. When soluble fluoride accumulate to appreciable quantities in the soil, a definite toxic effect is exerted on seed germination and plant growth. Therefore, mitigation of fluoride levels in soil is essential to ensure healthy growth of crops. Crop rotation experiments were hence conducted to test the capacity of these crops to overcome fluoride toxicity.

These studies were conducted with fluoride sensitive-sensitive crop (Methi-Guar) and sensitive-tolerant crop (Guar-Radish). Comparison was drawn between the yield of Guar rotated with the fluoride sensitive crop (Methi) and Fluoride tolerant crop (Radish).

In the sensitive-sensitive crop rotation trials, Methi was raised in three different levels of soil flyuoride-0.1, 0.2, and 0.4 g NaF/m^2. The linear growth of the Methi plants were reduced by 3.1 to 22 per cent at maturity, dry matter accumulation by 24.3 to 39.4 per cent (Figure 3.1) and the pod number decreased by 22.8 to 51.9 per cent and the seed number lowered by 29.2 to 57.1 per cent compared to the control (Figure 3.2). The proline accumulation was also high ranging from 93–107 per cent more than the control (Figures 3.2 and 3.4). On the contrary, the growth and yield of the fluoride tolerant radish was totally unaffected by fluoride in the soil. The linear growth was reduced marginally by 0.42 to 9.95 per cent at maturity while the dry matter accumulation in the shoot and root increased by 13.4 to 27.2 per cent compared to the control. However, the plants raised in 0.4 g NaF/m^2, registered a marginal reduction of 2.2 per cent in dry matter compared to the control (Figure 3.3). The proline accumulation was also marginal, *i.e.* 1 to 4 per cent more than the control at various concentrations of soil fluoride (Figure 3.4).

This very clearly indicates that Methi is sensitive to soil fluoride while Radish is tolerant to it.

The effectiveness of crop rotation trials was evident in the increase in the growth and yield of Guar (sensitive) rotated with Radish (tolerant) compared to Guar grown in rotation with Methi (sensitive). Guar when crop rotated with radish recorded growth improvement. The linear growth of Guar increased by 19 per cent, dry matter by 16 per cent, leaf yield by 11 per cent, pod number by 13.1 to 25.2 per cent and seed number by 6.5 to 35.5 per cent.

Guar when crop rotated with Methi (sensitive) showed reduction in growth parameters. The reduction in linear growth was 16 to 38 per cent, dry weight 19–63 per cent, Leaf number 28–60 per cent, pod number 31–45 per cent and seed number 6–54 per cent (Figures 3.4 and 3.5).

The proline accumulation in Methi was 107 per cent higher than control due to fluoride treatment when grown independently (Figure 3.2) while in radish, the fluoride tolerant crop the proline accumulation was only 4 per cent more than the control (Figure 3.6). This indicates that the fluoride sensitive crop when grown in soil with high fluoride content was under stress while the tolerant crop

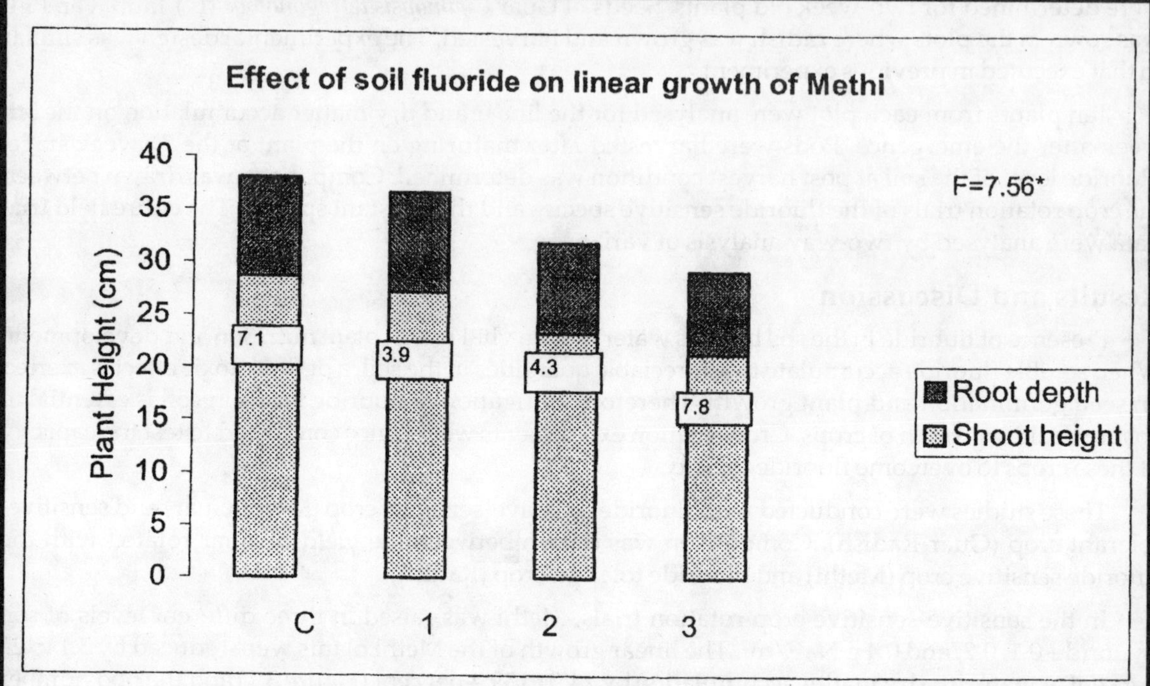

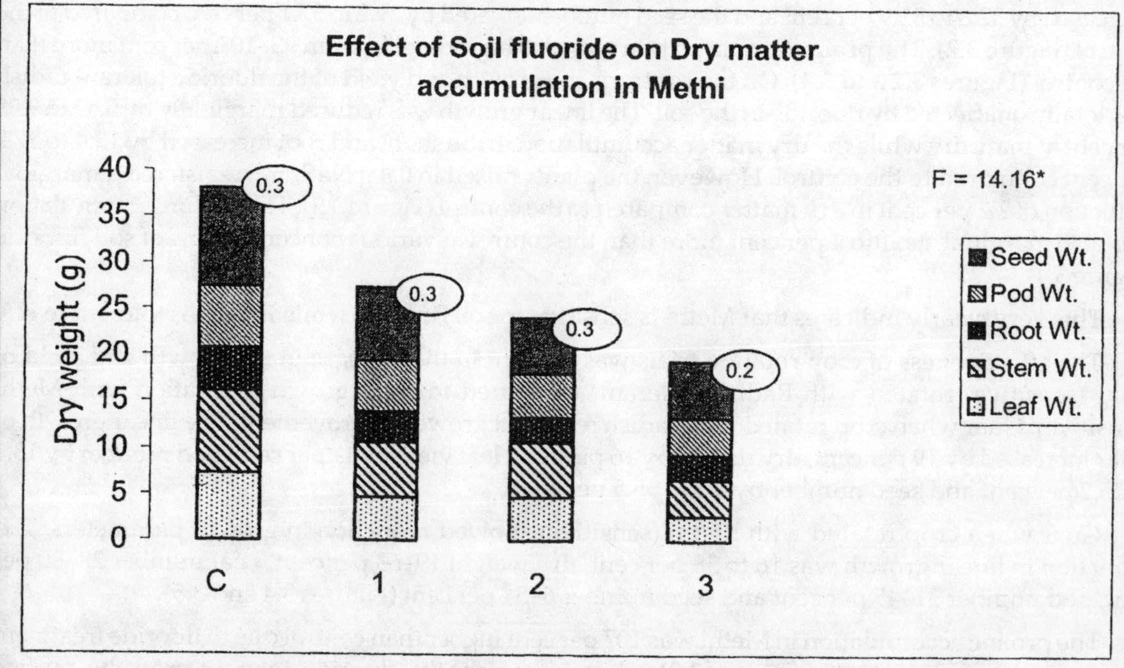

Figure 3.1: Growth Parameters in Methi (*Trigonella foenum graecum* L. var. Bangalore Local) Raised in 0.1gNaF/m² (1), 0.2 gNaF/m² (2) 0.4 gNaF/m² and Control (C)

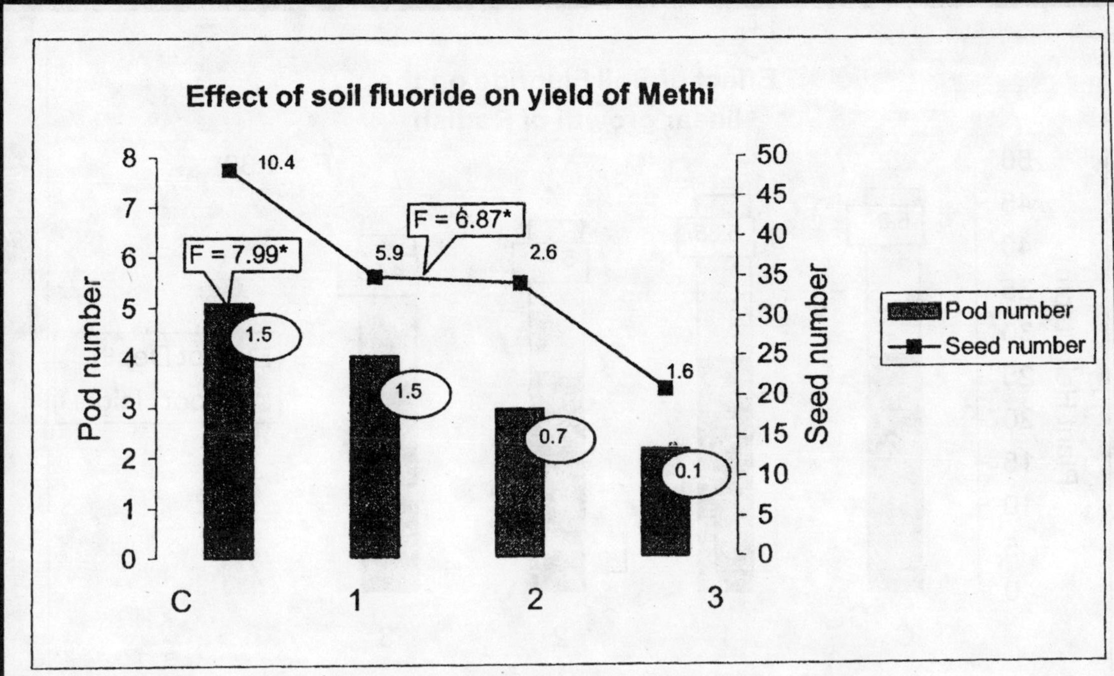

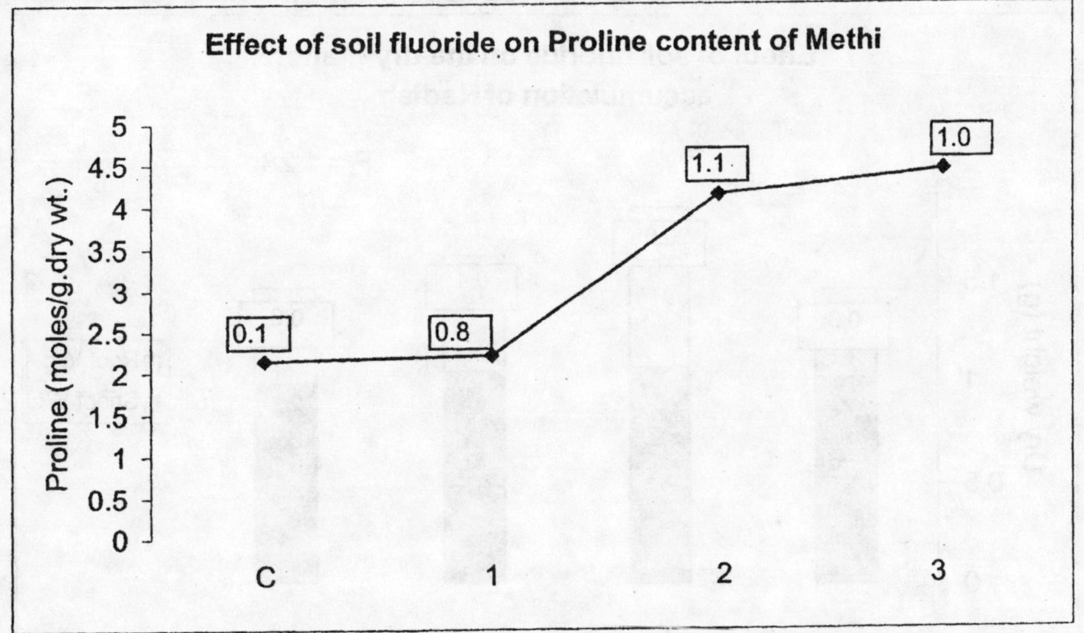

Figure 3.2: Yield Parameters (Pod and Seed Numbers) and Proline Accumulation in Methi (*Trigonella foenum graecum* L. var. Bangalore Local) Raised in 0.1 gNaF/m² (10, 0.2 g NaF/m² (2), 0.4 gNaF/m² (3) and Control (C)

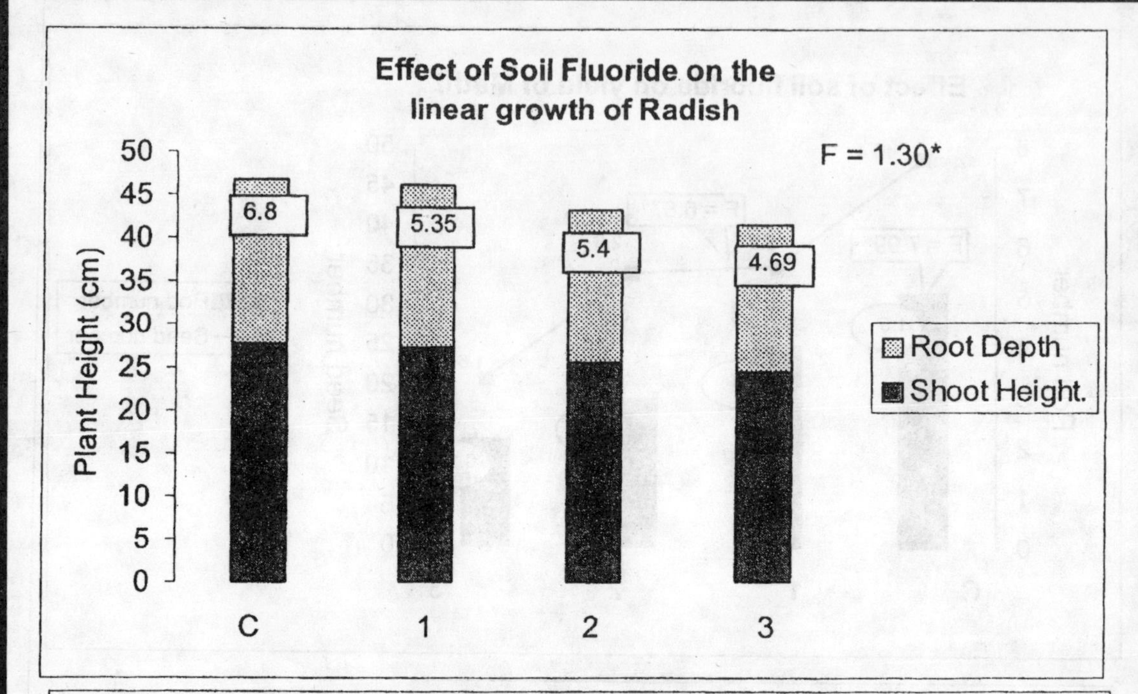

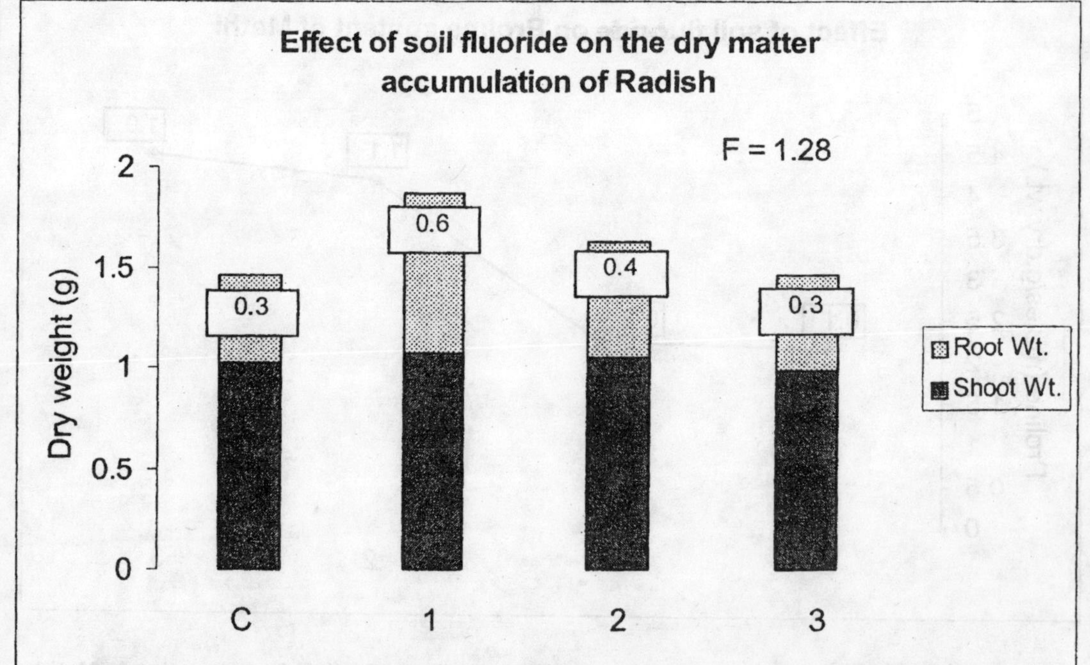

Figure 3.3: Growth Parameters in Radish (*Raphanus sativus* L. var. Pusa chetki) Raised in 0.1gNaF/m² (1), 0.2 gNaF/m² (2) 0.4 gNaF/m² (3) and Control (C)

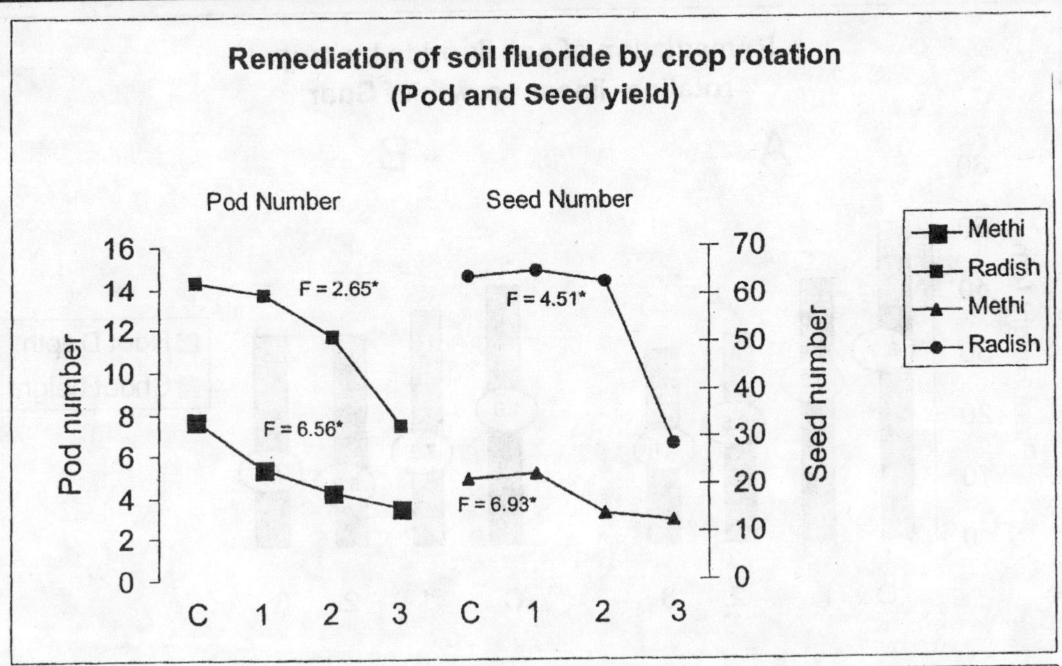

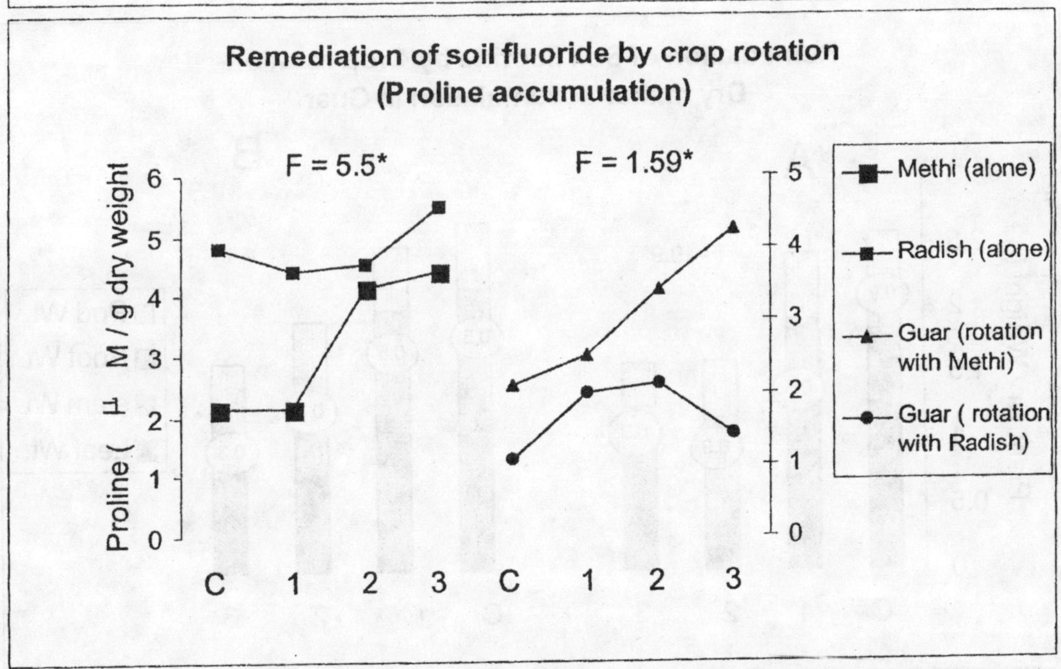

Figure 3.4: Proline Accumulation in Leaves of Methi (*Trigonella foenum graecum* L. var. Bangalore Local) and Radish (*Raphanus sativum* L. var. Pusa Chetki) when Grown Alone and in Guar [*Cyamopsis tetragonoloba* (Taub.) L. var. PNB] in Crop Rotation with Methi and Radish Raised in 0.1g NaF/m² (1), 0.2 g NaF/m² (2), 0.4 g NaF/m² (3) and Control (C)

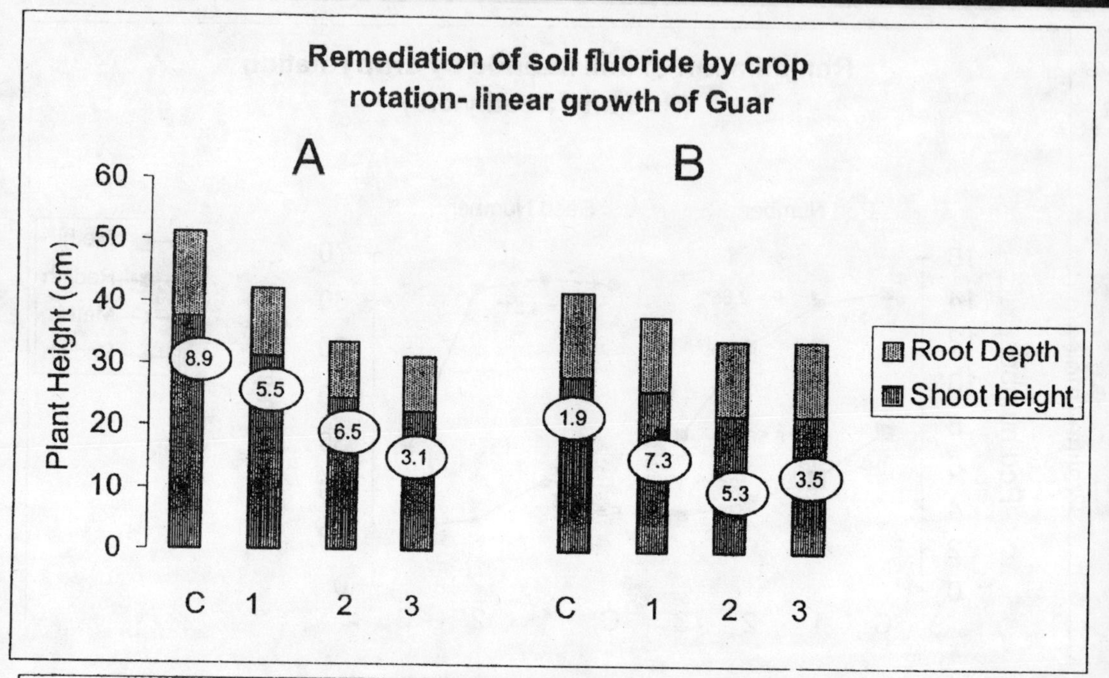

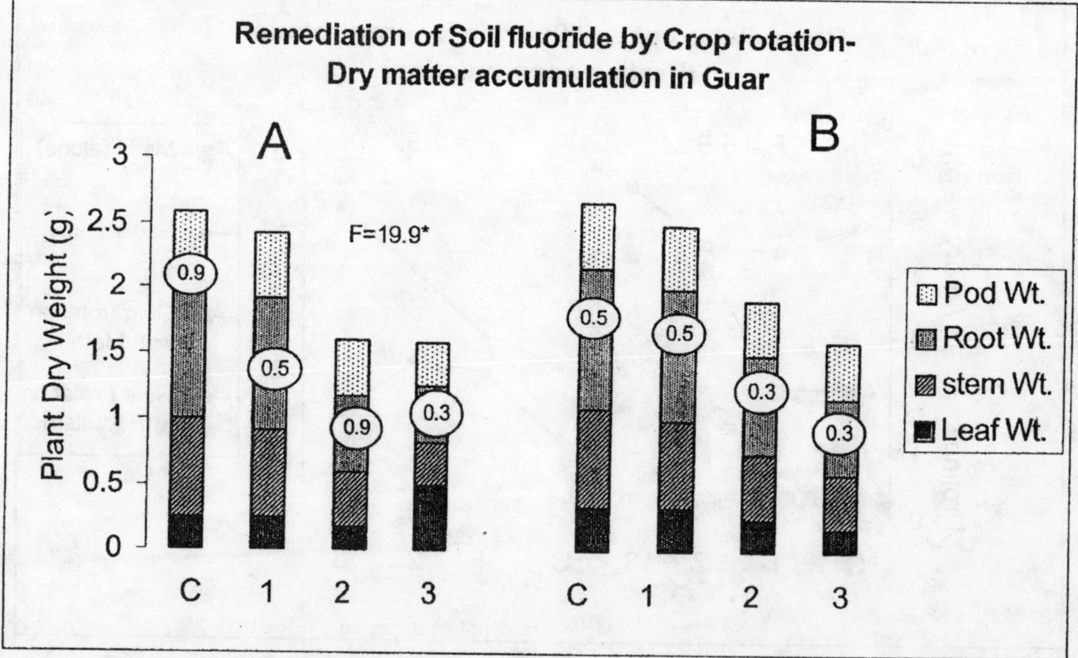

Figure 3.5: Growth Parameters in Guar (*Cyamopsis tetragonoloba* (L.) Taub. var. PNB) A and B–in Crop Rotation with Methi and Radish Raised in 0.1g of NaF/m² (1), 0.2 g NaF/m² (2), 0.4 g NaF/m² (3) and Control (C).

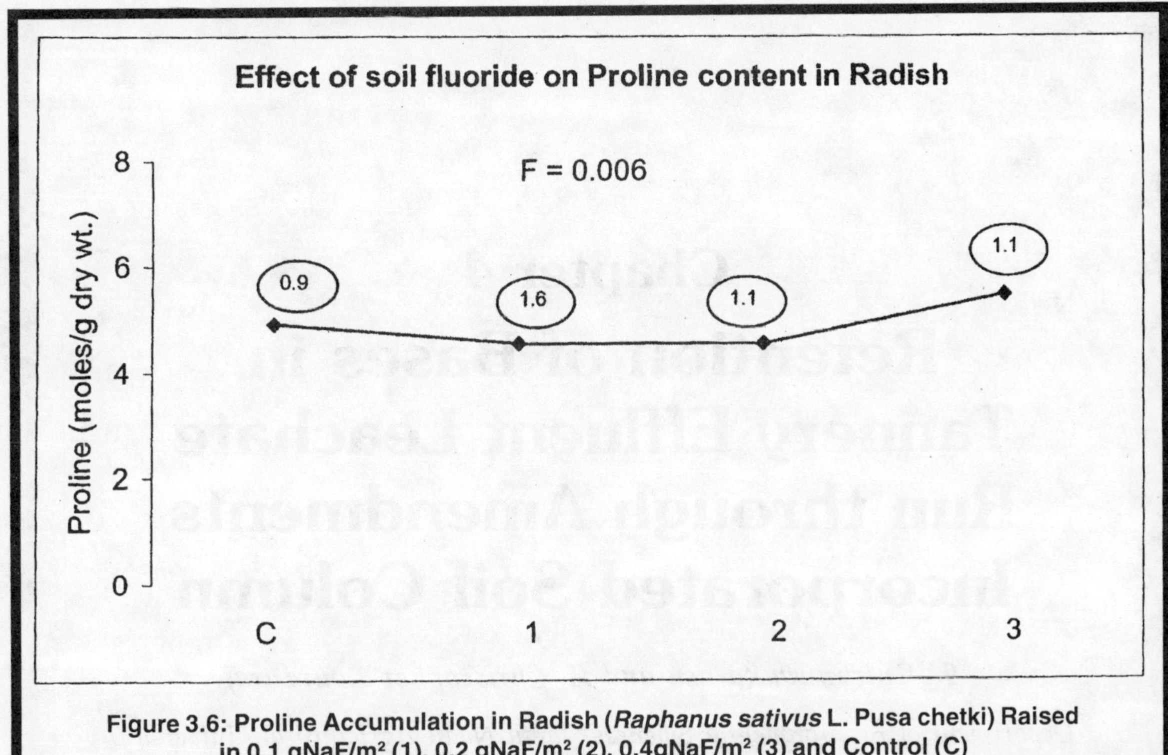

Figure 3.6: Proline Accumulation in Radish (*Raphanus sativus* L. Pusa chetki) Raised in 0.1 gNaF/m² (1), 0.2 gNaF/m² (2), 0.4gNaF/m² (3) and Control (C)

(Radish) was under little or no stress. In the presence of higher amount of soil fluoride, Methi crop registered 125 per cent more proline that the control (Figure 3.2). But in crop rotation studies, the accumulation of proline in the leaves was reduced considerably; 1 to 40 per cent (Figure 3.4).

Hence, the crop rotation technique was fairly effective in reducing the toxicity of fluoride to sensitive crops and can be recommended as a means of phytoremediating the soil contaminated with fluoride.

References

Hansen, D.E., Wiebe, H.H. and W. Thorne (1958). *J. Agron.* 50: 565.

Lavado, R.S. and N. Reinaudi (1979). *Fluoride.* 12: 28.

Mac Intire, W.H., Sterges, A.J. and W.M. Shaw (1955). *J. Agric. Food Chem.*, 3: 777.

Miller, G.W. (1993). The effect of fluoride on higher plants, with special emphasis on early physiological and biochemical disorders. *Fluoride*, 26(1): 3–22.

Miller, G.W. (1997). Fluoride: A toxic substance. *Fluoride*, 30(3): 141.

Preece, D.A. (1986). Some general principles of crop rotation experiments. *Expt. Agric.*, 22: 187–198.

Rathore, S. (1992). Effect of fluoride toxicity on leaf area, net assimilation rate, and relative growth rate of *Hordeum vulgare* and *Zea mays. Fluoride*, 25(4): 175–182.

Sarkar, R.K., Banerjee, A. and S. Mukherji (1982). Effect of toxic concentrations of fluoride on growth and enzyme activities of rice and jute seedlings. *Biol. Plant*, 24: 34–38.

Chapter 4

Retention of Bases in Tannery Effluent Leachate Run through Amendments Incorporated Soil Column

K. Thirunavukarasu and A. Christopher Lourduraj

Department of Environmental Sciences, Tamil Nadu Agricultural University, Coimbatore – 641 003

ABSTRACT

An experiment was conducted at the analytical laboratory of the Department of Environmental Science, Tamil Nadu Agricultural University, Coimbatore to study the impact of weekly application tannery effluent in the presence of various soil amendments on the quantum of bases collected in the leachate. The results of the study revealed that the amendment addition significantly influenced the retention of bases *viz.*, Ca^{++}, Mg^{++} and K^{+} in the soil column and decrease of these bases in leachate compared to control was observed.

Keywords: Tannery effluent, Amendments, Column, Leachate.

Introduction

Tannery effluent application as irrigation water to the agricultural crops has severe impact on the crop growth and to the soil fertility. This is due to presence of high load of harmful salts like Na^{+}, SO^{2-}_{4}, Cl^{-} and HCO_3^{-}. Apart from these harmful salts the effluent contains calcium, magnesium and potassium which improve the soil properties and good for crop. Hence with a view to quantify the extent of Ca^{++}, Mg^{++} and K^{+} retained in the presence of soil amendments, the present study was taken up.

Materials and Methods

Column Experiment Setup

Column experiments were conducted to evaluate the effectiveness of various absorbents in reducing the salt concentration from treated tannery effluent by simulating a field condition.

The column experiment was carried out using PVC pipes of 55 cm height and 6.1 cm internal diameter. Wire mesh, muslin cloth and filter paper of whatman no. 42 were placed at the bottom of each layer. The bottom of the column was closed with an end cap of PVC material. A hole was made in the endcap to collect the leachate. The leachate collected was analysed for Ca^{++}, Mg^{++} and K^{+}.

The column was divided into three layers of different bulk density of 15 cm height and separated each layer by muslin cloth (Figure 4.1).

The bottom layer was filled with 704 g of soil to maintain a bulk density of 1.6 g cc^{-1}, middle layer with 674 g of soil to maintain bulk density of 1.5 g cc^{-1}, the top layer was filed with mixture containing 613 g of soil and respective amendment at quantities as adopted in the field experiment. For each amendment three replications was carried out. Effluent was poured to every column at weekly interval and a constant head of upto 5 cm was maintained in each column. The leachate was collected upto 7 weeks and analysed for Ca^{++}, Mg^{++} and K^{+}. The tannery effluent used in the study were analysed prior to use and presented in Table 4.1.

Table 4.1: Characterisation of Tannery Effluent and Well Water

Properties	*Unit*	*Tannery Effluent*	*Well Water*
pH	–	7.91	6.98
EC	dS m^{-1}	15.1	11.4
BOD	mg l^{-1}	630	320
COD	mg l^{-1}	1244	870
Total solids	mg l^{-1}	14800	10060
Dissolved solids	mg l^{-1}	9700	7460
Suspended solids	mg l^{-1}	3500	1900
Water soluble sodium	me L^{-1}	153	109
Water soluble chloride	me L^{-1}	121	113
Water soluble calcium	me L^{-1}	27.24	20.16
Water soluble magnesium	me L^{-1}	7.93	6.32
Water soluble potassium	me L^{-1}	2.97	2.73
Water soluble sulphate	me L^{-1}	14.20	9.65
Water soluble bicarbonate	me L^{-1}	18.90	12.30
Water soluble carbonate	me L^{-1}	–	–
Sodium adsorption ratio	–	36.48	29.95
Residual sodium carbonate	me L^{-1}	– 16.27	– 14.18
Ion balance (anion excess)	me L^{-1}	44.8	2.612
Potential salinity	me L^{-1}	120.1	11.825

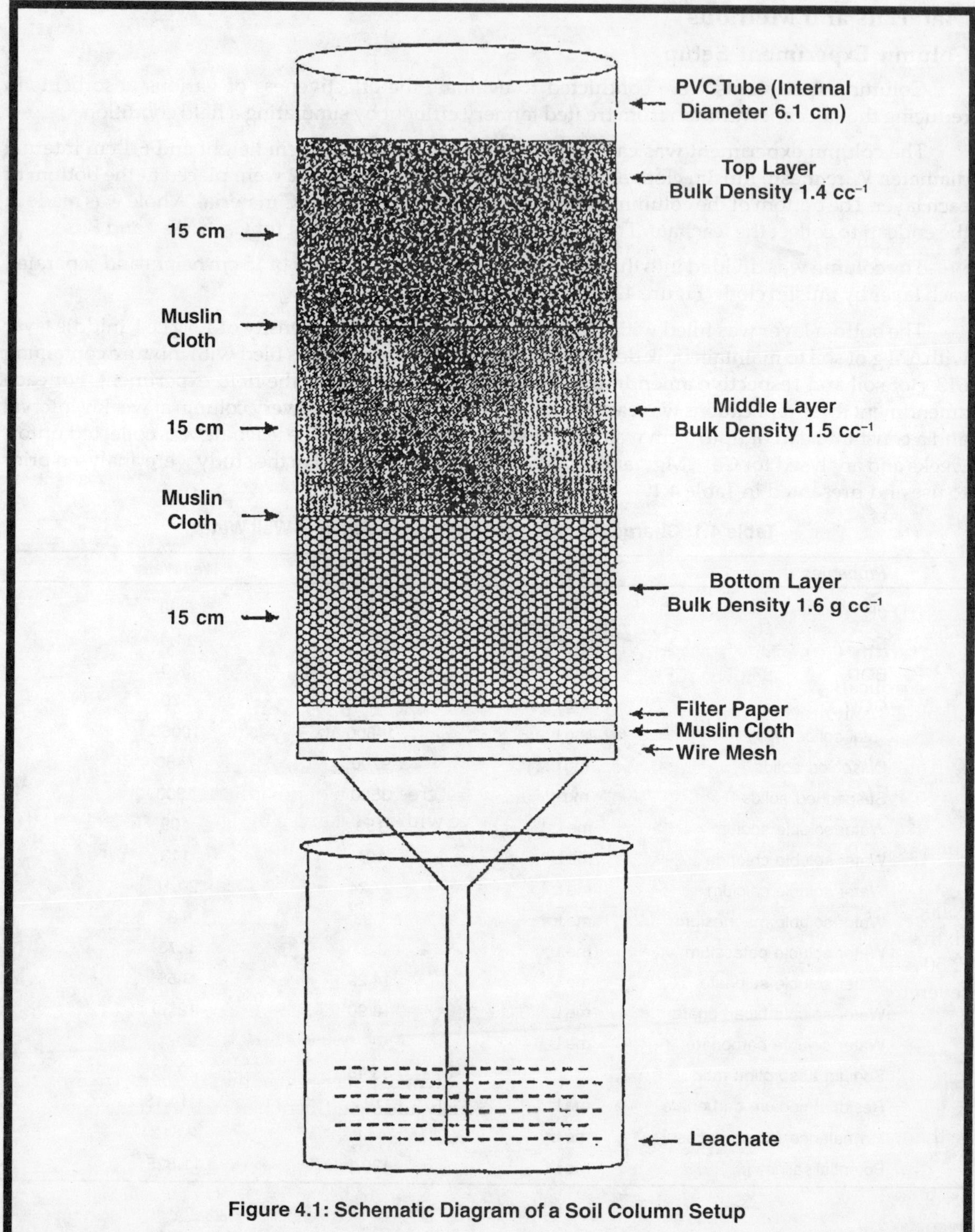

Figure 4.1: Schematic Diagram of a Soil Column Setup

Materials

A range of cheaply available biological wastes differing in their physical and chemical characteristics were used and examined for their efficiency in removing the salts (Na and Cl) from treated tannery effluent. Sawdust was collected from a saw mill in Coimbatore, composted coir pith and composted pressmud were collected from compost yard in central farmyard and poultry droppings from the TNAU poultry farm (the poultry droppings were kept heaped for 2 months before applying to the field). Vermiculite, a 2 : 1 (non-expanding) type alumino-silicate clay mineral of grade III was obtained from TAMIN (Tamil Nadu Minerals Limited), Chennai was used in the study.

The amendments used in the study were analysed for their nutrient content and the data is presented in Table 4.2

Table 4.2: Characterisation of Organic Amendments

Character	*Poultry Manure*	*Pressmud*	*Composted Coirpith*	*Sawdust*
Organic carbon (%)	20.1	20.69	28.2	21.06
N (%)	2.5	1.10	1.4	0.3
P (%)	1.3	1.85	0.38	0.02
K (%)	1.4	1.25	1.00	0.98
Ca (%)	1.39	3.05	0.46	–
Mg (%)	1.04	1.50	0.39	–

Results and Discussion

Potassium Content in the Tannery Effluent Leachate

Application of amendment significantly influenced the potassium level in the effluent leachate (Table 4.3). Significantly higher potassium level was recorded for control (T_1) (3.07 me L^{-1}) and was on par with vermiculite (T_7) (2.91 me L^{-1}). Lowest level of potassium was recorded with composted coir pith (T_3) (2.62 me L^{-1}). Composted coir pith addition had performed better than other amendments and control. Pressmud and gypsum addition also favoured reduction of potassium in leachate falling in line next to composted coir pith. This is in accordance with report given by Thabaraj *et al.* (1964) that composted coir pith was found to be best treatment for increasing the uptake of potassium nutrient.

Weekly intervals had significant influence on the potassium level in the effluent leachate. Leachate collected on seventh week (W_7) recorded higher level of potassium (3.07 me L^{-1}) and was on par with sixth week (W_1) (2.94 me L^{-1}). Leachate collected on first week (W_1) recorded lower level of potassium (2.50 me L^{-1}). Increasing trend was noted in the potassium level of the leachate from first week (W_1) to seventh week (W_7).

The continuous addition of effluent has saturated the adsorbing sites of the soil, hence the excess potassium could not be retained in the soil column and consequently increased the potassium concentration in the leachate. After every week, the content of potassium in the leachate increased due to maximum adsorption in the soil column. Continuously adding effluent had increased the potassium with increase in duration.

Interaction effect between amendment and weekly intervals significantly influenced the potassium level. Significantly higher level of potassium was recorded for the leachate collected on seventh week in control (T_1W_7) (3.37 me L^{-1}).

Table 4.3: Potassium Content in the Leachate Collected (me L^{-1})

Treatments	Period (Weeks)							Mean
	W_1	W_2	W_3	W_4	W_5	W_6	W_7	
T_1–control	2.65	2.79	2.97	3.16	3.24	3.28	3.37	3.07
T_2–gypsum	2.40	2.59	2.65	2.59	2.78	2.84	2.91	2.68
T_3–composted coir pith	2.36	2.48	2.56	2.63	2.68	2.79	2.85	2.62
T_4–poultry manure	2.61	2.71	2.76	2.78	2.84	3.01	3.14	2.83
T_5–pressmud	2.44	2.58	2.63	2.68	2.73	2.81	2.92	2.68
T_6–sawdust	2.57	2.74	2.82	2.84	2.86	2.86	3.10	2.83
T_7–vermiculite	2.48	2.90	2.92	2.95	2.93	2.99	3.19	2.91
Mean	2.50	2.69	2.76	2.80	2.86	2.94	3.07	

	SEd	CD (0.05)
T	0.064	0.127
W	0.064	0.127
TW	0.170	0.337

Calcium and Magnesium Content of Tannery Effluent Leachate

Application of amendments significantly influenced calcium and magnesium level of the tannery effluent leachate (Tables 4.4 and 4.5). Control treatment (T_1) recorded significantly higher calcium level (26.93 me L^{-1}), followed by vermiculite amendment (T_7) recording (25.91 me L^{-1}). T_1 and T_7 differed significantly. Lowest calcium level was recorded with application of composted coir pith (T_3) (23.79 me L^{-1}). This is in accordance with the report given by Ravikumar and Thygarajan (1980) about the beneficial effects of application of coir pith, pressmud and farmyard manure in improving the soil physical properties.

Table 4.4: Calcium (Ca) Content in the Leachate Collected (me L^{-1})

Treatments	Period (Weeks)							Mean
	W_1	W_2	W_3	W_4	W_5	W_6	W_7	
T_1–control	24.09	25.3	25.41	26.4	28.27	28.49	30.58	26.93
T_2–gypsum	21.78	23.54	24.09	23.54	25.19	25.74	26.4	24.33
T_3–composted coir pith	21.45	22.55	23.21	23.87	24.31	25.3	25.85	23.79
T_4–poultry manure	23.65	24.64	25.08	25.19	25.74	27.28	28.49	25.72
T_5–pressmud	22.11	23.43	23.87	24.31	24.75	25.52	26.51	24.36
T_6–sawdust	23.32	24.86	25.19	25.74	25.96	25.96	28.16	25.60
T_7-vermiculite	22.55	24.09	25.3	26.73	26.62	27.17	28.93	25.91
Mean	22.71	24.06	24.59	25.11	25.83	26.49	27.85	

	SEd	CD (0.05)
T	0.395	0.782
W	0.395	0.782
TW	1.05	2.07

Table 4.5: Magnesinm (Mg) Content in the Leachate Collected (me L^{-1})

Treatments	*Period (Weeks)*							*Mean*
	W_1	W_2	W_3	W_4	W_5	W_6	W_7	
T_1–control	5.02	6.42	6.48	7.42	7.88	8.24	8.32	7.11
T_2–gypsum	4.28	5.96	6.26	6.68	7.04	7.52	8.00	6.53
T_3–composted coir pith	3.94	5.46	5.92	6.56	6.78	7.38	7.78	6.26
T_4–poultry manure	4.38	6.22	6.74	6.78	7.82	8.18	8.26	6.91
T_5–pressmud	5.88	5.78	6.18	6.34	7.66	7.66	8.04	6.79
T_6–sawdust	4.80	6.12	6.40	6.56	7.80	7.86	8.12	6.81
T_7-vermiculite	4.74	6.28	6.58	6.62	7.58	7.94	8.02	6.82
Mean	4.72	6.03	6.37	6.71	7.51	7.83	8.08	

	SEd	*CD (0.05)*
T	0.133	0.264
W	0.133	0.264
TW	0.352	0.698

Significantly higher magnesium level was recorded in control (T_1) (7.11 me L^{-1}) followed by poultry manure application (T_4) (6.91 me L^{-1}). Composted coir pith (T_3) recorded lowest magnesium level (6.26 me L^{-1}) in the leachate.

Decomposition products of organic materials like composted coir pith and pressmud have very high exchange capacity. This is in accordance with Singaram (1996) report that application of 10t ha^{-1} either raw or composted coir pith had improving effect in tannery polluted soils. Their incorporation in soil irrigated by sodic water is associated with increased EC of soil. The colloids with higher CEC tend to accumulate more of divalent cations (Ca^{++}) than monovalent (Na^{+}) cations.

As the exchange sites in the soil have been saturated with calcium and magnesium from the previous effluent additions and the subsequent addition of effluent to the soil column has contributed to the marginal increase of calcium and magnesium in the leachate.

Weekly intervals significantly influenced the calcium and magnesium level of the leachate collected. Seventh week leachate (W_7) (27.85 me L^{-1}) recorded significantly higher level of calcium followed by sixth week leachate (W_6) (26.49 me L^{-1}) and fifth week leachate (W_5) (25.83 me L^{-1}). First week (W_1) leachate recorded lowest calcium level (22.71 me L^{-1}). Increasing trend was noted in the calcium level of the leachate from first week (W_1) to seventh week (W_7).

Significantly higher magnesium level was recorded in the leachate collected on seventh week (W_7) (8.08 me L^{-1}) and was on par with sixth week (W_6) (7.83 me L^{-1}) leachate. Magnesium level in the leachate collected on first week (W_1) recorded lowest magnesium level (4.72 me L^{-1}). Increasing trend is noted in the magnesium of the leachate from first week (W_1) to seventh week (W_7).

Interaction effect between amendment and weekly interval had significant influence on leachate calcium and magnesium level. Seventh week leachate in control (T_1W_7) had significantly higher level of calcium (30.58 me L^{-1}). Significantly higher magnesium level was recorded in the leachate collected in the seventh week in control (T_1W_7) (8.32 me L^{-1}).

References

Ravikumar, V. and T.M. Thyagarajan (1980). Soil physical problem in Coimbatore district. *Madras Agric. J.*, 67(4): 248–251.

Singaram, P. (1996). Characterisation of tannery effluent and K effect on soil physical properties. *Madras Agric. J.*, 83(3): 206–207.

Thabaraj, G.J., S.M. Bose and Y. Nayudamma (1964). Utilization of tannery effluent for agricultural purposes. *Indian J. Environ. Hlth.*, 6(1): 18–24.

Chapter 5

Study of Correlation of Coefficient of Physical, Chemical and Biological Characteristics with Catalse Activity in Industrially Polluted and Unpolluted Soils of Warangal (D.T.) A.P.

B. Lalitha Kumari and M.A. Sinrigara Charya

Department of Botany, Kakatiya University, Warangal – 506 009

ABSTRACT

Effect of various physical and chemical variables were studied in the production of catalse enzyme in different polluted and unpolluted soils. The correlation was positive with all parameters tried with 5 per cent, 1 per cent and 0.01 per cent level during 12 months of our observation (1995–96). The correlation was positive at 0.01 per cent level in sewage amended soils with variables such as pH, pore space, conductivity, calcium, manganese, iron, aluminium and silica. The highest significant ranges were recorded for dairy amended soils for catalase activity with calcium, magnesium, iron, phosphates, nitrates, organic matter and actinomycetes populations. Highest significance was noticed in Industrial estate soils, to catalse enzyme with bulk density, water holding capacity, manganese, aluminium, nitrates, nitrites, organic Matter and bacterial populations. Correlation coefficient at 0.01 per cent level was noticed in tannery industry soils with variables, potassium, sodium, nitrites and actinomycetes. Interestingly except Ammonia all other variables showed their significance of variation in unpolluted soils for catalase activity.

Keywords: Physicochemical, Correlation coefficient, Catalse.

Introduction

Enzyme activity of the soil is potentially a more specific tool for characterizing the biochemical capacities of soil than are microbial counts or respiration measurements. Soil enzyme measurements are often referred to soil type, vegetation, fertilization, microbial counts and respiration (Venkataraman and Rajya Laxmi, 1971; Tatabai, 1977; Yadav *et al.*, 1989; Evans, 1990; Hulugalle and Entwistla, 1997). Johnson and Temple (1964) presented correlation coefficients of three enzyme measurements and particle size, carbon content and nitrogen content for several soil types. Scheinost (1997) recommended catalase enzyme content as indicator of soil fertility. Scholes and Breemen (1997) in their studies on global change on tropical ecosystems determined the effects of edaphic factors on catalase measurements. Feller and Beare (1997) and Venkateswarlu *et al.* (1998) observed soil is just the sort of complex system in which catalase reaction rates might be confused by non-enzyme catalysis and by a variety of physicochemical factors. In the present study an important soil enzyme catalase was selected for this investigation and the effect of certain variables studied.

Material and Methods

No. of Study Sites: (1) Kumarpally Sewage; (2) Dairy Industry; (3) Industrial Estate; (4) Tannery Industry and (5) Kakatiya University (Unpolluted Site).

Hydrogen Peroxide Decomposition

2 groven dry weight soil sample was placed in a 125 ml Erlenmyer flask with 40 ml of distilled water and put on a rotary shaker. To this was added 5 ml of H_2O_2 solution (30 per cent H_2O_2 diluted 1 : 100) and the slurry was shaken for 20 minutes. The remaining peroxide was then stabilized by adding 5 ml 3 N H_2SO_4 and the contents of the flash filtered through Whatman filter Paper No. 1 and 25 ml of aliquot titrated with 0.1 N $KMnO_4$. The initial concentration of the peroxide used was determined by titration standardized against sodium oxalate. All titrations were corrected for a blank the soil slurry filtrate titration values were subtracted from the amount of permanganate needed to titrate the initial peroxide and calculated as ml of 0.1 N $KMnO_4$. Final results were expressed as ml of 0.1 N $KMnO_4$ equivalent to the peroxide decomposed per gram of even dry soil. All the soil variables were analysed in the soil samples as methods elaborated by Trivedy *et al.* (1987).

Results and Discussion

Correlation coefficient of 26 parameters were studied in four polluted soils and one unpolluted soil during 1995–1996 against to accumulation of catalase enzyme and presented in Table 5.1. From the table, it was evident that all variables were highly significant the rate of peroxide decomposition were effected by the soil collected from different polluted environments. The interactions between the soil microbes and catalase activities were highly significant.

pH showed its correlation at 0.01 level (0.9179) at Kumarpally sewage site. Significant correlation was recorded between bulk density of the soils with that of catalase activity in sewage amended soils (0.7837) ($P < 0.01$) and at Industrial Estate Soils (0.9094; $P < 0.001$). Similarly, pore space, water holding capacity and conductivity also recorded their significant levels of correlation with catalase production. Alkalinity and chlorides did not show much influence on the measurements of catalase in polluted and unpolluted soils. Calcium and Magnesium showed their significant levels of correlation with catalse activity in sewage soils (0.9676; $P < 0.001$) and Dairy effluents amended soils (0.9239; $P < 0.001$). Manganese, iron, aluminium and silica showed their highly significant results with catalase production in sewage soils. The eutrophication factors also influenced positively the soil enzyme

Table 5.1: Correlation Coefficient (r) Between Physical, Chemical and Biological Characteristics with Catalase Activity in Polluted and Unpolluted Sites of Warangal, Andhra Pradesh

	Polluted Sites				*Unpolluted Site*
	Kumarpally Sewage	*Dairy Industry*	*Industrial Estate*	*Tannery Industry*	*Kakatiya Univesity*
Physical					
pH	0.918***	0.585*	0.686*	0.528*	0.356
Bulk Density	0.784**	0.581*	0.909***	0.309	0.347
Pore Space	0.816**	0.562*	0.753**	0.229	0.497
Water Holding Capacity	0.567*	0.504*	0.946***	0.404	0.438
Conductivity	0.909***	0.445	0.568*	0.346	0.366
Permenant Wilting Coefficient	0.489	0.549*	0.581*	0.473	0.498
Chemical					
Alkalinity	0.591*	0.410	0.745**	0.594*	0.683*
Chlorides	0.565*	0.633*	0.240	0.235	0.451
Calcium	0.968***	0.924***	0.497	0.339	0.423
Magnesium	0.452	0.926***	0.369	0.227	0.655*
Calcium Carbonate	0.541*	0.454	0.463	0.560*	0.552*
Potassium	0.286	0.514*	0.178	0.901***	0.182
Sodium	0.558*	0.692*	0.621*	0.821**	0.441
Manganese	0.965***	0.579*	0.969***	0.411	0.609*
Iron	0.902***	0.919***	0.517*	0.398	0.363
Aluminium	0.906***	0.565*	0.927***	0.392	0.255
Silica	0.949***	0.641*	0.699*	0.261	0.424
Sulphates	0.732***	0.669*	0.458	0.208	0.473
Nitrates	0.590*	0.963***	0.947***	0.619*	0.219
Nitrites	0.809*	0.568*	0.975***	0.933***	0.487
Ammonia	0.621*	0.405	0.417	0.374	0.963***
Organic Matter	0.688*	0.921***	0.981***	0.088	0.459
Biological					
Fungi	0.789**	0.708*	0.487	0.986***	0.483
Bacteria	0.281	0.742**	0.973***	0.304	0.208
Actinomycetes	0.483	0.926***	0.374	0.963***	0.362

*P < 0.05; **P = 0.01; ***P < 0.01.

secretions. Phosphates showed their significant correlation levels (0.9353; P < 0.001) in dairy effluents amended soils while the other soil samples fail to correlate between phosphates and catalase enzyme. Among nitrogen substances nitrites showed a significant trend of correlation with catalase enzyme in sewage soils (0.8092; P < 0.01). The soils collected at industrial estate (0.9756) and tannery industry soils (0.9332) correlated even at 0.01 per cent. The organic matter content of the soils showed its correlations with highest significance in dairy industry amended soils (0.9203) and industrial estate soils (0.9813) at 0.0.1 per cent. The microbes, fungi, bacteria, actinomycetes increased substantially the activity of catalase enzyme in different soils under investigation. Fungi showed its positive significant correlation in sewage soils (0.7895; $P < 0.01$) and tannery industry amended soils (0.9862; $P < 0.001$). Bacteria also showed correlation coefficient values at 1 per cent level (0.7424) at dairy effluent amended soils and showed their highest significant in industrial estate soils (0.9739; $P < 0.001$). The actinomycetes population also could show their positive significance in dairy industry soils and tannery industry soils with coefficient values of (0.9267) and (0.9626) even at 0.01 per cent respectively. Similar to the present studies Broadman (1985) measured five soil types of catalase activity incubated with and without adding organic matter and bentonite. The measurement was effected by soil type, addition of organic matter, addition of bentonite and drying of the sample Brady (1990) characterized the soils in relation to catalase activity in presence of glucose, alfalfa, etc. and statistically proved the positive role. Nannipieri *et al.* (1980) studied the use of soils sterilents to limit the measurement to catalase in various soils studied by them. Recently Smith *et al.* (1997) while modelling procedures for nitrite accumulation in soils correlated positively nitrogen substances with catalase activity. Boarner *et al.* (1998) while studying the variation and spatial pattern in organic carbon content, total inorganic nitrogen and extractable inorganic phosphorous in soils among 6 sites in Ohio, USA clarified the role of catalase in different degradative processes.

Acknowledgements

Authors are thankful to Prof. S.M. Reddy, Head, Department of Botany, Kakatiya University for laboratory facilities and encouragement.

References

Boerner, R.E.J., Amy J. Scherzer Jennifer and A. Brinkman (1998). Spatial patterns of inorganic N, P availability and organic relation to soil disturbance: A chronosequence analysis. *Applied Soil Ecology*, 7(2): 159–177.

Brady, N.C. (1995). *The Nature and Properties of Soils, Organisms of the Soils*. Prentice Hall Pvt. Ltd., New Delhi, p. 253–277.

Broadman, J. (1985). *Soil and Quartenary Landscape Evolution*. Wiley Chiekester.

Evans, R. (1990). Soils at risk of accelerated erosion in England and Wales. *Soils, Use and Management*, 6: 125–131.

Feller, C. and M.H. Beare (1997). Physical control of soil organic matter dynamics in the tropics. *Geoderma*, 79(1–4): 69–116.

Hulugalle, N.R. and P. Entwistle (1997). Soil properties, nutrient uptake and crop growth in an irrigated vertisol after nine years of minimum tillage. *Soil and Tillage Research*, 42: 15–32.

Johnson, J.L. and K. Temple (1964). Some variables affecting the measurement of catalase activity in soil. *Soil Sci. Soc. Amer. Proc.*, 28: 207–209.

Mannipieri, P., B. Ceccanti, S. Cervelli and E. Matarese (1980). Extraction of phosphates, urease, protease, organic carbon and nitrogen from soil. *Soil. Sci. Soc. Am. J.*, 44: 1001–1016.

Scheinost, A.C., Sinowski, W. and K. Auerswald (1997). Regionalization of soil water retention curves in a highly variable soilscape, I: Developing a new pedotransfer function. *Geoderma*, 78(3–4): 129–143.

Scholes, R.J. and N. Van Breemen (1997). The effects of global change on tropical ecosystems. *Geoderma*, 79(1–4): 9–24.

Smith, R.V., Doyle, R.M., Burns, L.C. and R.J. Stenens (1997). A model for nitrite accumulations in soils. *Soil Biology and Biochemistry*, 29(8) 1241–1247.

Tatabai, M.A. (1977). Effect of trace elements on urease activity of soils. *Soil Biol. Biochem.*, 9: 9–73.

Trivedy, R.K., Goel, P.K. and C.C. Trisal (1987). *Practical Methods in Ecology and Environmental Science.* Environ. Media. Publications, KARAD (India).

Venkatraman, G.S. and Rajyalaxmi (1971). Interaction between pesticides and soil microorganisms. *Indian J. Exptl. Biol.*, 9: 521–522.

Venkateswarlu, B., Hari, K. and J.C. Katyal (1998). Influence of soil and crop factors on the native rhizobial population in soils under dryland farming. *Applied Soil Ecology*, 7(1): 1–10.

Yadav, K., Jha, K.K. and R. Jha (1989). Microbial decomposition of poultry manure and sewage sludge in soil. *Indian Soil Sci.*, 37: 301–305.

Chapter 6
Eco-toxicological Effects Caused by SWE of a Chlor-alkali Industry on the Biological Nitrogen Economy of Crop Fields

P.K. Pradhan, Alaka Sahu and A.K. Panigrahi

Environmental Science Research Center, Department of Botany, Berhampur University, Berhampur – 760 007, Orissa

ABSTRACT

The solid waste extract (SWE) of a chlor-alkali industry containing mercury was tested for possible use of BGA as a nitrogen fixer and an agent for decontaminating a contaminated environment (detoxifier). Significant increase in nitrogen fixation and removal of mercury by way of volatilisation by the blue-green alga was marked at sub-lethal or maximum allowable concentration of solid waste extract (SWE), showing stimulation. At higher concentration of the SWE both nitrogen fixation and removal of mercury from the contaminated environment was noticed. Hence, the SWE or solid waste of the chlor-alkali industry should be diluted to its sub-lethal dose in the crop fields, the tolerant BGA can be charged into the field and allowed to grow for 20 days and then any crop can be cultivated, which will show better yield/production. This alga can be used as a biofertiliser cum detoxifier in mercury contaminated environments.

Keywords: *Solid waste, Chlor-alkali industry, Mercury, BGA, Biofertiliser, Cellular nitrogen, Extra-cellular nitrogen, Detoxification.*

Introduction

Mercury pollution of the environment has created some serious hazards for mankind. As mercury has been used since ancient times, mercury poisoning also has a long history. The earliest case of industrial mercury poisoning was described by Paracelsus (1967) among miners. Ramazzini (1713) also made descriptions of the symptomatology observed in mercury miners (Singerman, 1976). But the most significant incidents of the toxicity of this metal from the scientific and epidemiological points of view have been those in Japan in Minamata (1953–1960) and Niigata (1965); these were caused by industrial release of mercury and its compounds into Minamata Bay and the Agano river, respectively (Fujuki, 1972; Tsubaki, 1971; Tsubaki and lrukayama, 1977). Alkyl-mercury fungicide used for seed dressings are important original sources of mercury in terrestrial food chains (WHO, 1989). Even today the dreadful repercussion of 1956 Minamata disease is still prevalent among the population (Davies, 1991).

Algal assays have proven to be very sensitive indicators of contaminant stress (Miller *et al.*, 1985). Claesson (1984) used mixed algal culture to evaluate the toxicity of industrial waste water. Tewari *et al.* (1990) studied the growth and biochemical composition of two marine macro algae exposed to chlor-alkali industry effluent. Gaur and Singh (1990) studied the growth, photosynthesis and nitrogen fixation of *Anabaena* exposed to Assam crude extract. Toxicity of textile mill effluents to freshwater and estuarine algae was studied by Walsh and Bahner (1980). Walsh and Alexander (1980) used marine algae as a biological tool for evaluating toxicity of industrial wastes. The role of blue-green algae as biological inputs in agriculture has been well documented and substantiated (Singh, 1961; Venkataraman, 1966 and Watanabe and Yamamato, 1970). A healthy aquatic system demands a balanced growth of the algal population. Contamination of the aquatic environment by anthropogenic heavy metals generated by industrial processes, agricultural run-off or domestic activities, and their disastrous impact on taxonomic diversity and phytoplankton production have been established (Leland *et al.*, 1979; and Wong *et al.*, 1979). The toxicity of heavy metals, pesticides, insecticides and other contaminants to algae have been reported by different workers like De Filippis and Pallaghy (1976a), Geike (1977), Prasad and Prasad (1982), Gupta *et al.* (1985), Jardim and Pearson (1985), Rath *et at.* (1983b), Rai *et al.* (1981a), Sahu (1987), Rath (1991), Wong and Chang (1991), Chawla *et al.* (1986), Sahu *et al.* (1986) and Adhikary (1989). Different heavy metals and physical and chemical factors of the environment have been found to interact among each other in their action towards growth of algae (Singh and Pandey, 1981b; Rai *et al.*, 1981b; Stratton, 1985). Industrial discharges occur both as effluents and solid wastes and cause hazardous effect on living organisms (Agarwal and Kumar, 1978). Although the toxic effect of the effluent have been studied on different algae (Palmer, 1983; Adhikary and Sahu, 1985; Tewari *et al.*, 1990; Gaur and Singh, 1990; Shaw, 1987 and Rath, 1991), much work has yet to be done on the effects of solid wastes and its leached chemicals on crop field inhabiting blue-green algae. The solid waste under present study contains a significant quantity of mercury. It has been reported that mercurial compounds were toxic to algae (den Doorendejong, 1965; Sakaguchi *et al.*, 1977), mercuric ion prolonged the lag phase of growth (Kleinen Hammans *et al.*, 1976) and at low concentrations of mercurial compounds stimulatory effects were reported (De Filippis and Pallaghy, 1976a; Rath, 1984; Shaw, 1987; Sahu, 1987 and Rath, 1991).

Nitrogen fixing blue-green algae make a major contribution to the fertility of paddy fields, became apparent following the work of De (1939). The atmospheric nitrogen fixed by the algae is released to the external medium as extra-cellular nitrogenous chemicals. Hence, the extra cellular and the cellular nitrogen content of algae, can be an index of nitrogen fixing ability of the organism. Rippka and Waterbury (1977) suggested that *Westiellopsis* can fix nitrogen equally well under aerobic, micro-aerobic and anaerobic conditions so that they can tolerate the very wide range of oxygen tensions found in rice fields. Therefore, any adverse effect on these organisms, caused by indiscriminate use of heavy metals or by discharge of industrial wastes might influence the total productivity and nitrogen fixing ability of the algae, which helps for increasing the fertility of the paddy fields. Some work has been done on the effect of pesticides, herbicides and insecticides (Da Silva *et al.*, 1975; Kar and Singh,

1978; Wurtsbaugh and Apperson, 1978; Rath *et al.*, 1986b and Sahu, 1987) on the nitrogen fixation capacity of blue-green algae. Shaw *et al.* (1989b) studied the effect of effluent of a chlor-alkali industry on the nitrogen fixation capability of the blue-green alga, *Westiellopsis prolifica*, Janet. The present piece of work was designed to study the effect of the leached chemicals of the solid waste of a chlor-alkali industry containing mercury on the same nitrogen fixing blue-green alga, abundantly available in nearby crop fields, the *Anabaena cylindrica*. The observed luxuriant growth of the alga in paddy fields might be used as a biological agent to detoxify the contaminated environments simultaneously fixing atmospheric nitrogen and acting as a biofertilizer.

Table 6.1: Physico-chemical Analysis of Solid Waste (SW) and Solid Waste Extract (SWE)

Physical Properties	*Solid Waste*	*Solid Waste Extract*
Texture	Clay	–
Colour	Greish white	Pale yellow
Temperature	26° ± 2°C	24° ± 2°C
Specific gravity	2.8	–
Water holding capacity	39% by volume	–
Air content	27% by volume	14% by volume
Chemical Properties	*Solid Waste (mg kg^{-1} dry wt.)*	*Solid Waste Extract (mg l^{-1})*
pH of solid waste	4.6 ± 0.4	4.8 ± 0.2
Phosphate	59.00 ± 10.93	42.6 ± 8.1
Chloride	11614 ± 165.2	18.2 ± 1.4
Calcium	126 ± 18.33	108 ± 26
Magnesium	75.00 ± 10.22	51.24 ± 16.87
Sodium	1157 ± 96.44	5.2 ± 0.8
Potassium	96.0 ± 10.25	122.4 ± 16.6
Total nitrogen	11.6 ± 4.5	0.24 ± 0.06
Mercury	1149 ± 248.5	12.86

(Values are mean of five samples ± standard deviation).

Materials and Methods

Anabaena cylindrica, Lemm. is photo-autotrophic, unbranched, filamentous, heterocystous, blue-green alga belonging to the family Nostocaceae. Allen and Arnon's (1955a) nitrogen free medium with trace elements of Fogg (1949) as modified by Pattnaik (1964) was most suitable for the organism. It was used as the basic culture solution in all the experiments.

Preparation of Solid Waste Extract (SWE)

A concentrated solid waste extract was prepared from the collected waste of the industry which was powdered, after being air dried in the laboratory. To five kilograms of this powdered solid waste, five litres of double distilled water was added and stirred continuously. Stirring and sedimentation continued for 3 months. After that, the extract was taken for experimental study. Estimation of cellular and extra-cellular nitrogen content: It was measured in terms of cellular and extra-cellular Kjeldahal nitrogen, which were estimated by Kjeldahal Nesslerization method of Herbert *et al.* (1971). The algal culture was centrifuged in a refrigerated centrifuge at 20°C and 5000 rpm for 10 minutes. From the supernatant the extra-cellular nitrogen and from the residue cellular nitrogen was extracted by digestion and was estimated. The data were expressed as µg of nitrogen/100 ml algal culture. Residual mercury was estimated by digesting the samples following Wanntorp and Dyfverman (1955) as modified by Sahu (1987) and measuring in a cold vapour atomic absorption spectrophotometer, Mercury Analyser (ECIL).

Results

The control set showed 100 per cent survival. The same data can also be interpreted as 10 per cent survival at 1.53 per cent, 50 per cent survival at 0.82 per cent, 90 per cent survival at 0.46 per cent, and 100 per cent survival at 0.21 per cent was marked. Out of the above concentrations, LC_{00} or PS_{100} as safe MAC value of 0.2 per cent was selected as 'A' and LC_{90} or PS_{10} value of 1.5 per cent was selected as 'B' for conducting future experiments.

In case of the control set, the cellular nitrogen content increased from 6.2 ± 0.8 µg/100 ml culture to 14.1 ± 2.6 µg/100 ml culture within 15 days of exposure. The cellular nitrogen content increased to 25.4 ± 4.8 µg/100 ml culture as recorded on 15th day of recovery. The cellular nitrogen increase showed a positive correlation with the exposure period (Table 6.2). In concentration 'A' (0.2 per cent SWE), significant higher values, when compared to control was recorded. The cellular nitrogen content increased from 6.2 ± 0.8 to 17.6 ± 2.2 µg/100 ml culture within 15th days of exposure. The cellular nitrogen content in concentration A was much higher than the control value at all exposure periods. When exposed alga was transferred to toxicant free medium, significant increase in cellular nitrogen content was recorded, where maximum value of 30.2 ± 3.6 µg/100 ml culture was recorded (Table 6.2). In concentration B (1.5 per cent SWE), the cellular nitrogen content increased insignificantly from 6.2 ± 0.8 to 7.1 ± 1.1 on 3rd day and 6.9 ± 0.9 µg/100 ml culture on 6th day of exposure. These values were less than the respective control values and concentration A set values. After 9th day of exposure, cellular nitrogen content significantly declined to 0.6 ± 0.1 µg/100 ml culture on 15th day of exposure. When the exposed alga was transferred to toxicant free medium, no recovery was marked. In case of concentration A, at all exposure periods, rise in cellular content level was marked, when compared to the control value. On 15th day of exposure, a maximum of 24.8 per cent increase over the control value was marked. In concentration B, significant depletion in cellular nitrogen content was recorded. A maximum of 95.74 per cent decrease over the control value was recorded on 15th day of exposure. In case of concentration A, significant recovery was marked and the 7th and 15th day recovery values were much more than the control value. But in case of concentration B, significant depression and no recovery was recorded. The correlation coefficient analysis between days of exposure and cellular nitrogen content shows the existence of a significant positive correlation in control ($r = 0.979$, $p \leq 0.001$) and in concentration A ($r = 0.969$, $p \leq 0.01$). A significant negative ($r = -0.905$, $p \leq 0.05$) correlation in concentration B was marked.

Table 6.2: Changes in Cellular Nitrogen Content (µg/100 ml culture) of Control and Exposed Blue-green Alga, at Different Exposure and Recovery Periods, and at Different Concentrations of the Toxicant. Data are mean of 5 samples ± standard deviation.

Concentration of the	*Exposure in Days*						*Recovery in Days*	
Toxicant (%) (v/v)	*0*	*3*	*6*	*9*	*12*	*15*	*7*	*15*
Control	6.2 ± 0.8	7.8 ± 1.1	8.5 ± 1.6	9.9 ± 0.6	11.2 ± 1.9	14.1 ± 2.6	18.6 ± 3.4	25.4 ± 4.8
A (0.2%)	6.2 ± 0.8	9.1 ± 0.4	9.8 ± 1.1	10.8 ± 0.7	13.9 ± 1.2	17.6 ± 2.2	24.2 ± 4.2	30.2 ± 3.6
B (1.5%)	6.2 ± 0.8	7.1 ± 1.1	6.9 ± 0.9	4.2 ± 0.4	2.1 ± 0.2	0.6 ± 0.1	0.8 ± 0.2	0.4 ± 0.1

The extra-cellular nitrogen was almost zero in the inoculation day, as the selected nutrient media is free from nitrogen which is more suitable for the growth of heterocystous, nitrogen fixing blue-green alga. The exuded extra-cellular nitrogen content in the control set showed an increasing trend, with the increase in exposure period. The value increased to 16.3 ± 2.4 on 3rd day, 31.6 ± 8.2 on 6th day, 54.3 ± 3.5 on 9th day, 86.7 ± 11.3 on 12th day and 122.7 ± 18.9 µg/100 ml culture on 15th day of exposure. The extra-cellular nitrogen content in the recovery flask, increased to 218.9 ± 16.8 µg/100 ml culture on 15th day of recovery. This value is the total amount of extra-cellular nitrogen fixed by the blue-green alga, which has been exceeded to the medium. In concentration A (0.2 per cent SWE), at all exposure periods, the extra-cellular exudate nitrogen content was more than the control value. The value increased from 22.8 ± 3.8 to 156.8 ± 18.8 µg/100 ml culture on 15th day of exposure and the value reached maximum to 254.2 ± 11.6 µg/100 ml culture on 15th day of recovery (Table 6.3). In case of concentration B (1.5 per cent SWE), an insignificant amount of nitrogen was fixed on 3rd day 10.8 ± 1.4, on 6th day 12.6 ± 2.2 and on 9th day 8.2 ± 1.1 µg/100 ml culture. These values were much less when compared to the control values. With the increase in exposure period, the extra-cellular nitrogen availability in the flask reduced and on 15th day of exposure, no extra-cellular nitrogen was recorded in the culture flask (Table 6.3). When the exposed alga was transferred to toxicant free medium for recovery, no recovery was marked (Table 6.3). In concentration A, higher fixation of nitrogen was marked on 3rd day. At all exposure periods, higher values recorded. But in the early phases, up to 9th day of exposure, maximum fixation was recorded. On 3rd day 38.18 per cent, 6th day 29.43 per cent, on 9th day 28.17 per cent, on 12th day 16.95 per cent and on 15th day 27.79 per cent increase over the control value was recorded. In case of concentration B, the extra-cellular nitrogen content decreased and the per cent decrease increased with the increase in exposure period. Hundred per cent inhibition was recorded on 15 day of exposure. When the exposed alga was transferred to toxicant free medium, no recovery was altogether marked. In concentration A, higher values and significant recovery was recorded. The obtained values were much more than the control values. But in concentration B, no recovery was marked. The correlation coefficient analysis between days of exposure and extra-cellular nitrogen content indicated the existence of a positive correlation in control ($r = 0.984$, $p \leq 0.001$) and in concentration A ($r = 0.980$, $p \leq 0.001$). A non-significant negative correlation ($r = -0.299$, p = NS) in concentration B was marked.

Table 6.3: Changes in Extra-cellular Nitrogen Content (µg/100 ml culture) of Control and Exposed Blue-green Alga at Different Exposure and Recovery Periods, and at Different Concentrations of the Toxicant. Data are mean of 5 samples ± standard deviation (NT–Not traceable).

Concentration of the Toxicant (%) (v/v)	*Exposure in Days*						*Recovery in Days*	
	0	*3*	*6*	*9*	*12*	*15*	*7*	*15*
Control	NT	16.5±2.4	31.6±8.2	54.3±3.5	86.7±11.3	122.7±18.9	194.3±14.2	218.9±16.8
A (0.2%)	NT	22.8±3.8	40.9±9.5	69.6±11.5	101.4±18.2	156.8±18.8	204.4±28.2	254.2±11.6
B (1.5%)	NT	10.8±1.4	12.6±2.2	8.2±1.1	1.4±0.2	NT	NT	0.8±0.1

Table 6.4 shows the residual mercury accumulation in SWE exposed blue-green alga at different exposure period and recovery period. In concentration A (0.2 per cent SWE), the exposed alga could accumulate 0.682 ± 0.065 µg of Hg/100 ml culture within 3 days of exposure. With the increase in exposure period, the residual mercury accumulation increased showing a positive correlation. The alga could accumulate 1.324 ± 0.065 µg of Hg/100 ml culture within 15th days of exposure. When the exposed alga was transferred to toxicant free medium, on 7th day, 1.106 ± 0.066 µg of Hg/100 ml culture was noted and on 15th day 0.864 ± 0.071 µg of Hg/100 ml culture was recorded. Within a

period of 15 days during recovery period, 0.46 µg of Hg/100 ml culture was excreted from the exposed alga. Within 7 days, 0.218 µg of Hg/100 ml culture excreted from the exposed alga. In concentration B (1.5 per cent SWE), the exposed alga could accumulate 0.914 ± 0.084 µg of Hg/100 ml culture within 3 days of exposure and the residual accumulation of mercury increased significantly and a maximum of 1.336 ± 0.114 µg of Hg/100 ml culture was recorded on 15th day of exposure. When the exposed alga, was transferred to toxicant free medium, on 7th day 1.208 ± 0.092 µg of Hg/100 ml culture was recorded and on 15th day 1.106 µg of Hg/100 ml culture was recorded. When the exposed alga was transferred to toxicant free medium, within 7th day, 0.128 µg of Hg/100 ml culture has been excreted and within 15 days of recovery 0.23 µg of Hg/100 ml culture was excreted. A maximum of 16.47 per cent and 34.74 per cent residual mercury got excreted during recovery period in concentration and 9.58 per cent of residual mercury and 17.22 per cent residual mercury got excreted from the exposed alga in concentration B set.

Table 6.4: Showing the Mercury Content in the Medium, Accumulation in the Alga after 15 days of Exposure, Amount of Volatilised Mercury within 15 days of Exposure (µg/100ml Culture), Total Mercury Estimated Budget and Unseen/Non-recordable Mercury in the Exposed systems. (Sensitivity–88.65 per cent, Accuracy–99.8 per cent).

Concentration of SWE in (%) (V/V)	*Hg Content in the Medium*	*Residual Hg in the BGA After 15 Days of Exposure*	*Hg in the Medium After 15 Days of Exposure*	*Volatised Hg within 15 Days of Exposure*	*Total Hg Removal from the Medium*	*Total Hg Recorded in the Experiment*
	A	B	C	D	E (B+D)	F (C+E)
A (0.2%)	3.065± 0.002	1.324	0.005	1.732	3.056	3.061
B (1.5%)	6.122 ± 0.041	1.336	0.165	4.702	6.038	6.203

Discussion

Algae, the most important primary producers of the aquatic environments and crop fields have received, least attention. Very few references are available particularly on the toxicity effects and physiological changes induced by heavy metals on algae. The review made by Whitton (1970), Gadd and Griffiths (1978) and Sorentino (1979) on impact and effect of heavy metals on algae added a lot of information to the literature of algal toxicology. Algae have been shown to concentrate heavy metals to a larger extent (Mclean and Jones, 1975; Silverberg, 1975; Jannett and Wixson, 1975; Trollope and Evans, 1976; and Say *et al.*, I and II, 1977). So more detailed knowledge is needed if we are to understand the interactions between algae and heavy metals. Information's are available pertaining to the toxicity of mercury in the form of metal, mercury based pesticides, industrial wastes containing mercury etc. on fresh water blue-green algae (Shaw, 1987; Sabu, 1987; Rath, 1991; Mohapatra, 1992 and Sahu, 2000). But reports on the toxicity of solid waste of the chlor-alkali industry on the fresh water blue-green algae is scanty.

Agarwal and Kumar (1978) showed decrease in growth of *Chlorella* sp when exposed to mercurial effluent and solid wastes indicating toxic nature of mercury on the organism. A liquid industrial waste may affect the algal growth in any of three ways: stimulation, inhibition and stimulation at lower concentrations but inhibition at higher concentrations (Walsh and Alexander, 1980; Sabu, 1987; and Rath, 1991). The enhancement of growth, heterocyst frequency and nitrogen fixation at lower doses of furadon (0.75 µg/ml of carbofuran) have also been reported (Kar and Singh, 1978). Some suggested uptake and metabolisation of the constituent as the probable mechanism for growth stimulation (O'Brien and Dixon, 1976). Prasad and Prasad (1982) observed stimulation in the algal growth at low concentrations of Cd, Pb and Ni. Neither Cd, Pb and Ni have been reported to be essential micronutrients for algae (O'Kelley, 1968, 1974) nor the pure solution of these heavy metals are expected to contain any growth regulator. Thus, an ideal explanation for stimulation of growth at lower concentrations of the toxicant is yet to be ascertained. The solid waste extract under present investigation contains huge amount of mercury. Compounds based on mercury are toxic to algal organisms (Rath *et al.*, 1983b; Roederev, 1983). Organic forms of Hg is more toxic than inorganic form (Rath, 1984 and Sahu, 1987). Further inorganic forms are active only when these are present in free ionic state (Spencer and Nicholas, 1983; Jardim and Pearson, 1985). Rana and Kumar (1974) and Say *et al.* (1977) reported that the toxicity of zinc to green and blue-green algae could be reduced by supplementing the culture medium with phosphate, calcium and magnesium. Possible formation of complex of heavy metals with calcium and phosphate ions might be responsible for reduction in the toxicity, because of the complex formation they might not be getting an entry into the cell either through membrane transport or by surface adsorption which are the two known mechanisms for uptake of the substances (Rai and Khatonair, 1980; Rai and Kumar, 1980 and Rai *et al.*, 1981b). Further, different heavy meals interact among themselves either to reduce or to increase the toxicity of each other (Singh and Pandey, 1981b).

Misra *et al.* (1985) reported a decrease in pigment content of algae, cultured in solid waste from a chlor-alkali industry, with a crop plant, with the increase in concentration of the waste soil and exposure period which was also significantly correlated with mercury uptake by algae. The interesting findings observed here like stimulation at 0.2 per cent of SWE concentration at all exposure periods and significant drastic depletion at 0.82 per cent SWE concentration at all exposure periods, indicated the nature and effect of the solid waste of a chlor-alkali industry. Rath (1984), Shaw (1987) and Mohapatra (1992) reported the dual behaviour of mercurial compounds on the growth of BGA and confirmed the dichotomous behaviour of the toxicants on living systems (Sahu, 1987 and Rath, 1991). However, this piece of work strongly agrees with the findings of the above authors. Probably the result indicated a new line of thinking, which can become a possibility in case of heterogeneous toxicants, where synergistic and antagonistic effects were expected. Here, it can be presumed that chemicals present other than mercury probably act as a masking agent on mercury, reducing the toxicity in turn, showing variation in the observed data. More work is essential on different live systems to confirm the synergistic and antagonistic characteristic features of the mixture toxicants.

The toxicant used in this investigation is not a single unit toxicant rather a complex heterogeneous chemical, and contains other chemicals along with mercury. Hence, the effect observed in higher concentrations may not only be attributed to mercury, but to other compounds present in the heterogeneous media of solid waste extract. The correlation values will give a clear insight to the mercurial poisoning in the system wider study, which needs further confirmation. Stimulation in nitrogen fixing ability at low concentrations of various toxicants have been reported by several authors. Wurtsbaugh and Apperson (1978) reported an increase in nitrogen fixation with some insecticides. Henriksson and Da Silva (1978) observed a species dependent resistance towards heavy metals with respect to nitrogen fixation. Da Silva *et al.* (1975) studied the effects of different pesticides on eight asymbiotic cyanophyceae species and a blue-green phycobiont of a lichen which has nitrogen fixing capacity and observed two patterns of response: *viz.* (a) in some cases an initial period of depression was succeeded by increased activity in nitrogen fixing ability; and (b) an initial decrease was observed in other cases with a subsequent decrease in nitrogen fixation as a function of time. Shaw *et al.* (1989) reported a decrease in nitrogen fixation capacity of BGA, *W. prolifica*, Janet, when exposed to the effluent of a chlor-alkali industry. Rath (1984) described a decrease in the cellular Kjeldhal nitrogen level with an increase in the concentration of $HgCl_2$. Mishra (1986) and Mishra *et al.* (1985a) reported a decrease in algal nitrogen content with the increase in waste soil concentrations from a chlor-alkali industry and exposure period. The solid waste is loaded with pollutants like Hg, Na^+ and Cl^- in very high concentrations. Stratton *et al.* (1979) stated that mercuric ion was toxic towards *Anabaena inaqualis's* growth, photosynthesis and nitrogenase activity to be the primary site of toxicity (Kamp-Nielson, 1971 and Passow *et al.*, 1961).

In the present study, however, mercury with other chemicals does not appear to play any significant role in its highest concentration applied, to inhibit the nitrogen fixation capacity of the BGA, though, in lower concentrations it showed a stimulatory effect which was correlated with the amount of mercury present in the medium. The increase in the amount of extra-cellular and total nitrogen in lower concentrations of the. toxicant was also related significantly with the amount of mercury present in the medium, whereas the change in "B" concentration showing a negative significant correlation with residual mercury accumulation, indicated that SWE, at higher concentration is toxic. The solid waste is a heterogeneous compound and its extract containing an objectionable amount of mercury is deadly toxic in nature. This toxicity may be due to the presence of mercury or due to the presence of some other substances; the derogatory effect of which on living organisms in the ecosystem cannot be overruled, because this solid waste extract may show antagonistic or synergistic effect in field conditions rather than in laboratory controlled conditions. On the basis of experimental evidences for the existence of various Hg species in the environment (Nriagu, 1979; Brosset, 1981; Slemr *et al.*, 1985; Kim and Fitzgerald, 1988), elemental Hg and methylmercury (dimethyl and monomethyl) are the principal candidates for volatilisation (emission and/or re-emission) into the atmosphere. Total vapour-phase Hg fluxes from the agricultural and forest soils were found to be smaller than those from the surface of a lake (Schroeder *et al.*, 1989). Volatilisation and methylation of Hg appear to occur at the highest rates under aerobic conditions (Hamdy and Noyes, 1975; Olson and Cooper, 1976). Under aerobic conditions the rate of volatilisation by Hg reducing strains has been shown to be reduced, but methylation takes place in a natural way as such (Olson, 1978; Barkey *et al.*, 1979). In the present investigation, it was observed that the alga *Anabaena cylindrica* can reduce the toxicity of the mercury contained solid waste extract by way of volatilisation of elemental mercury from the medium because the alga did not contain methyl mercury (Sahu and Panigrahi, 2002) even though their total residual mercury levels were high (Holm and Cox, 1974). In the lower concentration of mercury in the medium, higher rate of detoxification was marked which strongly indicated the dilution of the toxicant to a particular level, where volatilisation and absorption become faster (Sahu and Panigrahi, 2002). Sahu (2000) reported a similar finding, while working with the liquid effluent of a chlor-alkali industry containing mercury either in the dissolved form, or in the form microsize elemental mercury. The same author reported that *Anabaena cylindrica* could remove a substantial amount of mercury from the effluent medium. The only question, he raised was the concentration of the effluent. He suggested that the effluent should be diluted sufficiently and a base nutrient medium should be provided, where the BGA can grow. Once the BGA grows, at lower concentration, the alga can remove a lot amount of mercury by way of volatilisation. We are of the opinion that the solid waste should be diluted proportionately and the extract is also to be diluted to get a value of 0.2 per cent dilution, so that the alga can grow profusely fix higher amount of atmospheric nitrogen and exude higher amount of extra-cellular nitrogen and simultaneously detoxify the environment by way of removal of mercury from the environment, may be by a way of volatilisation. Zingmark and Miller (1975) reported considerable loss of mercury from culture of dinoflagellates and opined that loss of mercury was probably due to chemical volatilisation, although some might be due to the biological process. We also agree with the remarks of the above authors.

Anabaena cylindrica, Lemm, the heterocystous blue-green alga is more sensitive and less tolerant to the solid waste extract (SWE) of the chlor-alkali industry, when compared to *Westiellopsis prolifica*, Janet, which is less sensitive and more tolerant (Sahu and Panigrahi, 2002). The alga *Anabaena* was exposed to SWE at two different concentrations *i.e.* 0.2 per cent and 1.5 per cent many interesting results were obtained. The alga could accumulate maximwn residual mercury within 1.5 days of exposure. A maximum of 1.324 ± 0.065 and 1.336 ± 0.114 µg of Hg/100 ml culture was recorded at 0.2 per cent (A) and 1.5 per cent (B) SWE concentration. When the exposed alga was transferred to toxicant free medium for recovery studies only 24.79 per cent and 17.22 per cent of residual mercury was excreted from their body leaving 0.864 ± 0.011 and 1.106 ± 0.082 µg of Hg/100 ml culture in concentration A and B, respectively.

The only hint at one point that low doses of mercury in the medium can stimulate growth and nitrogen fixation. We do not agree with the concluding remarks given by Mohapatra (1992) that SWE contains other chemicals and ions, in addition to mercury, which showed a combined effect and increased toxicity. At 0.2 per cent SWE, the extract is diluted to a greater extent, where these ions can never play a crucial role to increase toxicity. We agree with the findings of Rath (1984), Shaw (1987), Sahu (1987), Rath (1991) and strongly disagree with the findings and remarks made by Mohapatra (1992). Probably, the same author mainly depended upon theoretical presumptions rather than experimental facts. The residual mercury played the key role for the reduction in growth parameters including nitrogen fixation at higher concentrations in the medium and in the alga. The observed trend and results in concentration A contradicts the trend and results observed in concentration B. In both concentrations A and B, the same solid waste extract was used. But the only difference between two sets was the dose/concentration of the SWE and the main factor was mercury. Hence, it can be concluded that mercury at low concentrations acts as a growth stimulant and fixes higher amount of atmospheric nitrogen and the same metal mercury at higher concentration is growth inhibitor and reduces atmospheric nitrogen fixation. This indicated again that mercury played a dual role depending on the concentration of the chemical.

These industrial chemicals have brought immense benefit to the society, but they have also brought new dangers largely through the wastes generated in their manufacture. One of the most worrying features of the problem is that very little is known about the long term consequences of exposure to the chemicals. The situations made even more difficult because, once they are in the environment, chemicals spread in a very complex way and may be converted into other substances which have different effects. The industrial pollution has spoiled the water, air and soil. The effluents and wastes have caused havoc for all living organisms. One such microorganism is blue-green algae inhabiting crop fields whose importance as biofertiliser in rice cultivation is undisputed and well documented. A rice field with a healthy growth of BGA would require no input of nitrogenous fertiliser. India being predominantly an agriculture based country and the agriculture is totally based on the nitrogen economy, any effect on the blue-green algae will affect the nitrogen budget of the paddy fields. The hazards of uncontrolled release of certain industrial wastes and by products have been revealed through several incidents of mercury poisoning, the most notorious was that of the "Minamata Disease" in Japan.

A thorough knowledge of basic mechanisms by which toxic agents impair organisms are fundamental to any eco-toxicological approach. The toxicity of the metal was probably detoxified due to the presence of the masking agents in the SWE, or the nutrients present in the medium or the exudates of the blue-green alga. The initial depression of the parameters of the exposed blue-green alga was probably due to the stress or sudden change of the environment, which probably changed slowly due to the modification of the environment either due to formation of complexes in the medium or the chelating effect caused due to interaction or might be due to the non-availability of toxic chemicals, indicating either acclimatization or adaptation of the blue-green alga to the changed medium. The variance ratio test, pertaining to the parameters linked to the nitrogen fixation and from the correlation matrix, it can be hinted that the extra-cellular products of the blue-green alga probably helped to change the nature and toxicity of the solid waste extract. Detoxification mechanisms can involve storage of metals at inactive sites within organisms on a temporary or more permanent basis. Temporary storage is generally by binding of metals to proteins, polysaccharides and amino acids in soft tissues or in body fluids. This particular alga detoxified the medium by the process of volatilisation, which showed that like bacteria, this BGA (Cyanobacteria) has the capacity to volatilise mercury from the medium, the rate of which was higher in lower concentrations of solid waste extract. So it can be suggested that the waste of this chlor-alkali industry should be diluted sufficiently, before its discharge to outside environment either to be used in the crop fields for irrigation purpose or for any other purpose. The solid waste being a heterogeneous medium, containing different soluble and insoluble chemicals and compounds like PO_4^{3-}, Cl^{3-}, Ca^{++}, Mg^{++}, Na^{+}, and K^{+} etc. along with mercury may be showing antagonistic or synergistic effects on target organism the BGA, *Anabaena cylindrical,* Lemm. It was clearly observed that though the alga accumulated a high amount of mercury from the environment, it did not show any inhibitory effects but showed growth regulatory effect at lower concentrations of the toxicant but at higher concentrations of the SWE drastic changes in different parameters of growth was marked. But in a long run, the amount of accumulated mercury will be very high in the BGA, which will bioconcentrate and biomagnify through different trophic levels and will represent a potential hazard ultimately to man. In addition, the alga, *Anabaena cylindrica* can be used as an agent, which can remove mercury from a mercury contaminated area as a detoxifier and simultaneously can act as a biofertiliser.

Acknowledgements

The authors wish to thank the Head, Department of Botany for providing laboratory facilities. The financial support of DOD, Government of India, New Delhii and OSTC, Berhampur University in the form of a project is sincerely acknowledged.

References

Adhikary, S.P. and J. Sahu (1985). *Environ. and Ecol.* 3(4): 580–590.

Agarwal, M. and H.D. Kumar (1978). *Ind. J. Environ. Hlth.*, 20: 141–155.

Allen, M.B. and D.I. Arnon (1955a). *Pl. Physiol.*, Lancaster, 30: 366–372.

Bakir, F., S.F. Damluji, L. Amin-Zaki, M. Murtadha, A. Khalidi, N.Y. Al-Rawi, S.T. Kriti, H.I. Dhahir, T.W. Charkson, J.C. Smith and R.A. Doherty (1973). *Science*, 181: 230–241.

Barkay, T., B. Bolson and R.R. Coldwell (1979). In: *Management and Control of Heavy Metals in the Environment*. CEP Consultants Ltd., Edinburgh. U.K., p. 356–363.

Brosset, C. (1981). *Water, Air and Soil Pollut.*, 16: 253.

Chawla, G., P.M. Viswanathan and S. Devi (1986). *Environ. and Expt. Bot.*, 26: 39–51.

Claesson, A. (1984). *Ecotoxicol. and Environ. Safety*, 8: 80.

Da Silva, E.J., L.E. Henrikson and E. Henrikson (1975). *Arch. Environ. Contam. Toxicol.*, 3: 193–204.

Davies, F.C.W. (1991). *Environ. Geochem. and Hlth.*, 13: 35.

De, P.K. (1939). *Proc. R. Soc. B.*, 127: 121–139.

De Filippis, L.F. and C.K. Pallaghy (1976a). *Z. Pflanzen-physiol. Bd.*, 78: 197–207.

Den Dooren deJong, L.E. (1965). Antonie Vane Leeuwenhoek. *J. Microbiol. Serol.*, 31: 301–313.

ECIL (Electronic Corporation of India Limited) (1981). Analytical methods for determination of mercury with mercury analyser, MA 5800 A.

Fogg, G.E. (1949). *Ann. Bot. N.S.*, 13: 241–259.

Fujuki, M. (1972). 6th Int. Conf. Water Pollut. Res. Paper No. 12.

Gadd, G.M. and A.J. Griffith (1978). *Microbial Ecol.*, 4: 303–317.

Gaur, J.P. and A.K. Singh (1990). *Bull. Environ. Contam. Toxicol.*, 44: 494–500.

Geike, F. (1977). *J. Plant, Diseases and Protection*, 84: 84–94.

Gordson, D.C. Jr. and N.J. Prouse (1973). *Mar. Biol.*, 22: 329–333.

Gupta, S.L., A.K. Kashyap and A.P. Singh (1985). *Indian J. Environ. Hlth.*, 27: 224–229.

Hamdy, M.K. and O.R. Noyes (1975). *Appl. Microbiol.*, 30: 424–432.

Herbert, D., P.J. Phipps and R.E. Strange (1971). Chemical analysis of microbial cells. In: *Methods of Microbiology*, (Eds.) J.R. Norris and D.W. Ribbons. Academic Press, N.Y., London, p. 210–344.

Holm, H.W. and M.F. Cox (1974). In: *Mercury in Aquatic Systems: Methylation, Oxidation-Reduction, and Bioaccumulation*, (Ecological Research Series), U. S. Environmental Protection Agency, (E.P.A.) 660/3–74–021, p. 39.

Hufford, G.L. (1971). The biological response of oil in the marine environment: A review. Background report, Washington, D.C., U.S. Coast Guard Oceanographic Unit, Office of Research and Development, U.S. Coast Guard Headquarters.

Jardim, W.F. and H. W. Pearson (1985). *Microb. Eco.*, 11: 139–148.

Jennett, J.C. and B.C. Wixson (1975): In: *Proc. 30th Purdue Ind. Waste Cong.*, Ann Arbor, Mi: Ann. Arbor Science Publishers, Inc., p. 1173–1180.

Kamp-Nielsen, L. (1971). *Plant Physiology*, 24: 556–561.

Kar, S. and P.K. Singh (1978). *Bull. Environ. Contam. Toxicol.*, 20: 707–714.

Kim, J.P. and W.F. Fitzgerald (1988). *Geophys. Res. Lett.*, 15: 40.

Kleinen Hammans, J.W., L.P.L.M. Rabou and H.Q. Platersen (1976). *Photosynthetica*, 10: 440–445.

Leland, H.V., S.N. Luoma and J.M. Fielden (1979). *J. Wat. Pollut. Contr. Fed.*, 51: 1592–1616.

McLean, R.O. and A.K. Jones (1975). *Freshwater Biol.*, 5: 431–444.

Miller, W.E., S.A. Peterson, J.C. Greene and C.A. Callahan (1985). *J. Environ. Qual.*, 14: 569–574.

Mishra, B.B., D.R. Nanda and B.N. Misra (1985a). *Environ. Pollut.*, Ser. A, 37: 97–104.

Mishra, B.B. (1986). Role of blue-green algae in biodegradation of chlor-alkali industrial wastes (soil) with special reference to rice cultivation. *Ph.D. Thesis*, Berhampur University, Orissa, India.

Mohapatra, A. (1992). Eco-physiology, resistance and ecological implications of a mercurial compound on a blue-green alga. *Ph.D. Thesis*, Berhampur University.

Nriagu, J.O. (Ed.) (1979). *The Biogeochemistry of Mercury in the Environment*, Elsevier/North-Holland Biomedical Press, Amsterdam.

O'Brien, P.Y. and P.S. Dixon (1976). *Br. Phycol. J.*, 11: 115–142.

O'Kelley, J.C. (1968). *Ann. Rev. Plant Physiol.*, 19: 89.

O'Kelley, J.C. (1974). Organic nutrients. In: *Algal Physiology and Biochemistry,* (Ed.) W.D.P. Stewart, Blackwell Scientific Publications, Oxford, p. 610–635.

Olsen, B.H. and R.C. Cooper (1976). Comparison of aerobic and anaerobic methylation of mercuric chloride by San Francisco Bay Sediments. *Water Res.*, 10: 113–116.

Olson, B.H. (1978). In: *Microbial Ecology*, (Eds.) Loutit, M.W. and J.A.R. Miles. Springer-Verlag, Berlin, Heidelberg and New York, p. 416–422.

Palmer, C.M. (1983). *Ann. N.Y. Acad. Sci.*, 108: 389–395.

Paracelsus (1967). In: *Four-treaties of Theophrastus Von Hohenheim called Paracelsus*, (Ed.) H.E. Sigerist. Johns Hopkins Press, Baltimore, 1941.

Passow, H., A. Rothstein and T.W. Clarkson (1961). *Pharmacol. Rev.*, 13: 185–223.

Pattnaik, H. (1964). Studies on nitrogen fixation by *Westiellopsis prolifica*, Janet. *Ph.D. Thesis*, University of London.

Prasad, P.V.D. and P.S.D. Prasad (1982). *Water, Air and Soil Pollut.*, 17: 263–268.

Rai, L.C. and N. Khatoniar (1980). *Indian J. Environ. Hlth.*, 22: 113–123.

Rai. L.C. and A. Kumar (1980). *Microb. Lett.*, 13: 79–84.

Rai, L.C., J.P. Gaur and H.D. Kumar (1981a). *Biol. Rev.*, 56: 99–151.

Rai, L.C., J.P. Gaur and H.D. Kumar (1981b). *Environmental Research*, 25: 250–259.

Ramazzini, B. (1713). De Morbis Artificum (Diseases of Workers). W.C. Right (trans.) Hafner Publishing Company, New York, 1964.

Rana, B.C. and H.D. Kumar (1974). *Phykos.*, 13: 60–66.

Rath, P. (1984). Toxicological effects of pesticides on a blue-green algae, *Westiellopsis prolifica* Janet. *Ph.D. Thesis,* Berhampur University, India.

Rath, P., A.K. Panigrahi and B.N. Misra (1983b). *J. Environ. Biol.*, 4: 103–109.

Rath, P., B.N. Misra and A.K. Panigrahi (1986b). *Microb. Lett.*, 31: 15–20.

Rath, S.C. (1991). Toxicological effects of a mercury contained toxicant on *Anabaena cylindrica*, L. and its ecological implications. *Ph.D. Thesis*, Berhampur University, India.

Rippka, R. and J.B. Waterbury (1977). *FEMS Lett.*, 2: 83–86.

Roederev, G. (1983). *Aquat. Toxicol.*, 3(1): 23–24.

Sahu, A. (1987). Toxicological effects of a pesticide on a blue-green alga: III. Effect of PMA on a blue-green alga, *Westiellopsis prolifica* Janet. and its ecological implications. *Ph.D. Thesis*, Berhampur University, India.

Sahu, A. (2000). Eco-toxicological effects of mercury contained waste of a chlor-alkali industry on a blue-green alga, *Westiellopsis prolifica*, Janet. *D.Sc. Thesis*, Berhampur University, India.

Sahu, A. and A.K. Panigrahi (2002). Eco-toxicological effects of a chlor-alkali industry effluent on a cyanobacterium and its possible detoxification. In: *Algological Research in India*, (Ed.) N. Anand. Bishen Singh Mahendra Pal Singh, Dehra Dun, India, p. 351–430.

Sahu, A., B.P. Shaw, A.K. Panigrahi and B.N. Misra (1986). *Microb. Lett.*, 33: 45–50.

Sakaguchi. T., T. Horikoshi and A. Nakajima (1977). *Agric. Chem. Soc. Jpn. J.*, 5: 497–505.

Say, P.J., B.M. Diaz and B.A. Whitton (1977a). *Freshwater Biol.*, 7: 357–376.

Say, P.J., B.M. Diaz and B.A. Whitton (1977b). *Freshwater Biol.*, 7: 377–384.

Schroeder, W.H., J. Munthe and O. Lindqvist (1989). *Water, Air and Soil Pollut.*, 48. 337–347.

Shaw, B.P. (1987). Eco-physiological studies of industrial effluent of a chlor-alkali factory on biosystems. *Ph.D. Thesis*, Berhampur University, Orissa, India.

Shaw, B.P., A. Sahu and A.K. Panigrahi (1989a). *Bull. Environ. Contam. Toxicol.*, 43: 618–626.

Silverberg, B.A. (1975). *Phycol.*, 14: 265–274.

Singerman, A. (1976). Clinical signs versus biochemical effects for toxic metals. In: *Effects and Dose-Response Relationships of Toxic Metals*, (Ed) G.F. Nordberg. Elsevier Scientific Publishing Company, Amsterdam, Oxford, New York.

Singh, R.N. (1961). *Role of Blue-green Algae in Nitrogen Economy of Indian Agriculture*. ICAR, New Delhi.

Singh, S.P. and A.K. Pandey (1981b). *Acta. Bot. Ind.*, 9: 290–296.

Slemer, F., G. Schuster and W. Seiler (1985). *J. Atmos. Chem.*, 3: 407.

Sorentino, C. (1979). *Phykos.*, 18: 149–161.

Stratton, G.W., A.L. Huber and C.T. Carke (1979). *Appl. Environ. Microbiol.*, 38: 537–543.

Statton, G.W. (1985). *Bull. Environ. Contam. Toxicol.*, 34: 676–683.

Tewari, A., S. Thampan and H.V. Joshi (1990). Effect of chlor-alkali industry effluent on the growth and biochemical composition of two marine macroalgae. *Mar. Pollut. Bull.*, 21(1): 33–38.

Tezuka, T. and K. Tonomura (1976). *J. Biochem.*, Tokio, 80: 79.

Trollope, D.K. and B. Evans (1976). *Environ. Pollut.*, 11: 109–116.

Tsubaki, T. (1.971). *Proceedings of the Symposium on Mercury in Man's Environment*. 15–16th February, 1971, Ottawa, Canada. The Royal Society of Canada, p. 131–136.

Tsubaki, T. and K. Irukayama (Eds.) (1977). *Minamata Disease* (Methyl mercury poisoning in Minamata and Nigata, Japan). Elsevier Scientific Publishing Company, " Kodansha" Ltd., New York.

Venkataraman, G.S. (1966). *Phykos*, 5: 164–174.

Walsh, G.E. and S.V. Alexander (1980). *Water, Air Soil Pollution*, 13: 45–55.

Wanntorp, H. and A. Dyfverman (1955). *Arkiv. for Kemi.*, 9(2): 7.

Watanabe, A. and Y. Yamamato (1970). *Proc. Second Symp. on Nitrogen Fixation and Nitrogen Cycle*, Sendai, Japan, p. 22–28.

Whitton, B.A. (1970). *Phykos*, 9: 116–125.

W.H.O. (World Health Organisation) (1989). *Mercury-Environmental Aspects*, IPCS, Environmental Health Criteria–86.

Wong, P.T.S., G. Burmison and Y.K. Chau (1979). *Bull. Environ. Contam. Toxicol.*, 23: 487–490.

Wong, P.K. and L. Chang (1991). *Environ. Pollut.*, 72: 127–139.

Wurtsbaugh, W.A. and C.S. Anderson (1978). *Bull. Environ. Contam. Toxicol.*, 19: 641–646.

Zingmark, R.C. and T.G. Miller (1975). The effects of mercury on the photosynthesis and growth of estuarine and oceanic phytoplankton. *Belle W. Baruch Libr. Marine Science*, 3: 45–57.

Chapter 7
Use of Modified *Ficus religiosa* Bark for Scavenging Zinc Ions from Industrial Wastewater

M.N. Gourkar, P.U. Meshram** and P.V. Patil****

**Lecturer in Chemistry, Mohota Science College, Nagpur – 440 009*
*** Head of Department in Environmental Science,*
Sevadal Mahila Mahavidhyalaya, Nagpur – 440 009
****Ex. Director and Emeritus Professor, Laxminarayan Institute of Technology, Nagpur*

ABSTRACT

Use of *Ficus religiosa* bark substrate for the removal and recovery of Zinc Ions from industrial wastewater is discussed here. The dried bark powder is contacted with acidified formaldehyde and resin product so obtained, is highly efficient in removing Zn^{2+} ions from the solution. The metal ion removal increases with increasing pH values. It is also observed that more than 84 per cent of Zn^{2+} ions is removal by substrate from solution instantaneously. By using packed columns of the substrate, the metal ion concentration from wastewaters can be reduced to very low levels conforming to the acceptable water quality standards.

Keywords: *Adsorption, Zinc (II), Ficus religiosa (Pimpal).*

Introduction

The treatment of Zinc bearing effluent has been reported by several methods such as ion exchange, evaporation, reverse osmosis and adsorption on activated carbon. The high capital cost and recurring expenses of most of these method do not suit the small scale industries. Studies on the treatment of effluent bearing heavy metal have revealed adsorption to be highly effective, cheap and an easy method among the physico-chemical treatment processes owing to the high cost and difficult in procurement of activated carbon. Effects are being directed towards finding efficient and also low adsorption material like fly ash [Adriano *et al.* (1980)], wood charcoal [Deepak *et al.* (1991)], Bituminous Coal Nagesh *et al.* (1989)], Lignite Coal [Kannan *et al.* (1991)], Rice husk [Shrinivasa *et al.* (1998)] and saw dust [Flynn *et al.* (1980)].

The present study is undertaken with a view to assess the suitability of.natural available adsorbents for the removal of Zn (II) from wastewater. The study also includes the scope positive effect of both batch study and column study.

Material and Methods

Dried bark of the plant *Ficus religiosa* was procured locally and crushed to small size, soaked overnight in demineralized water, wash several times and dried at room temperature. It was further powder and sieved to get suitable particular size.

The dried powder was further treated with 20 parts of 0.2N H_2SO_4 and 5 parts of 39 per cent formaldehyde (HCHO). It was kept in a water bath at 50°C for 6 hours and occasionally stirred. The powder was washed with double distilled water for several times and dried at 60°C. The dried powder was used in the experiment.

All reagents used in the experiment were of analytical grade. All glass-wares used were leached with 10 per cent nitric acid, washed with distilled water and dried in oven. The initial stock Zn (II) solution was prepared by dissolving oven dried zinc acetate in concentrated nitric acid and distilled water. Standard solution of Zn (II) were prepared by taking different aliquonts from stock solution with subsequent diluted with distilled water.

Atomic Absorption Spectrophotometer (AAS, GBC 932 dobule beam) was used for the estimation of Zinc. The instrument was operated at 357–9 nm. wave length, 0.2 slit width (spectral band pass) and 6.0 m A lamp current with nitrous oxide acetylene flame. For pH adjustment, pH meter (Model No. 111 E) used and standardized by using different pH buffers.

Experimental

Batch experiments were performed for the optimization of pH, contact time, dosages, temperature, anion, light metal ions. The experiments were carried out at room temperature.

Result and Discussion

Effect of pH

The results of the experiments conducted to evaluate the ability of *Ficus religiosa* bark substrate to adsorb zinc from aqueous solution at different pH are presented in Figure 7.1. It is observed from the figure that the maximum adsorption efficiency of *Ficus religiosa* was found to be 10, which is evident from the studies carried out earlier by Pawan Kumar *et al.* (1979).

Effect of Contact time

It was found that 78 per cent of the Zn^{2+} ions removal from the solution occurring with in 10 minutes, showing a very fast metal ions removal on the substrate. The Zn^{2+} ions removal from solution recorded a value of 85 per cent after a contact time of 1 hours and the value remained same even after a contact time of 24 hours (Figure 7.2).

Effect of Dosages

The effect of bark dosages for the adsorption of Zinc was studied and result are presented in Figure 7.3. It was observed from the figure that the removal increases with in dosages at them optimum

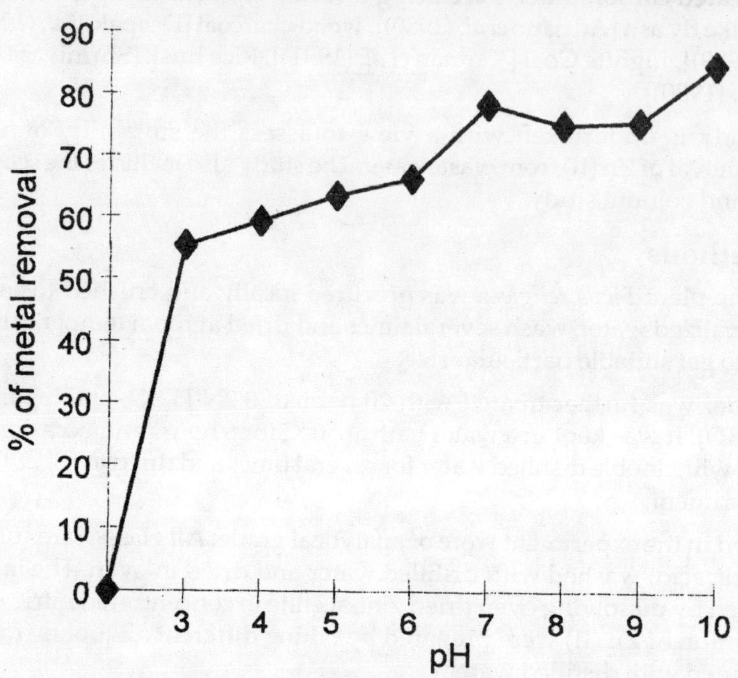

Figure 7.1: Effect of pH for the Adsorption of Zinc on *Ficus religiosa* Bark Substrate

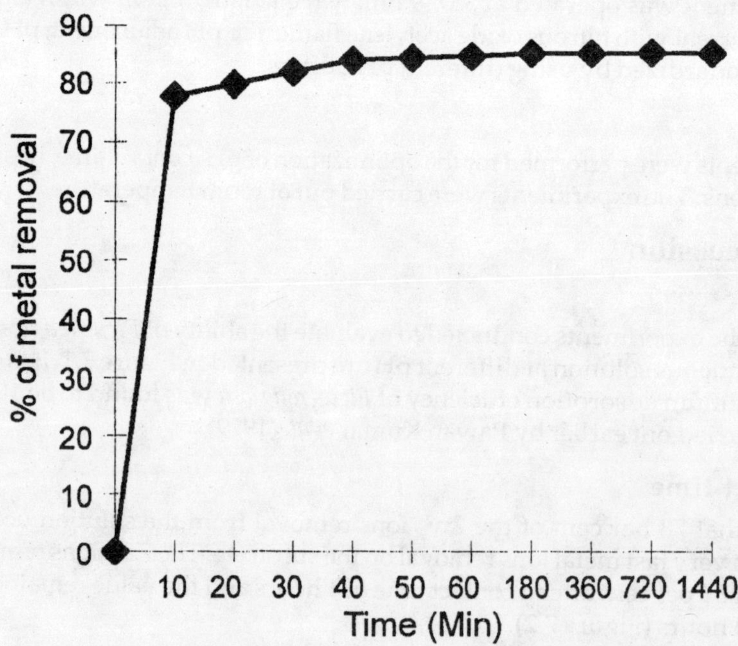

Figure 7.2: Effect of Time for the Adsorption of Zinc on *Ficus religiosa* Bark Substrate

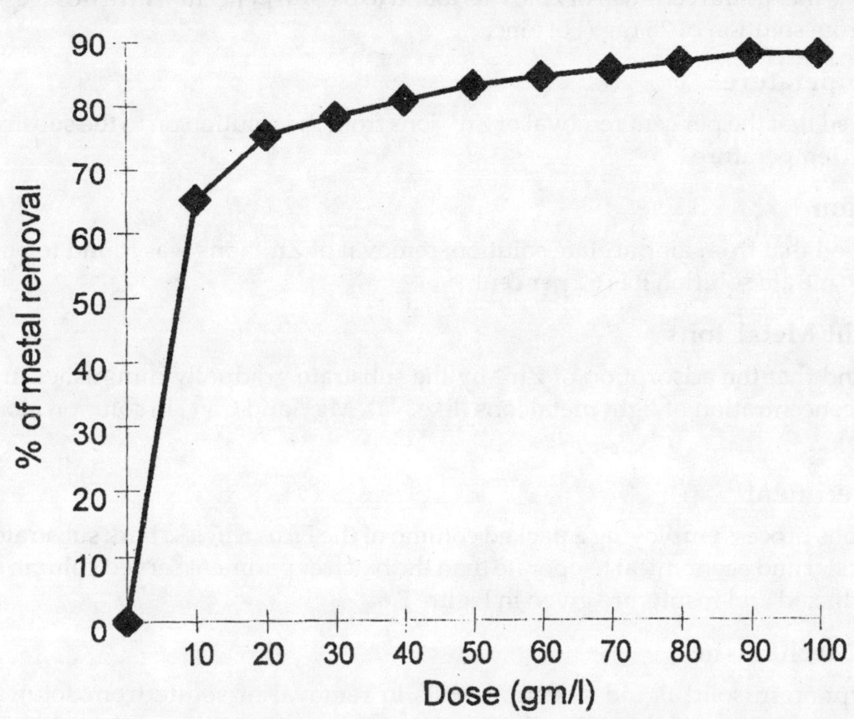

Figure 7.3: Effect of Dosage for the Adsorption of Zinc on *Ficus religiosa* Bark Substrate

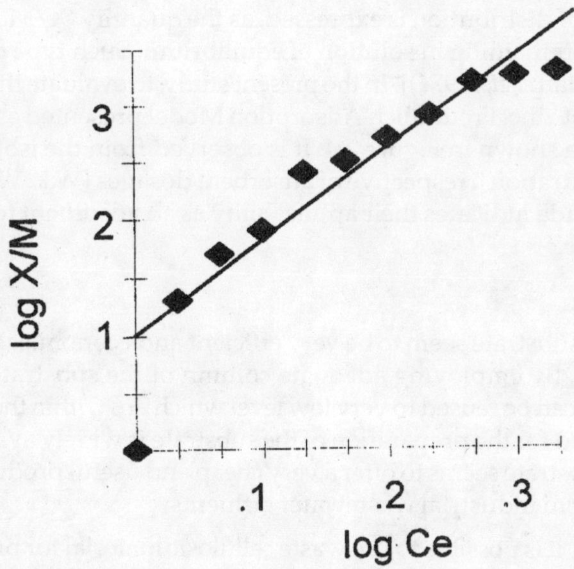

Figure 7.4: Freundlich Adsorption Isotherm for Zinc by Using *Ficus religiosa* Bark Substrate

pH and time. The maximum removal of zinc was found to be 87.8 per cent with a dosages of 90 gms of *Ficus religiosa* from solution of 25 mg/l of zinc.

Effect of Temperature

It is observed that the per cent removal of Zn^{2+} ions from the solution on to the substrate decrease with increase in temperature.

Effect of Anion

It is observed that from zinc acetate solution, removal of Zn^{2+} ions was found to be 82 per cent, while from zinc nitrate solution it is 62 per cent.

Effect of Light Metal Ions

It was found that the adsorption of Zn^{2+} by the substrate gradually diminishes in presence of increasing the concentration of light metal ions (like Na^{+}, Mg^{2+} and Ca^{2+}) in solution along with Zn^{2+} ions (Figure 7.5).

Column Experiment

A continuous process employing a packed column of the *Ficus religiosa* bark substrate is expected to be more efficient and economical to operate than the batch experiment served column experiments have been conducted and results are given in Figure 7.6.

Adsorption Mechanism

The adsorption in solid liquid system, results in removal of solute from solution and their concentration of the surface of the solid, till such a time the concentration of the solute remaining in solution is a dynamic equilibrium with that of the surface. At this equilibrium position, there is a definite distribution of adsorbent where as the quantity Ce is the amount of solute between the liquid and solid phases. This type of distribution is expressed, as the quantity 'X/M' is the amount of solute adsorbed /gm of the solute remaining in solution of equilibrium. Such type of expression is term as 'Adsorption Isotherm' [Randall *et al.* (1974)]. In the present study, to evaluate the efficiency of the barks being used for the removal of zinc. Freundlich Adsorption Model presented as $X/M = KC^{1/n}$ was used and the isotherms draws are shown in Figure 7.4. It is observed from the isotherms that adsorption applicable for higher concentration irrespective of adsorbent dosages [W.L. Webber (1972)]. The shift of isotherms towards right side indicates their applicability as an adsorbent for higher concentration of adsorbate (zinc).

Conclusion

The *Ficus religiosa* bark substrate seem to be very efficient and economical for removing Zn^{2+} ions from industrial wastewater. By employing adequate column of the substrate, the residual Zn^{2+} ion concentration in the effluent can be reused to very low level which are within their acceptable discharge limits. The raw materials used for the preparation of the substrate is widely available and inexpensive. Hence *Ficus religiosa* bark substrate seems to offer a very cheap and useful product for effective removal and recovery of Zn^{2+} ions from industrial wastewater effluents.

By using such substrate, it is possible to use waste cellulosic material for preparing a very efficient cation exchange, which is cheaper than the cation exchange resins available in the market. The metal ions can be removed and reused, these by solving the problem of toxic effects in wastewater on living organism. They also helps to solve the water pollution problem.

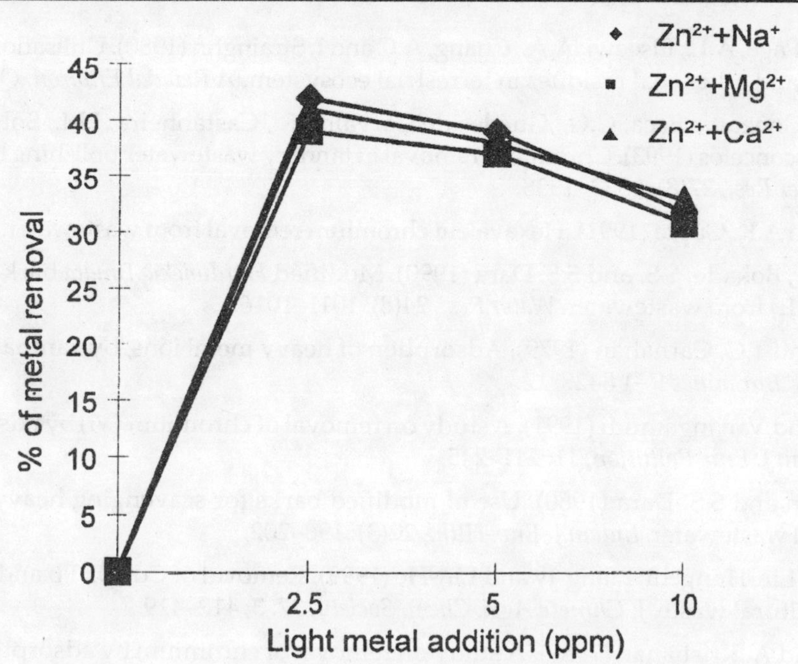

Figure 7.5: Effect of Light Metal Ion Concentration on Zn^{2+} Adsorption Using *Ficus religiosa* Bark Substrate

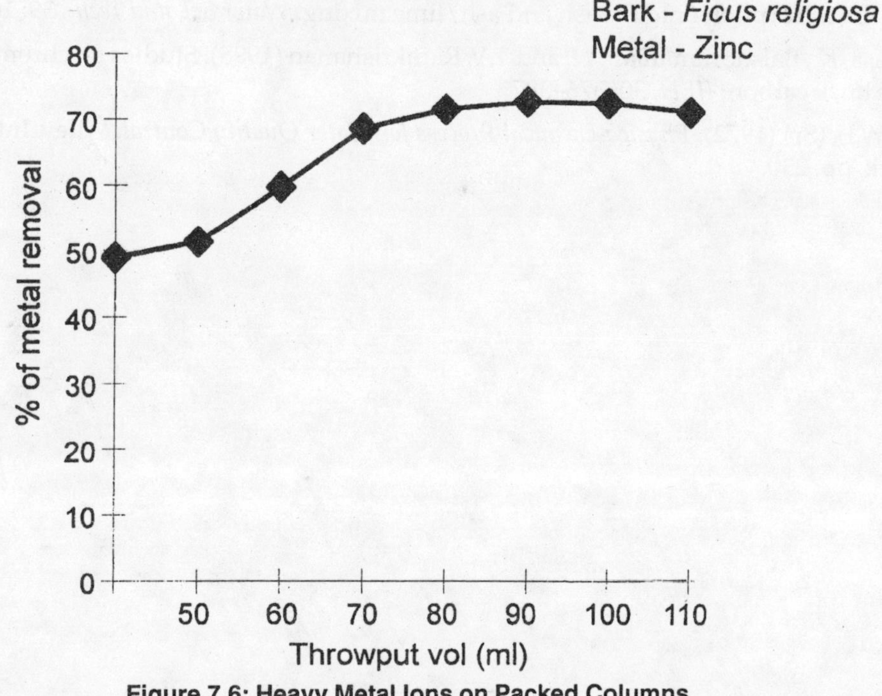

Figure 7.6: Heavy Metal Ions on Packed Columns

References

Adriano, D.C, Page, A.L., Elseewi, A.A., Chang, A.C and I. Strainghn (1980). Utilisation Kumar disposal of fly ash and other coal residues in terrestrial ecosystem. *A Revidal Environ. Qual.*, 9: 333–334.

Alves, M.M., Gonzalez Beca, C.G., Guedes de Carvalho R., Castanheira, J.M., Sol Pereira, Mc. and L.A.T. Vasconcelos (1993). Chromium removal in tannery wastewater polishing by *pinus sylverstris* bark. *Water Res.*, 27(8): 1333–1338.

Deepak, D. and A.K. Gupta (1991). Hexavalent chromium removal from wastewater. *IJEP*, 33: 297–305.

Deshkar, A.M., Bokade, S.S. and S.S. Dara (1990). Modified *Hardwickia Binata* bark for adsorption of mercury (II) from wastewater. *Water Res.*, 24(8): 1011–1016.

Flynn, C.M. and T.G. Carnahan (1979). Adsorption of heavy metal ions, by xanthated saw dust. *Rep Invest U.S. Bur mines* R–1 8427: 12.

Kannam, N. and Vanangamudi (1991). A study on removal of chromium (Vi) by adsorption on lignite coal. *Indian J. Env. Pollution*, 11: 241–245.

Kumar, Pawan and S.S. Dara (1980). Use of modified barks for scavenging heavy metal ions from industrial wastewater. *Indian J. Env. Hlth.*, 22(3): 196–202.

Liang-Jiyuan, Lin-Hongchi, Liang-Jy and Lin-Hc (1999). Removal of Cd, Ni, Pb and Zn from solution by agricultural waste. *J. Chinese Agri. Chem. Society*, 37: 3; 412–419.

Nagesh, N. and A. Krishanan (1989). A study on removal of chromium by adsorption of bituminous coal from synthetic effluents. *IJEN*, 31: 304–308.

Ricou, P., Lecuyer, I., Cloirec, P., Englande (Jr.) (ed) A.J. and N. Syholm (1997). Removal of Cu^{2+}, Zn^{2+} and Pb^{2+} by adsorption onto fly-ash and ash/lime mixing. *Water Sci. and Tech.*, 39: 10–11, 239–247.

Shrinivasa, K., Balsubramanian, N. and T.V. Ramkrishanan (1998). Studies on chromium removal by rice husk carbon. *IJEH*, 30: 376–387.

Weber, W.L. (Sr) (1972). *Physico-chemical Process for Water Quality Control*. Wiley, Inter Science, New York, pp. 230.

Chapter 8

Utilisation of Municipal Wastewater in Aerobic Composting of Solid Organic Waste of Bhubaneswar City

S.P. Panda, C.S.K. Mishra** and D.K. Behera**

*Orissa Pollution Control Board, A/118, Nilakanthanagar, Bhubaneswar – 751 012

**Orissa University of Agriculture and Technology,
College of Basic Sciences, Bhubaneswar – 751 003

ABSTRACT

Approximately 600 m tons of solid organic wastes and a huge bulk of sewage water are generated in Bhubaneswar city daily endangering the urban environment. Solid wastes in windrows with sewage water inoculum decomposed faster compared to cowdung slurry and plain water treated wastes C : N and C : P ratios declined relatively faster in sewage water treated windrows than cowdung slurry and water treated windrows during the experimental period of 45 days. The results indicated that the Bhubaneswar Municipal sewage water would be successfully utilised in solid waste composting.

Keywords: Municipal solid wastes (MSW), Sewage water, Aerobic composting, Windrows, C : N ratio.

Introduction

Enormous increase in the quantum and diversity of waste materials generated in urban areas and their detrimental effects on the environment demand urgent need for adoption of feasible scientific methods for their disposal and utilisation (Williams, 1994). In recent years a number of techniques have been developed to help reduce the quantity of wastes, convert the wastes into organic manures and generate substantial decentralised energy (Chalapathi *et al.*, 2000). Among the most popular means for utilisation of organic wastes, aerobic composting has proved to be most efficient and cost effective (Saha *et al.*, 2000; Hajra *et al.*, 2000). Aerobic composting is the biological decomposition process in which organic matter is decomposed to carbon dioxide and water while stabilised products principally humic substances are synthesised (Brunt *et al.*, 1985).

Literatures are available on survey and study of physico-chemical characteristics of municipal solid wastes and its qualitative improvement (Nanda *et al.*, 2000a; Nanda *et al.*, 2000b and Rao and Radhakrishna, 2001). Bhubaneswar, the capital city of Orissa State has witnessed a rapid growth in the last decade and presently has a population of approximately 6.7 lakh. A huge quantity of solid organic wastes amounting to approximately 600 m tons is produced daily which are disposed of in open dumps. Further, a huge volume of sewage water produced in the city is allowed to accumulate in small lakes which are connected to the rivers Daya and Kuakhai. No systematic study has been done on the municipal solid and liquid wastes of Bhubaneswar city and methods of its utilisation.

Hence the present study was conducted to assess the feasibility of utilising the municipal sewage water in the aerobic composting of solid organic wastes of Bhubaneswar city and its effect on per cent N, P, K and C. C/N ratio and C/P ratio of the decomposing wastes.

Materials and Methods

Municipal solid waste was collected from local urban areas of Bhubaneswar. Sampling points were selected from all the 31 wards of the city. Sampling sites represented all types of localities, such as residential, commercial, slums and markets. The wastes were mixed thoroughly before collection from the sampling sites. Further, the waste samples collected from different wards were mixed thoroughly before the experiment. per cent C, P, N and K in the waste were measured as per Baruah and Barthakur (1997).

The sewage water sample was collected from the main city drain of Bhubaneswar Municipal Corporation which opens into the Vani Vihar lake. Physical and chemical parameters such as temperature, pH, DO, BOD, COD, NO_3, PO_4, chloride, alkalinity, total hardness, turbidity, SO_4, Na, K, TDS, TSS, TPS, TC, fe conductivity of the wastewater were measured following standard procedures (APHA, 1985).

The aerobic composting of the MSW was done by windrow method (Brunt *et al.*, 1985). Three windrows of the size (20 × 3 × 5 ft) were constructed on a concrete floor. Each windrow contained 3 tons of wastes for experimental purpose. The wind row was designed as follows. There were three layers inside each windrow. In first windrow the first layer contained 1 ton of garbage which was soacked with 80 lit. of municipal sewage water. In the second layer 1 ton of garbage was soaked with 80 lts of municipal sewage water and it was spread above the 1st layer. The top or, third layer of 1 ton garbage was also soaked with 80 litres of sewage water and it was spread above the 2nd layer of garbage.

The second windrow (20 × 3 × 5 ft) contained 3 tons of garbage, which was soaked with 80 lt. of cowdung slurry (2 lit water: 1 kg cowdung) in subsequent layers. The third windrow (20 × 3 × 5 ft) of 1 ton garbage was treated as control which was treated with equal volume of plain water. The moisture level in the windrow was maintained at 50–60 per cent and temperature of the windrow was maintained 44°C–50°C which is the optimal temperature for thermophillic bacteria, which accelerate the composting process. The turning of the organic waste materials were done at regular intervals for forced aeration, since oxygen is required for enhanced activity of the aerobic microbes. The windrows were covered with black polythene sheets to accelerate the thermophillic biodegradation. The per cent of N, P, K of all the windrows were measured at an interval of 15 days for a period of 45 days. The sample size was kept constant at 50 for both control and experimental windrows. The experiment was conducted during the months of May and June of the year 2002.

Results and Discussion

The per cent nitrogen, phosphorus, potassium and carbon of solid waste samples have been depicted in Table 8.1. Table 8.2 represents the physico-chemical characteristics of the municipal wastewater. Figure 8.1 shows a gradual increase in percent-nitrogen from the day 1 to 45th day in all the windrows. The rate of increase in per cent nitrogen was highest in sewage water treated windrow compared to cowdung treated and negligible increase in control. This corroborates the earlier observations by Horwath *et al.*, 1995 on grass straw decomposition that the per cent nitrogen increased steadily from the first to 30th day in aerobic composting. The increase in per cent nitrogen may indicate an accumulation of microbial byproducts with decline in microbial carbon. The minimum per cent of nitrogen was observed in the control.

Table 8.1: Changes in Certain Composting Parameters During Decomposition of Solid Organic Wastes of Bhubaneswar City

Days	*N*	*P*	*K*	*C*	*C : N*	*C : P*
			W_1			
0	0.5 ± 0.3	0.1 ± 0.01	0.4 ± 0.01	12 ± 3.1	24 ± 4.61	120 ± 8.45
15	0.53 ± 0.01	0.2 ± 0.01	0.51 ± 0.11	11.90 ± 2.1	22.45 ± 3.91	59.5 ± 3.36
30	0.56 ± 0.02	0.28 ± 0.03	0.58 ± 0.01	11.65 ± 2.4	20.80 ± 4.12	41.60 ± 2.99
45	0.6 ± 0.02	0.3 ± 0.01	0.9 ± 0.02	11.35 ± 2.7	18.91 ± 3.13	37.8 ± 2.35
			W_2			
0	0.5 ± 0.02	0.1 ± 0.01	0.4 ± 0.01	12 ± 2.1	24 ± 4.51	120 ± 8.4
15	0.52 ± 0.01	0.15 ± 0.01	0.5 ± 0.011	11.97 ± 2.2	23.01 ± 3.5	79.8 ± 3.5
30	0.54 ± 0.02	0.2 ± 0.01	0.51 ± 0.01	11.96 ± 2.2	22.14 ± 4.2	59.8 ± 2.99
45	0.55 ± 0.02	0.28 ± 0.02	0.7 ± 0.021	11.94 ± 2.4	21.70 ± 3.9	42.64 ± 2.34
			W_3			
0	0.5 ± 0.01	0.1 ± 0.01	0.4 ± 0.02	12 ± 2.4	24 ± 4.21	120 ± 8.34
15	0.51 ± 0.02	0.11 ± 0.01	0.4 ± 0.02	12 ± 2.41	23.98 ± 3.9	109 ± 7.89
30	0.51 ± 0.03	0.11 ± 0.01	0.41 ± 0.01	12 ± 2.4	23.98 ± 4.1	109 ± 7.89

W_1 = Windrow with sewage water treatment;

W_2 = Windrow with cowdung slurry treatment;

W_3 = Windrow with water treatment.

The per cent carbon of the windrows declined in all the windrows from the 1st to 45th day (Figure 8.2). The highest decline was observed in windrow treated with wastewater in comparison to cowdung treated and control windrows. The decrease in per cent carbon in the windrows might be due to utilisation of carbon by aerobic microbes and subsequent oxidation of carbon-to-carbon dioxide (CO_2). Similar findings have been reported by Horwath *et al.*, 1995 during aerobic decomposition of grass straw.

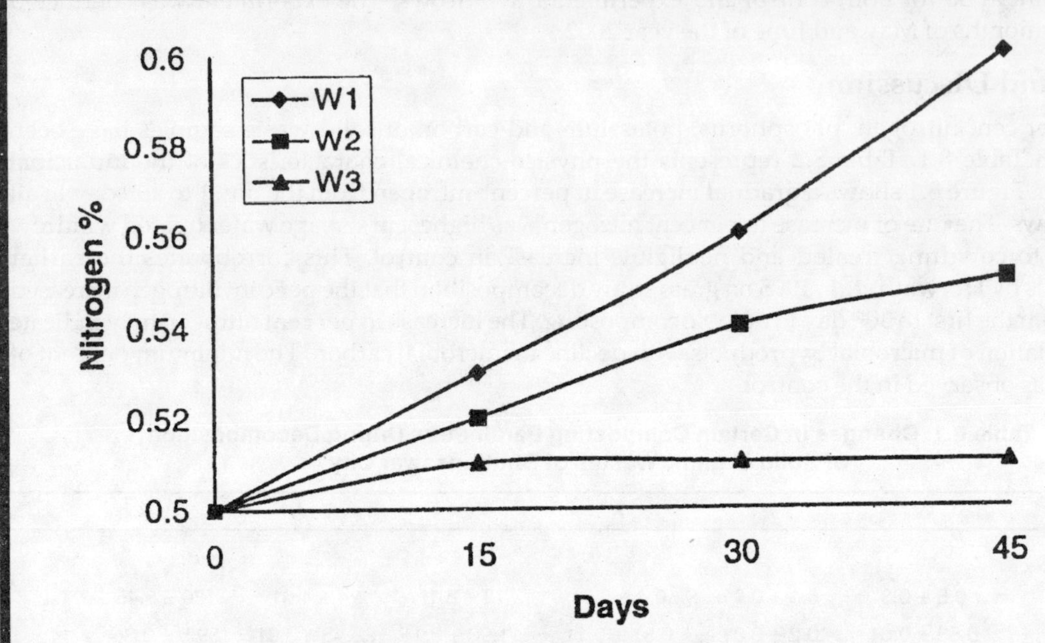

Figure 8.1: Increase in Per cent Nitrogen During 45 Days of Composting

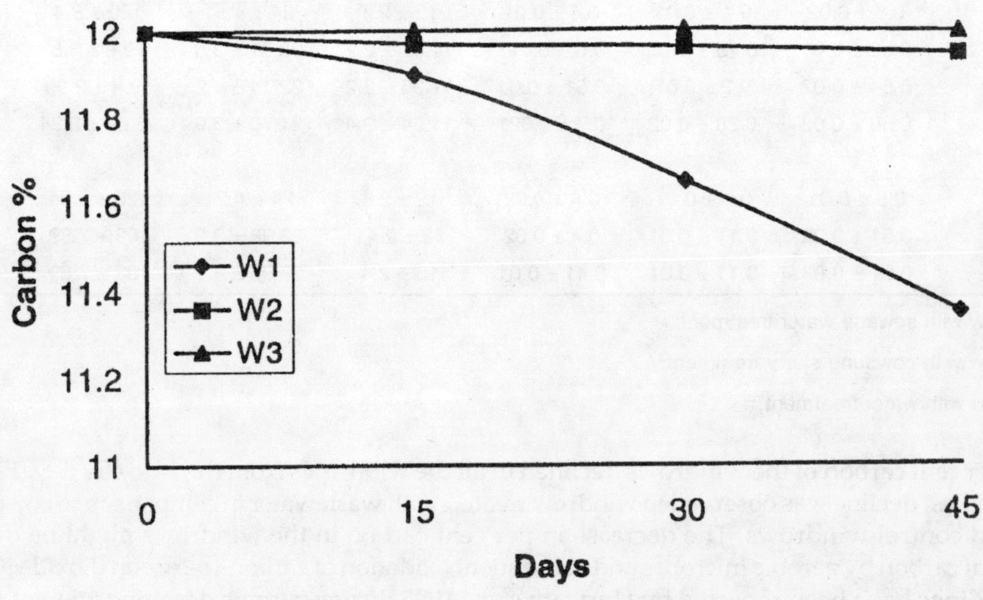

Figure 8.2: Decrease in Per cent Carbon During 45 Days of Composting

Table 8.2: Certain Physico-chemical Parameters of the Sewage Water

Parameters	$\overline{X} \pm$ *S.D.*
Temperature °C	35.4 ± 3.1
pH	7.41 ± 2.1
DO mg/l	4.43 ± 1.1
BOD mg/l	26.6 ± 2.1
COD mg/l	76 ± 4.1
NO_3 mg/l	0.595 ± 0.02
PO_4 mg/l	1.83 ± 0.2
CL mg/l	74.74 ± 4.1
Alkalinity	76 ± 4.1
Total hardness	40 ± 3.2
Ca Hardness	32 ± 2.2
Mg Hardness	8 ± 0.4
Turbidity (NTU)	12.6 ± 1.2
Sulphate mg/l'	12.85 ± 1.1
Na	21 ± 3.1
K	5.33 ± 0.2
TDS mg/l	266.3 ± 15.9
TSS	53.66 ± 3.1
TFS mg/l	317.33 ± 19.4
TC MPn/100 ml	1,60,000 ± 500
FC MPn/100 ml	70,000 ± 200
Conductivity µmho/cm	507 ± 50.8
Ammonical-nitrogen	17.33 ± 02.2
Total fe. mg/l	0.508 ± 0.01

The C : N ratio declined from initial 24 per cent to 23.98 per cent in the control windrow in 45 days (Figure 8.3). In the windrow with cowdung slurry inoculation the C : N ratio declined from 24 per cent to 21.7 per cent. The highest rate of decline in the C : N ratio was observed in the windrow with municipal sewage water inoculation. The C : N ratio here declined from 24 per cent to 18.91 per cent. This indicates relatively faster decomposition process in the sewage water treated wastes in comparison to cowdung slurry treated and plain water treated wastes. This corroborates earlier reports on decomposition of organic wastes (Reinhart and Trainor, 1995; Horwath *et al.*, 1995).

There was negligible increase in the per cent phosphorus (Figure 8.4) and potassium (Figure 8.5) in the control windrow in 45 days and in the cowdung treated windrow the per cent phosphorus increased from 0.1 to 0.28 and per cent potassium increased from 0.4 to 0.7. Similarly in the wastewater treated windrow the per cent phosphorus increased 0.1 to 0.3 and per cent potassium increased from 0.4 to 0.9.

The highest per cent phosphorus was recorded in the wastewater treated windrow on the 45th day (Figure 8.4). Brown *et al.* (1998) have reported during composting of MSW that biological activity

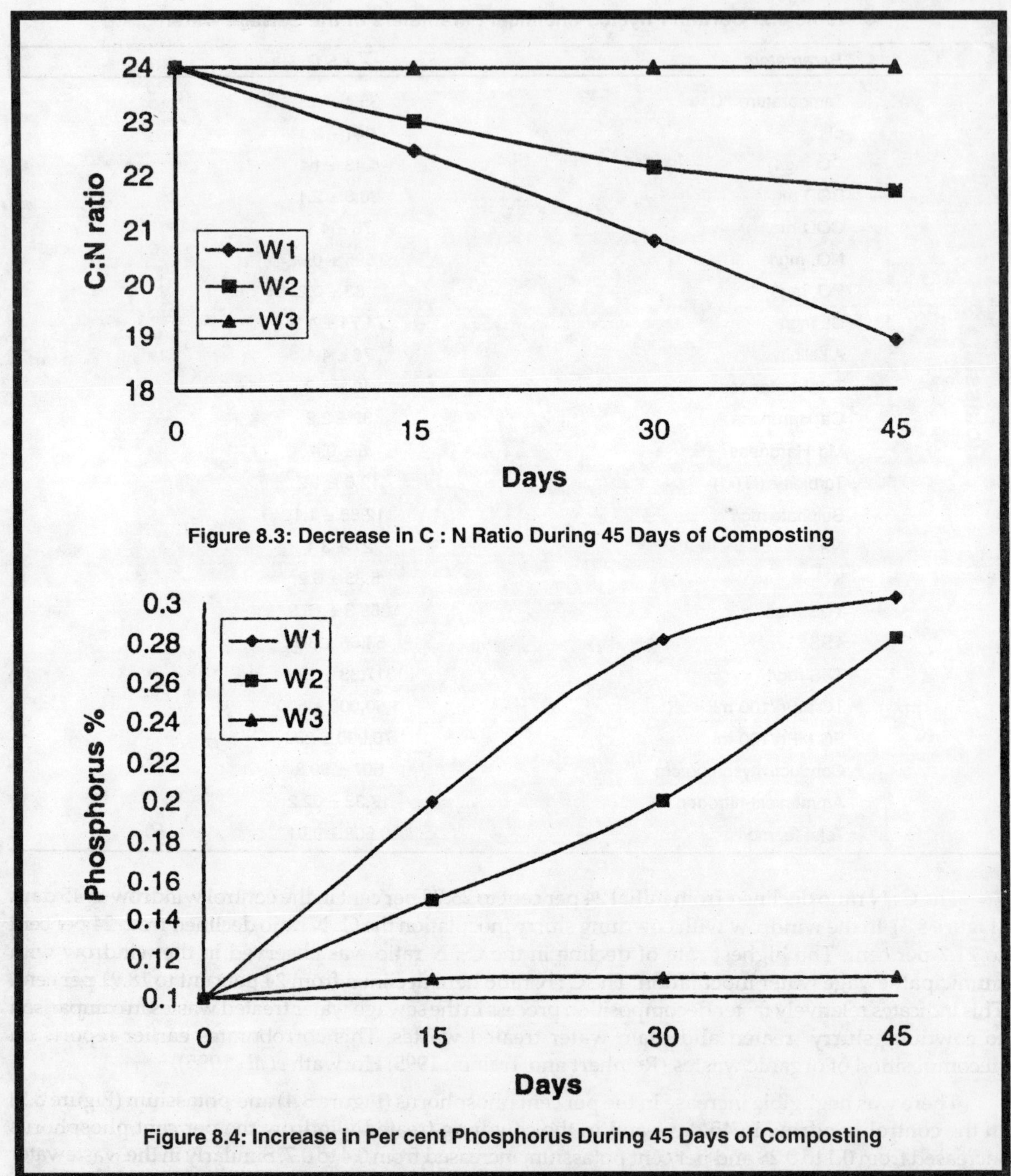

Figure 8.3: Decrease in C : N Ratio During 45 Days of Composting

Figure 8.4: Increase in Per cent Phosphorus During 45 Days of Composting

of microbes is dependent on availability of phosphorus. An increase in the biological activity has been reported with higher percentage of available phosphorus. The C : P ratio in all the windrows declined during 45 days of composting (Figure 8.6). The maximum decline was observed in wastewater treated windrow and minimum in the control. A relatively low C : P ratio has been reported to be congenial for faster decomposition due to enhanced activity of decomposing microbes (Brown *et al.*, 1998). The present finding supports the above observation.

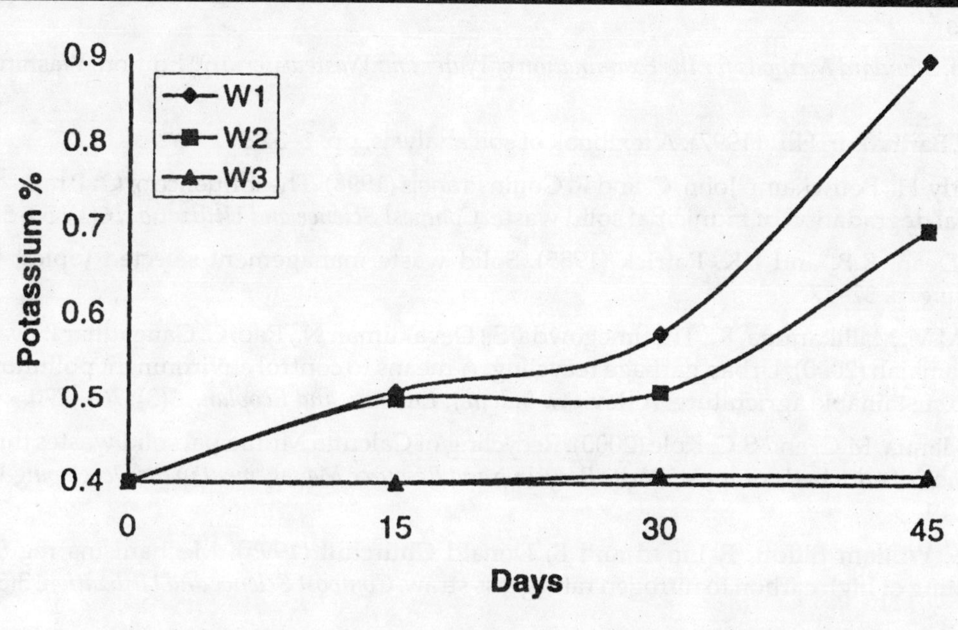

Figure 8.5: Increase in Per cent Potassium During 45 Days of Composting

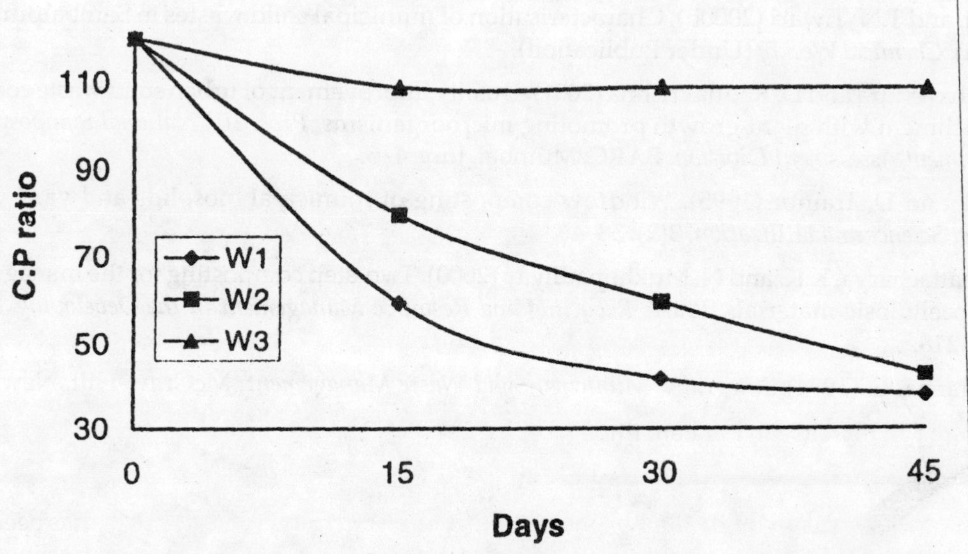

Figure 8.6: Decrease in C : P Ratio During 45 Days of Composting

The above findings clearly indicate that municipality wastewater inoculum accelerates the rate of decomposition processes in municipality solid wastes in comparison to cowdung and water. Hence the municipal wastewater from Bhubaneswar City can be utilized for large scale composting. This obviously will minimise water pollution problems and help in urban solid waste management.

References

APHA (1985). *Standard Methods for the Examination of Water and Wastewater*, 16th Edition. Washington, D.C.

Baruah, T.C., Barthakur, H.P. (1997). A textbook of soil analysis, pp. 1–322.

Brain Kimberly, H., Bouwkamp John, C. and R. Gouin Francis (1998). The influence of C : P ratio on the biological degradation of municipal solid waste. *Compost Science and Utilization*, 6(1): 53–58.

Brunt, L.P., Dean, R.B. and P.K. Patrick (1985). Solid waste management selected topics. *WHO Composting*, p. 37–77.

Chalapathi, M.V., Mallikarjuna, K., Thimmegowda, S., Devakumar, N., Rao, G. Gangadhar Eswar and R. Jayaramaiah (2000). Urban garbage recycling: A means to control environment pollution and a way to sustainable agriculture: A Review. *Indian J. Environ. and Ecoplan.*, 3(3): 761–770.

Hajra, J.N., Manna, M.C. and S.C. Kole (2000). Recycling of Calcutta Municipal solid wastes through production of enriched compost. *Waste Recycling and Resource Management in the Developing World*, p. 197–201.

Horwath, R., William Elliott, F. Lioyd and B. Donald Churchill (1995). Mechanisms regulating composting of high carbon to nitrogen ratio grass straw. *Compost Science and Utilization*, 3(3): 22–30.

Nanda, S.N. and T.N. Tiwari (2000a). Physico-chemical characteristics of the municipal solid wastes of Sambalpur (India). *Indian J. Env. Prot.* (Under Publication).

Nanda, S.N. and T.N. Tiwari (2000b). Characterisation of municipal solid wastes in Sambalpur town (Orissa). *Chemical Weekly* (Under Publication).

Rao, B., Raghavendra and D. Radhakrishna (2001). Quality improvement of urban solid waste compost by enrichment with plant growth promoting microorganisms. *Proc. 10th National Symposium on Environment Assessment Division*, BARC, Mumbai, June 4–6.

Reinhart, R. and D. Trainor (1995). Windrow composting of municipal biosolids and yard waste. *Compost Science and Utilization*, 3(2): 38–46.

Saha, N., Bhattacharya, K.K. and N. Mukhopadhyay (2000). Two-step composting for the management of lignocellulosic materials. *Waste Recycling and Resource Management in the Developing World*, p. 213–218.

Williams Marcia, E. (1994). *Integrated Municipal Solid Waste Management*. McGraw-Hill, New York, Ch. 2.

Chapter 9
Dust Accumulation Studies at Hyderabad

Nirmala Babu Rao, I. Sobha Rani and Amena

Department of Botany, Osmania University College For Women, Koti, Hyderabad

ABSTRACT

Dust accumulation studies show that there was more dust accumulation at Railway Station and MC EME Mess samples than Indira Park, Women's College and Saroor Nagar. A continuous accumulation of dust particles from 1st to 30th day was noticed with an increase in size and weight. There was more accumulation of living aggregates on the slides in summer season than in winter and rainy seasons. Rains cleared the atmosphere and the load of dust aggregates were less in the air. Electrical conductivity was less in rainy season than in winter and summer seasons. This type of studies help to monitor the dust pollution levels of a place.

Keywords: Dust, Pollution, Aero-allergies, Aggregates.

Introduction

Due to dust accumulation air loses its freshness and a continuous encounter with dust causes damage to lungs, bronchus, eyes and skin often beyond repair (Misra, 1978). The dust fall in Chembur suburbs in Bombay was found to be 25 tons/month/km^3 which was close to that in New York or Tokyo (Sunavala, 1978). Calcutta air becomes more loaded with dust in the winter. Madras and Hyderabad seem to be in a better position having particulate matter below 150 μg/m^3 (Seth, 1976). Due to non-availability of information on dust pollution in Hyderabad the present investigation was taken up to estimate the quantum of dust fall, concentration of living and non-living aerosols, size of the particulate matter, with their pH and conductivity at different localities exhibiting Urban, Suburban, Rural, and protected environment in and around Hyderabad.

Materials and Methods

Dust is collected by glass plate method where 10 cm^2 plates were employed for the measurement of the particulate matter. Glass plates were washed, dried and weighed and then were coated with grease and again weighed. Glass plates were exposed over a height of 30 feet at Secunderabad Railway Station, MC EME Mess, Saroor Nagar, Women's College and Indira Park. The glass plates were exposed horizontally by placing them on a tripod stand. Nine glass plates were exposed at each place and the first set of 3 slides were removed after 24 hours, next 3 after 15 days and the last 3 on 30th day of exposure to air in each season. The experiment was similarly performed at all the localities. The slides were brought back to the laboratory and weighed to know the amount of dust accumulated per unit area per unit time (Chaphekar *et al.*, 1980). pH was found with the pH meter and the conductivity with the help of conductivity meter.

Result and Discussion

Weight of dust particles per unit area (g/cm^2) were presented in Table 9.1. Greater amount of dust has accumulated on the slides kept over Secunderabad Railway Station. There was more smoke at the Railway Station resulting in greater dust accumulation over a given period of time per unit area. Lowest quantity of dust was accumulated on Women's College and Indira Park slides, which were regarded as protected areas having a plant cover of several types of trees. Saroor Nagar provided a rural atmosphere with less dust accumulation. Further the barrier of trees in the College Campus and Indira Park, Saroor Nagar intercepted the dust particles carried by air. There was approximately twenty times increase in the dust accumulation values from 1 to 30 days interval in all the localities similar to the observations of Chaphekar *et al.*, 1980. But the quantum increase was not the same as observed by Kayastha and Kumar (1978) and Chaphekar *et al.* (1980) at Kanpur and Bombay conditions respectively. Hyderabad is also more populated and polluted area as far as dust pollution is concerned. In general the accumulation of dust particles in rainy season was less than that was observed in summer and winter season. However, the rate of increase was higher initially followed by a steep fall, except in the rainy season.

Table 9.1: Weight of the Dust Particles per unit area (g/m^2) Collected from the Air Sample from the Height of 30 Feet above the Ground Level

Locality	*Dust Accumulated in*								
	Winter			*Summer*			*Rains*		
	1 day	*15 days*	*30 days*	*1 day*	*15 days*	*30 days*	*1 day*	*15 days*	*30 days*
Railway Station	0.23	3.10	5.10	0.27	3.40	5.40	0.25	0.50	3.06
(Urban) Rate	–	2.87	1.93	–	3.13	1.00	–	0.25	2.56
MC EME Mess	0.13	1.83	2.59	0.15	2.00	2.90	0.00	0.30	1.86
(Sub Urban) Rate	–	1.70	0.76	–	1.85	0.90	–	0.30	1.56
Saroor Nagar	0.08	0.92	1.21	0.10	1.10	1.41	0.00	0.25	1.16
(Rural) Rate	–	0.84	0.29	–	1.00	0.31	–	0.25	0.91
Women's College	0.05	0.52	0.99	0.08	0.80	1.06	0.02	0.08	0.86
(Protected) Rate	–	0.47	0.47	–	0.72	0.26	–	0.06	0.78
Indira Park	0.10	1.00	1.95	0.12	1.22	2.03	0.11	0.18	1.40
(Protected) Rate	–	0.90	0.90	–	1.10	0.81	–	0.07	1.22

Table 9.2: List of Trees and Shrubs from Twin Cities of Hyderabad

Railway Station (Urban)	*MC EME Mess (Sub-urban)*	*Saroor Nagar (Rural)*	*Women's College (Protected)*	*Indira Park (Protected)*
Annona squamosa, Linn.	*Annona squamosa*, Linn.	*Annona squamosa*, Linn.	*Annona squamosa*, Linn.	*Annona squamosa*, Linn.
Citrus limonum, Linn.	*Citrus limonum*, Linn.	*Citrus limonum*, Linn.	*Citrus limonum*, Linn.	*Citrus limonum*, Linn.
Zizyphus jujuba, Lamk.	*Zizyphus jujuba*, Lamk.	*Zizyphus jujuba*, Lamk.	*Zizyphus jujuba*, Lamk.	*Zizyphus jujuba*, Lamk.
Ipomea carnea, Jacq.	*Ipomea carnea*, Jacq.	*Ipomea carnea*, Jacq.	*Ipomea carnea*, Jacq.	*Ipomea carnea*, Jacq.
Lantana camera, Linn.	*Lantana camera*, Linn.	*Lantana camera*, Linn.	*Lantana camera*, Linn.	*Lantana camera*, Linn.
Cassia occidentalis, Linn.	*Cassia occidentalis*, Linn.	*Cassia occidentalis*, Linn.	*Cassia occidentalis*, Linn.	*Cassia occidentalis*, Linn.
Hyptis, Jack.	*Hyptis*, Jack.	*Hyptis*, Jack.	*Hyptis*, Jack.	*Hyptis*, Jack.
Bauhinia variegata, Linn.	*Bauhinia variegata*, Linn.	*Bauhinia variegata*, Linn.	*Bauhinia variegata*, Linn.	*Bauhinia variegata*, Linn.
Tamarindus indica, Linn.	*Tamarindus indica*, Linn.	*Tamarindus indica*, Linn.	*Tamarindus indica*, Linn.	*Tamarindus indica*, Linn.
Albizzia lebbek, Benth.	*Albizzia lebbek*, Benth.	*Albizzia lebbek*, Benth.	*Albizzia lebbek*, Benth.	*Albizzia lebbek*, Benth.
Delonix iegia, Ref.	*Delonix iegia*, Ref.	*Delonix iegia*, Ref.	*Delonix ieqia*, Ref.	*Delonix iegia*, Ref.
Saraca indica, Linn.	*Saraca indica*, Linn.	*Saraca indica*, Linn.	*Saraca indica*, Linn.	*Saraca indica*, Linn.
Ficus religiosa, Linn.	*Ficus religiosa*, Linn.	*Ficus religiosa*, Linn.	*Ficus religiosa*, Linn.	*Ficus religiosa*, Linn.
Syzigium jambolanum, D.C.	*Syzigium jambolanum*, D.C.	*Syzigium jambolanum*, D.C.	*Syzigium jambolanum*, D.C.	*Syzigium jambolanum*, D.C.
Mangifera indica, Linn.	*Mangifera indica*, Linn.	*Mangifera indica*, Linn.	*Mangifera indica*, Linn.	*Mangifera indica*, Linn.
Melia azedarach, Linn.	*Melia azedarach*, Linn.	*Melia azedarach*, Linn.	*Melia azedarach*, Linn.	*Melia azedarach*, Linn.
	Calotropis gigantea, R.B.r.	*Calotropis gigantea*, R.B.r.	*Calotropis gigantea*, R.B.r.	*Calotropis gigantea*, R.B.r.
	Hibiscus rosa sinensis, Linn.	*Hibiscus rosa sinensis*, Linn.	*Hibiscus rosa sinensis*, Linn.	*Hibiscus rosa sinensis*, Linn.
	Hibiscus mutabilis, Linn.	*Hibiscus mutabilis*, Linn.	*Hibiscus mutabilis*, Linn.	*Hibiscus mutabilis*, Linn.
	Murraya koenogii, Spr.	*Murraya koenogii*, Spr.	*Murraya koenogii*, Spr.	*Murraya koenogii*, Spr.
	Pongamia glabra, Vent.	*Pongamia glabra*, Vent.	*Pongamia glabra*, Vent.	*Pongamia glabra*, Vent.
	Ficus benghlensis, Linn.	*Ficus benghlensis*, Linn.	*Ficus benghlensis*, Linn.	*Ficus benghlensis*, Linn.
	Artocarpus integrifolia, Linn.	*Artocarpus integrifolia*, Linn.	*Artocarpus integrifolia*, Linn.	*Artocarpus integrifolia*, Linn.
	Cocos nucifera, Linn.	*Cocos nucifera*, Linn.	*Cocos nucifera*, Linn.	*Cocos nucifera*, Linn.
	Borassus, Linn.	*Borassus*, Linn.	*Borassus*, Linn.	*Borassus*, Linn.
	Rosa, Linn.	*Rosa*, Linn.	*Rosa*, Linn.	*Rosa*, Linn.
		Nerium odorum, Soland	*Nerium odorum*, Soland	*Nerium odorum*, Soland
			Polyalthia longifolia H k.f &T.	*Polyalthia !ongifolia* H k.f &T.
			Jatropha pandurifolla, Linn.	*Jatropha pandurifolla*, Linn.

Contd...

Table 9.2–Contd...

Railway Station (Urban)	*MC EME Mess (Sub-urban)*	*Saroor Nagar (Rural)*	*Women's College (Protected)*	*Indira Park (Protected)*
			Bombax malabaricum, D.C.	*Bombax malabaricum*, D.C.
			Aegle marmelos, Corr.	*Aegle marmelos*, Corr.
			Caesalpinia	*Caesalpinia*
			Pulcherrina, Sw.	*Pulcherrina*, Sw.
			Ceasalpinia bonducella, Flem.	*Ceasalpinia bonducella*, Flem.
			Cassia fistula, Linn.	*Cassia fistula*, Linn.
			Cassia glauca, Lam.	*Cassia glauca*, Lam.
			Acacia arabica, Wild.	*Acacia arabica*, Wild.
			Eucalyptus globulus, Labill.	*Eucalyptus globulus*, Labill.
			Adhatoda vasica, Ness.	*Adhatoda vasica*, Ness.
			Clerodendron inerme, Gaertn.	*Clerodendron inerme*, Gaertn.
			Minusops, Linn.	*Minusops*, Linn.
			Spathlobus, Hassk.	*Spathlobus*, Hassk.
			Guazuma tomentosa, L.	*Guazuma tomentosa*, L.
			Boerhaavia diffusa, L.	*Boerhaavia diffusa*, L.
			Tinospora cordifolia, L.	*Tinospora cordifolia*, L.
			Thespesia populenea	*Thespesia populenea*
			Sida acuta	*Sida acuta*
			Sida rhombifolia	*Sida rhombifolia*
			Argemone maxicana	*Argemone maxicana*
			Malva sylvestris	*Malva sylvestris*
			Thuja occidentalis	*Thuja occidentalis*
			Arucaria spp.	*Arucaria* spp.
			Pongamia glabra	*Pongamia glabra*
			Santalum album	*Santalum album*
			Millingtonia hortensis	*Millingtonia hortensis*
			Sida acuta	*Sida acuta*
			Corchorus acutangularis	*Corchorus acutangularis*
			Klenohaavia hospata	*Klenohaavia hospata*
			Abutilon indicum	*Abutilon indicum*
			Commelina benghalensis	*Commelina benghalensis*
			Cleome viscose	*Cleome viscose*
			Gynandropsis gynandra	*Gynandropsis gynandra*
			Tribulus terrestris	*Tribulus terrestris*

The list of trees and shrubs in different localities is given in Table 9.2. When compared with Bombay 25 tons/month/km^3 and Madras and Hyderabad had less particulate matter (Seth, 1976). However, a minimum of dust particles (10 tons/month/km^3) in air acts as condensing nuclei for water vapour, which gives rise to rain and this confines to the primary standards of the dust fall (EPA, 1971). However, continuous dust monitoring is to be done to check pollution levels of environment.

Number and the percentage composition of living and non-living matter accumulated during the study period were given in Table 9.3. It increased during 1 to 30 days exposure. Up to 19 to 50 per cent increase in living matter was observed between 15 to 30 days of exposure in summer than in winter and rainy season. This is due to the quantitative reduction in the dust aggregates in the air due to rain. In 30 days samples due to polarity repulsion reduction in the number of particulate matter was observed. The settlement of the particulate matter was higher up to 15 days and declines there after in the slides kept for 30 days.

Table 9.3: Percentage Composition of Living Matter in the Air Dust Samples Collected Over a Height of 30 Feet from Different Localities of Twin Cities of Hyderabad

Localities		*Accumulaiion on Slides Exposed for 1 Day*		*Accumulation on Slides Exposed for 15 Days*		*Accumulation on Slides Exposed for 30 Days*	
		Living	*Non-living*	*Living*	*Non-living*	*Living*	*Non-living*
Railway	W	15.9	94.1	17.2	82.8	17.6	82.4
Station	S	16.9	83.1	19.2	80.8	33.9	66.1
(Urban)	R	5.5	94.5	15.0	85.0	9.7	90.3
MC EME	W	8.1	91.9	17.1	82.9	24.0	76.0
Mess	S	8.0	92.0	19.2	80.8	41.0	59.0
(Sub-urban)	R	–	–	15.2	84.9	5.9	94.1
Saroor	W	2.4	97.6	24.8	75.2	36.5	63.5
Nagar	S	11.1	88.9	55.1	44.9	44.8	55.2
(Rural)	R	–	–	5.8	94.2	2.7	97.3
Women's	W	1.3	98.7	7.3	92.7	6.6	93.7
College	S	9.2	90.8	41.5	58.5	42.3	57.7
(Protected)	R	1.5	98.5	3.3	96.7	3.4	96.6
Indira	W	1.6	98.4	7.6	92.4	19.3	80.8
Park	S	15.2	84.8	48.4	51.6	40.0	60.0
(Protected)	R	1.2	98.8	6.7	93.3	5.9	94.1

W = Winter; S = Summer; R = Rainy.

Size of the particulate matter settled per unit area on the glass plate per unit time were given in the Table 9.4. It was observed that there was no uniform pattern in the number of particles collected on the slides. An increase in the size of the particles was observed in the samples accumulated over a period of 30 days due to aggregation of dust particles resulting in the increase of their size. Increase in the size of the aggregates has made the number compact and less.

Table 9.4: Size in Microns of Aggregate Particles Collected from the Air Samples Over a Height of 30 Feet in Different Localities of Twin Cities of Hyderabad

Localities	Winter			Summer			Rains		
	1 day	15 days	30 days	1 day	15 days	30 days	1 day	15 days	30 days
Railway Station (Urban)	9.7	21.3	48.2	10.7	13.2	49.4	9.1	9.4	80.6
MC EME Mess (Sub-urban)	7.7	17.3	23.3	8.7	9.1	9.8	0.0	25.1	90.6
Saroor Nagar (Rural)	6.6	14.5	20.5	7.45	7.4	22.6	0.0	10.1	69.3
Women's College (Protected)	5.5	9.4	16.7	6.4	6.7	17.8	3.2	6.4	88.2
Indira Park (Protected)	7.2	15.3	30.9	8.2	8.3	33.1	7.6	8.7	19.2

Table 9.5 shows electrical conductivity measured in micro ohms / per cm. Electrical conductivity (EC) was higher in summer and winter than in rainy season. EC was very much reduced in rainy season probably due to variation in the types of particulate matter that settle in various seasons.

Table 9.5: Electrical Conductivity of Dust Pollutants in Different Areas of Twin Cities of Hyderabad Placed 30 Feet Above the Ground

Localities	1 day			5 days			30 days		
	Winter	Summer	Rains	Winter	Summer	Rains	Winter	Summer	Rains
Railway Station (Urban)	0.7 × 10 70	0.75 × 10 75	9.5 × 10 50	1.1 × 10 110	1.25 × 10 125	0.6 × 10 60	1.5 × 10 150	1.6 × 10 160	1.25 × 10 125
MC EME Mess (Sub-urban)	0.15 × 10 15	0.2 × 10 20	0 0	0.5 × 10 50	0.6 × 10 60	0.55 × 10 55	1.0 × 10 100	0.05 × 10 105	0.9 × 10 90
Saroor Nagar (Rural)	0.25 × 10 25	0.35 × 10 35	0 0	0.65 × 10 65	0.75 × 10 75	0.45 × 10 45	0.7 × 10 70	0.8 × 10 80	0.65 × 10 65
Women's College (Protected)	0.1 × 10 10	0.15 × 10 15	0.05 × 10 5	0.4 × 10 40	0.5 × 10 50	0.4 × 10 40	0.5 × 10 50	0.60 × 10 60	0.55 × 10 55
Indira Park (Protected)	0.6 × 10 60	0.65 × 10 65	0.4 × 10 40	1.05 × 10 105	1.15 × 10 115	0.3 × 10 30	1.2 × 10 120	1.25 × 10 125	1.15 × 10 115

Table 9.6: pH of Dust Pollutants in Urban, Sub-urban, Rural and Protected Areas at Different Place in Twin Cities of Hyderabad Placed at 30 Feet Above the Ground

Localities	1 Day			15 Days			30 Days			Average
	Winter	Summer	Rains	Winter	Summer	Rains	Winter	Summer	Rains	
Railway Station (Urban)	7.0	7.5	7.0	7.5	8.5	7.5	9.0	8.5	6.5	7.6
MC EME Mess (Sub-urban)	5.5	6.0	7.0	5.5	6.5	6.5	6.5	7.0	6.0	6.2
Saroor nagar(Rural)	6.0	6.5	7.0	6.0	7.0	6.0	6.5	7.0	6.0	6.3
Women's College (Protected)	6.5	7.0	7.0	7.0.	8.0	7.0	7.5	7.5	7.0	7.1
Indira Park (Protected)	5.5	6.0	7.0	5.5	6.5	7.0	6.5	6.5	6.5	6.3

Table 9.6 represented the data on pH of the dust particles. Slightly acidic range in the air dust samples were observed at and near Indira Park, Saroor Nagar and MC EME Mess due to small scale industries *viz.*, iron melters, hand made paper mills, cement dust etc. At Saroor Nagar which has chemical industries also show acidity of air dust samples. Due to coal and wood burning at the MC EME Mess there was accumulation of smoke and the samples showed acidic pH. Women's College dust environment was neutral.

References

Chaphekar, S.B., Boralkar, D.B. and R.P. Shetye (1980). Effects industrial air pollutants on plants. *Final Reports of the Sponsored Research Project* No. F. 23–170/75 (SR. 11), pp. 83

E.P.A. (1977). National air quality standard. *J. Air Poll. Control Assoc.*, 21: 149.

Kayastha, S.C., and V.K. Kumar (1978). Air pollution problem in Kanpur city. *Indian J. Air Poll. Control*, 1: 38–42.

Misra, R.C. (1978). Hazards of dust pollution. *Indian J. Air Poll. Control*, 1: 32–34.

Seth, G.K. (1976). Know your environment. *Science Reporter*, 13: 7–11.

Sunavala, P.D. (1977). Chemical processing and engineering [Misra, R.C. (1978) Hazards of dust pollution]. *Indian J. Air Poll. Control*, 1: 32–34.

Chapter 10
Maturation Biology and Spawning Ecology of *Schizothorax plagiostomus* (Pisces: Cyprinidae) from River Pinder, Uttaranchal

K.L. Bisht, Anoop K. Dobriyal*#, H.K. Joshi**, P.K. Bahuguna** and H.R. Singh****

**Department of Zoology, Government P.G. College, Kotdwara, Garhwal, Uttaranchal*
***Department of Zoology, H.N.B. Garhwal University Campus, Pauri Garhwal, Uttaranchal – 246 001*
****Department of Zoology, University of Allahabad, U.P.*
#E-mail: anoopkdobriyal@rediffmail.com

ABSTRACT

Schizothorax plagiostomus (Heckel) is an important snowtrout of Garhwal Himalaya, which form a major part of the fishery. The present communication deals with the macroscopic and microscopic observations on the gonads of *S. plagiostomus* to determine its maturation process and ecology of spawning behaviour in the high altitude river Pinder in the Garhwal Himalaya, Uttaranchal. The male fish attains maturity little earlier (33.8 cm) than the female fish (35.2 cm) and spawning is twice a year during April–May and October–November.

Introduction

Schizothoracid is a group of snowtrout species, which show a great diversity in their spawning behaviour in different ecological conditions around the entire Himalayan regions of the country. The breeding bio-ecology of different schizothoracid species have been studied by different workers

(Malhotra, 1971; Raina, 1977; Sunder, 1984 in Jammu; Sehgal *et al.*, 1971 in Himachal Pradesh; Bhatnagar, 1964 in Punjab; Bisht, 1974 in Kumaun; Misra, 1982 and Agarwal *et al.*, 2004 in Garhwal). However, the biology of snowtrouts in the river Pinder has not been studied so far, which was once considered as an abode for the endemic schizothoracids and the exotic rainbow trouts by the Britishers during pre-independence period. The study was undertaken to fill up this lacuna and to provide a platform for further conservation and management of the species in near future.

Materials and Methods

The fish were collected from river Pinder for a period of two years during 1984–86 at a stretch of about twenty kilometers. After taking morphometric measurements, the fish were preserved in 5 per cent formalin for further study. The gonads were removed when they were hardened enough and sex, physical appearance, length and weight of gonads were noted down. The Gonado-Somatic Index (GSI), which is an important macroscopic observation, was calculated as GSI = Weight of gonads/ Weight of body multiplied by 100.

For microscopic observations, ovaries alone were taken into consideration. Three samples of ova were collected from different regions of ovary and were measured by means of an ocular micrometer. At least 100 ova from each ovary were measured. The maturity stages were determined on the basis of rate of ova development throughout the year. The different maturity stages were determined according to the ICES scale (Wood, 1930). The frequency of spawning and the spawning season were studied by tabulation of percentage occurrence of fish in various stages of maturity month-wise and size-wise and also by the Ova diameter frequency polygons.

For the determination of size at first maturity, the percentage occurrence of mature fish (both sex) in various size groups during pre-spawning and spawning time was taken into consideration. 50 per cent level in the maturity was considered as size at first maturity (Dobriyal and Singh, 1987, 89, 93).

Observations

Gonado-Somatic Index

The Gonado-Somatic Index (GSI) was calculated for each fish and finally an average value was calculated for each month. The values for both the sexes are presented separately in Figures 10.1 and 10.2. The peak GSI values were obtained as 12.1 and 13.0 during second fortnight of April and October respectively for the male fish and 15.8 and 16.0 in the first fortnight of April and October respectively for the female fish. The minimum GSI values were observed as 1.0 for males and 1.2 for females in the month of January.

Maturation Biology

The ova diameters of each female fish were recorded month-wise round the year and finally on the basis of growth of ova, the maturity stages were determined. The classified maturity stages are presented in Table 10.1. The stage I–(Immature 1st) of maturity was represented by the ova ranging from 0.1 to 0.3 mm with the peak at 0.1 mm. Stage II–(Immature 2nd) was close to stage I and was represented by the ova ranging from 0.1 to 0.4 mm with two peaks at 0.2 and 0.4 mm. In the III stage of maturity (Maturing 1st, the two peaks were separated at 0.6 and 1.0 mm with an overall range of 0.2 to 1.0 mm. These two peaks were continuous till the seventh stage of maturity. In the IV stage (Maturing 2nd) the ova range was broaden to 0.5 to 1.75 mm. A much faster growth of ova was recorded during V stage of maturity (Mature) when the peaks were shifted to 1.0 and 2.0 mm respectively. In spawning stage (Stage VI), there was the highest range of ova (0.5 to 3.0 mm) with two clear peaks showing

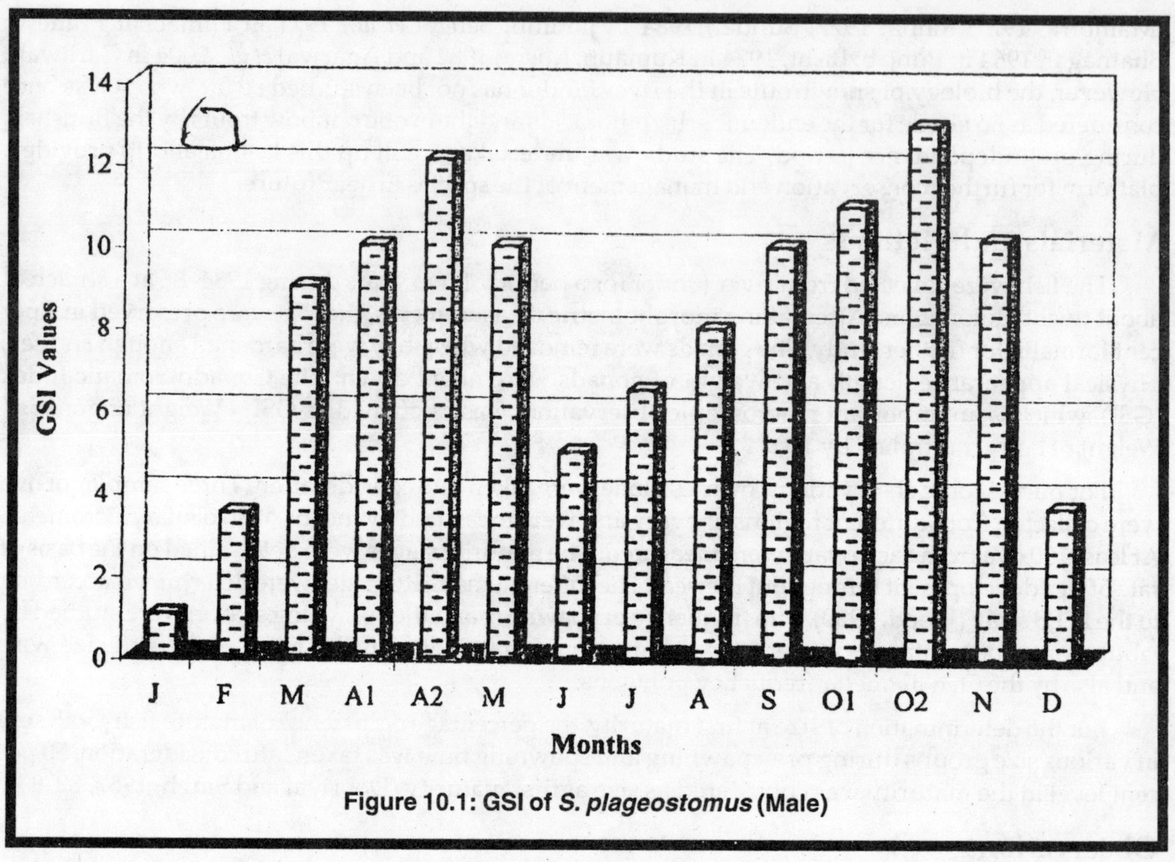

Figure 10.1: GSI of *S. plageostomus* (Male)

double frequency of spawning. In spent stage (Stage VII) two types of eggs were observed, *viz.*, either they were un-spawned mature eggs or were in the second stock of eggs to be laid within few months (Figure 10.3).

Table 10.1: Classification of Maturity Stages of *Schizothorax plageostomus* from River Pinder

Maturity Stages	*Ova Diameter (mm)*	*Peaks*
I–Immature 1st	0.1–0.3	0.1
II–Immature 2nd	0.1–0.4	0.2, 0.4
III–Maturity 1st	0.2–1.0	0.6, 1.0
IV–Maturity 2nd	0.5-1.75	0.5, 1.25
V–Mature	0.5–2.5	1.0, 2.0
VI–Spawning	0.5–3.0	1.5, 2.5
VII–Spent	0.5–3.0	0.5, 1.5

Size at First Maturity

For the determination of size at first maturity the percentage occurrence of male and female fish in

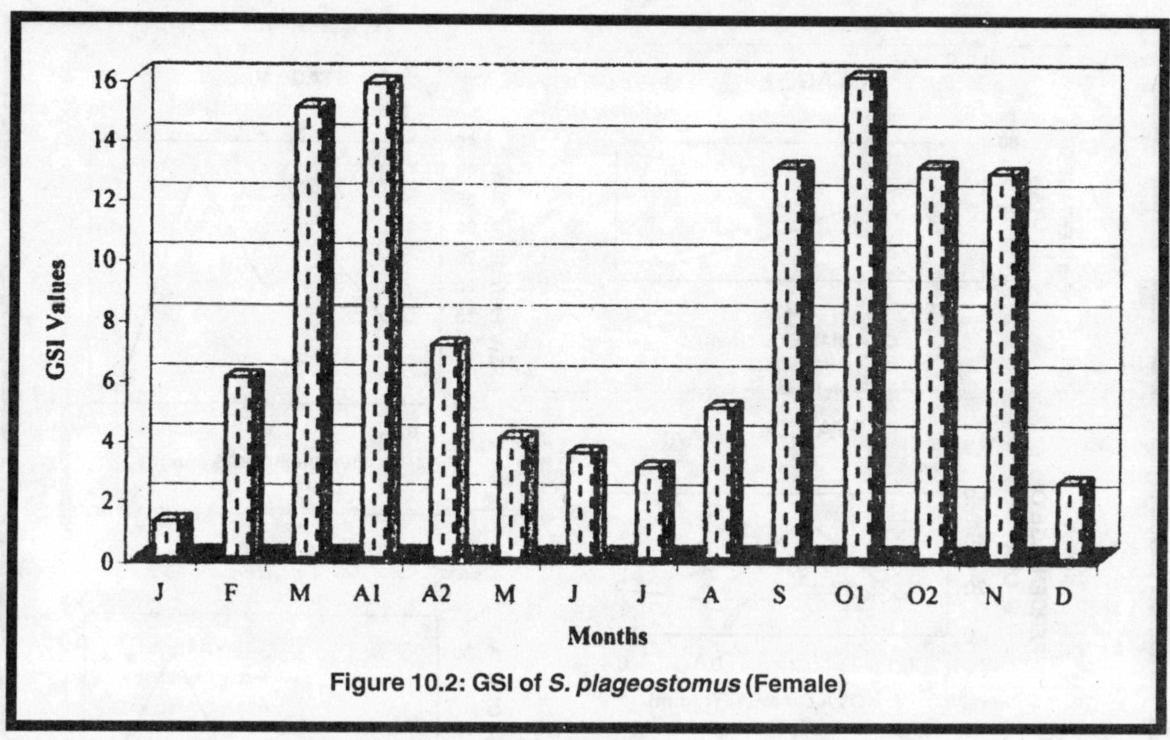

Figure 10.2: GSI of *S. plageostomus* (Female)

various size groups during pre-spawning and spawning time were tabulated. 50 per cent level of maturity after interpolation was considered the size at first maturity (Figure 10.4). The size at first maturity obtained for male fish was 33.8 cm and for female fish 35.2 cm.

Spawning Ecology

The percentage occurrence of fish of various maturity stages round the year are presented in the Figure 10.5. It showed that the mature fish were available during March–April and September–October. The spent fish were recorded during April–May and again in October–November. It clearly indicated two spawning frequency during late spring and late autumn for *S. plagiostomus.* Pebbly substratum enriched with algal carpet in the shallow areas were recognized as spawning grounds of the fish. The range of water temperature of these spawning grounds was observed as 12.5 to 16.6°C and that of pH and dissolved oxygen was 7.8–8.3 and 10–13 ppm respectively.

Discussion

Pioneering work on the spawning behaviour in fish has been conducted by Clark (1934), and Hickling and Rutenberg (1936) based on the size distribution of intra-ovarian eggs in different fishes. A majority of teleost fishes all over the world are seasonal breeder and in the Indian subcontinent a vast majority of freshwater fishes breed during the monsoon months of heavy rainfall (Jhingran, 1982). However, schizothoracid is the only inter-specific pisces group, which has been reported to perform varied kind of spawning behaviour in different ecological characteristics. Present study indicated that *Schizothorax plagiostomus* (Heckel) is a double spawner, which spawns during April–May and October–November in the river pinder.

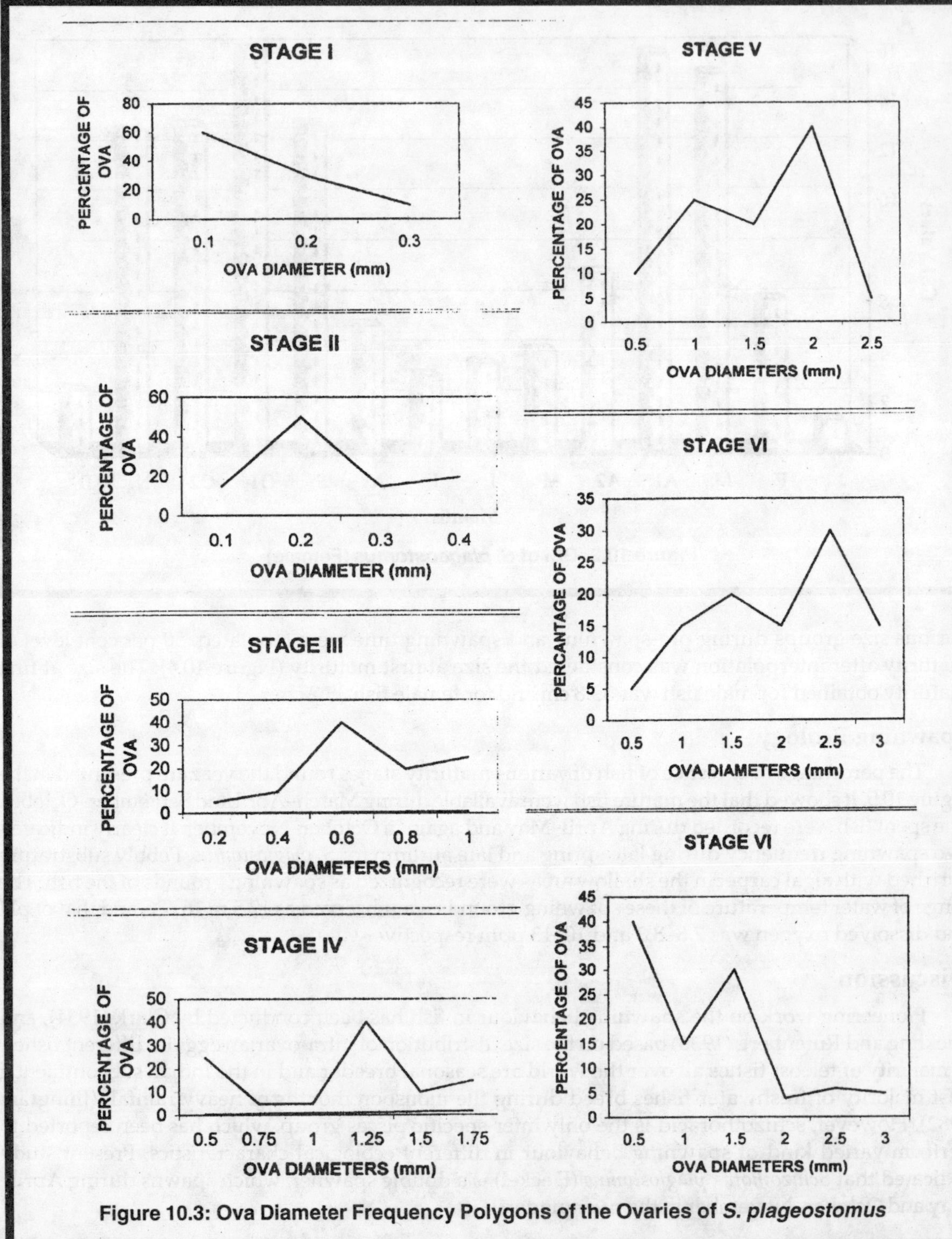

Figure 10.3: Ova Diameter Frequency Polygons of the Ovaries of *S. plageostomus*

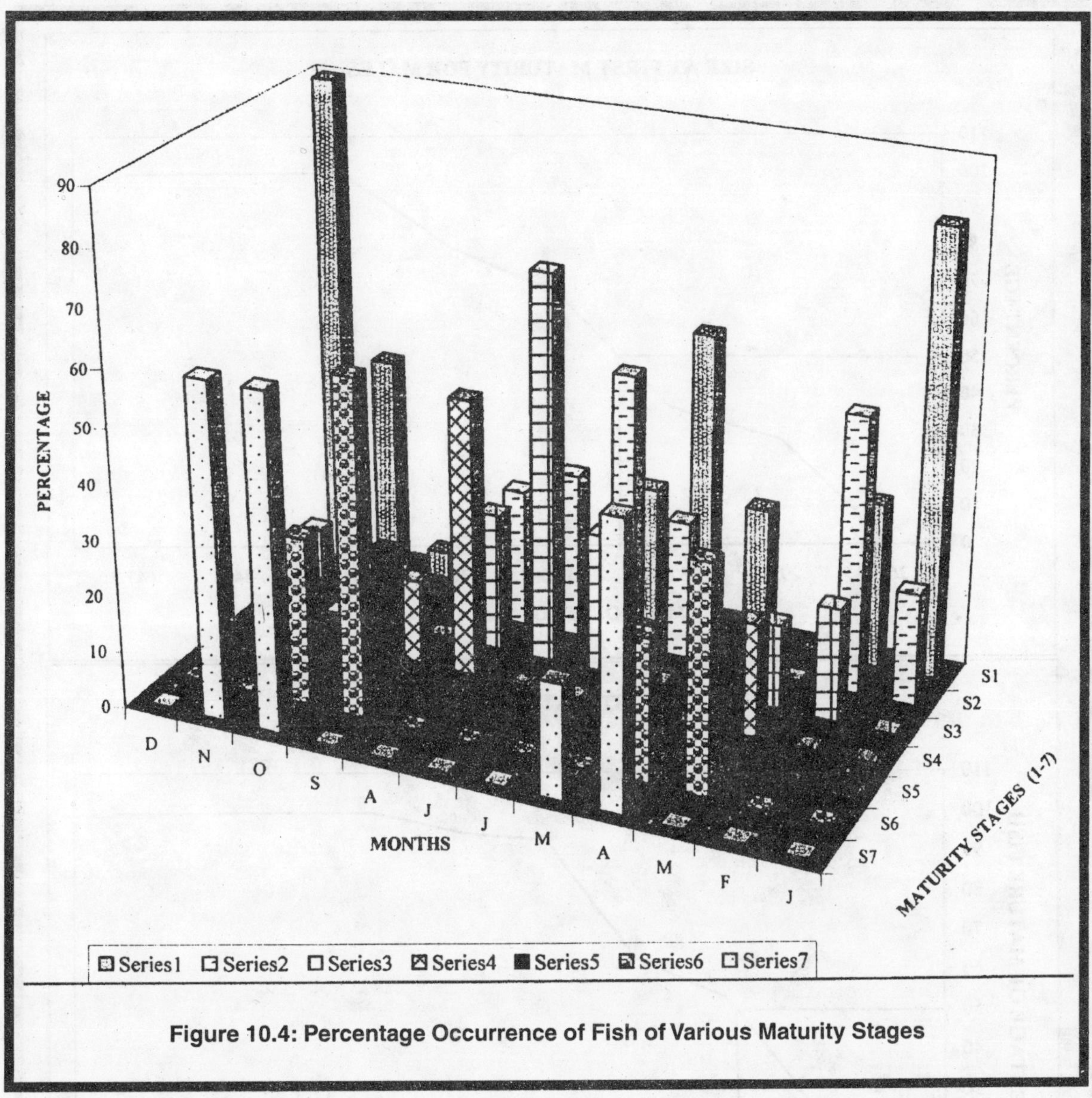

Figure 10.4: Percentage Occurrence of Fish of Various Maturity Stages

The determination of maturity stages in fish becomes easy when ova diameter frequency polygons are plotted for each fish during each month of investigation. Most of the fishery scientists follow the trend set in ICES scale (Wood, 1930). Some of the investigators suggested some modifications depending upon the maturity condition of an individual fish. In the present investigation seven maturity stages were determined. The stages were classified as immature I, immature II, maturing I, maturing II, mature, spawning and spent. Observations of the ova diameter frequency polygons showed that there were two continuous batches of eggs throughout that spawned in two batches during two different spawning periods.

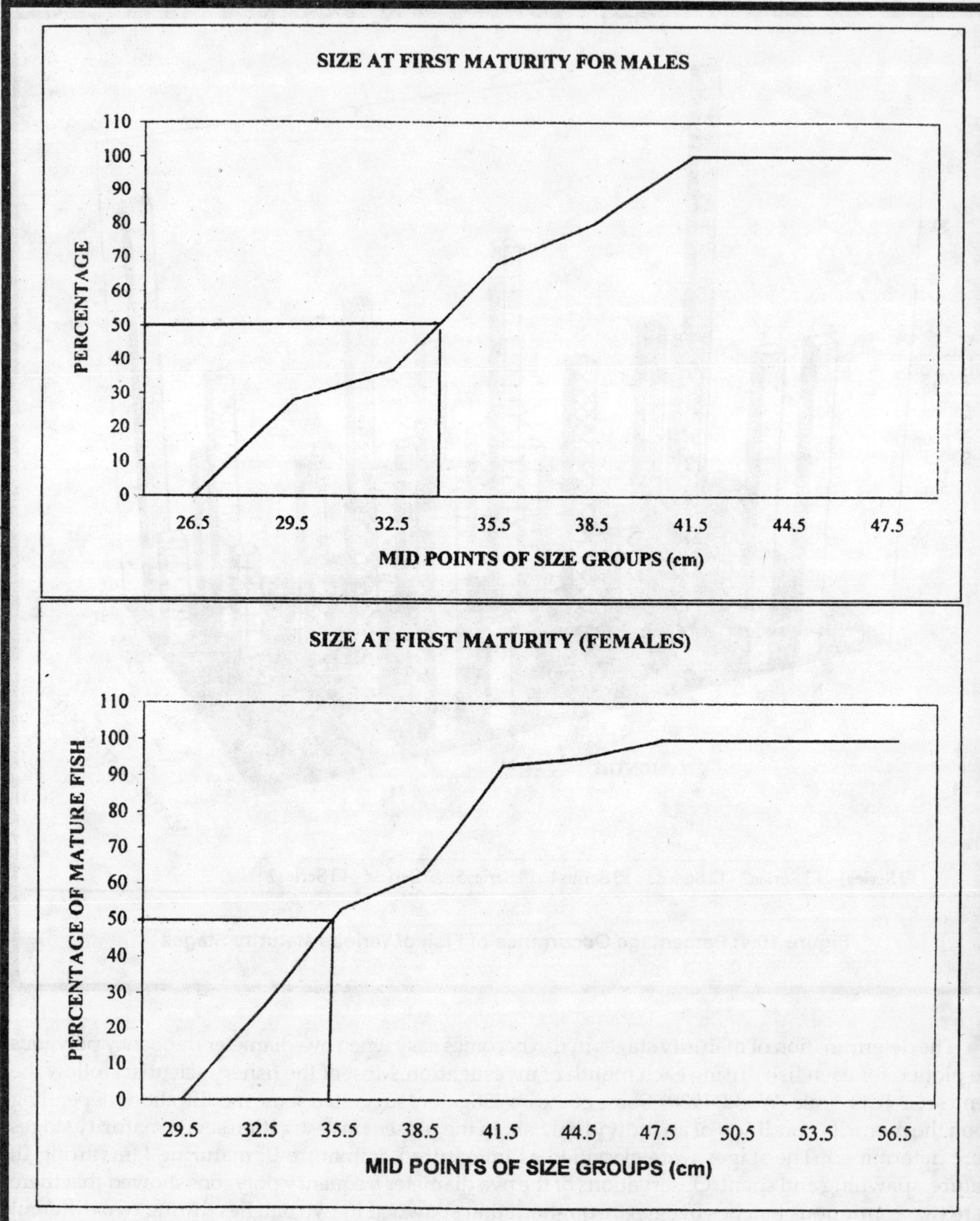

Figure 10.5: Size at First Maturity of *S. plageostomus*

Several previous workers also determined maturity stages. Five maturity stages were reported for *S. niger* by Sunder *et al.* (1979), for *S. esocinus* by Raina (1976) from Kashmir valley and for *S. plagiostomus* by Agarwal (1996) from Alaknanda river of Garhwal. Six stages of maturity were reported for *S. richardsonii* by Agarwal (2004) from Bhagirathi river in Garhwal and also by Qadri *et al.* (1983) from Kashmir waters. Our attempt of reporting seven maturity stages corroborates with that of Dobriyal and Singh (1987) and Negi and Dobriyal (1997) who reported seven maturity stages in *Barilius bendelisis* and *Crossocheilus latius latius* respectively. The maximum GSI values for both the sexes of *S. plagiostomus* were noted in the months of April and October, which show the highest level of maturity in these months. The first fall in the value was observed during second fortnight of same months, which indicated towards start of spawning act.

The 50 per cent level in maturity, which has been taken to represent the mean length at which maturity was obtained, were 33.8 cm and 35.2 cm for the male and female fish respectively. Studies conducted on this aspect on the hillstream fishes of the Garhwal Himalaya indicated that *Glyptothorax telchitta* matures at a size 98 mm (female) and 112 mm (male). In *Garra lamta* this size was 99 mm for male and 115 mm for the females (Rautela, 1999). Dhasmana (1990) reported this size as 154 mm for males and 174 mm for females for *G. gotyla gotyla.* Dobriyal and Singh (1989) reported it as 125 mm (male) and 121 mm (female) for *Glyptothorax pectinopterus.* Dobriyal and Singh (1987) reported it as 104 mm for male and 83 mm for the female *Barilius bendelisis.* Among schizothoracids, the reported size at first maturity was 25 cm for *S. curvifrons* (Sunder, 1984), 33 cm for *S. plagiostomus* of Alaknanda river (Agrawal, 1996), 40 cm for female *S. richardsonii* (Misra, 1982).

A diverse spawning time has been reported in the schizothoracine fishes all over India. In present study it was observed that *S. plagiostomus* spawns during April–May and October–November in the river Pinder. Similar bimodal spawning has been reported in *S. plagiostomus* from Bhakra reservoir by Bhatnagar (1964) but it was during July–August and December–January. For same species, Agarwal (1996) reported a prolonged spawning from August to November in river Alaknanda. Kashmir snowtrouts (*S. niger* and *S. esocinus*) have been reported to breed during April–June by Malhotra (1966) and Raina (1977) respectively.

In present study it was noticed that the fish spawns in the shallow areas of side water on the pebbly substratum that is rich in algal carpets and prefers a temperature range between 12.5 to 16.6°C, pH range of 7.8-8.3 and higher dissolved oxygen contents (10–13 ppm). Temperature has been recognized as stimulating factor in this case as rising temperature during late spring and falling temperature during late autumn triggered spawning. Khan (1945), and Nautiyal (1984) observed the thermal profile during the breeding season and recorded the range of temperature 21 to 25°C for the breeding of Himalayan mahseer. Desai (1973) considers the range of 19.9 to 28.4°C to be optimum for *T. tor* in Narmada river.

References

Agrawal, N.K. (1996). *Fish Reproduction.* APH Publishing Corporation 5, Darya Ganj, Ansari Road, New Delhi.

Agrawal, N.K., Rawat, U.S., Thapliyal, B.L. and S.K. Raghuvanshi (2004). Breeding biology of snowtrout *Schizothorax richardsonii*, inhabiting the Bhagirathi river of Garhwal Himalayas. *J. Inland Fish. Soc. India*, 36: 1–8.

Bhatnagar, G.K. (1964). Spawning and fecundity of Bhakra reservoir fishes. *Indian J. Fish*, 11: 485–502.

Bisht, J.S. (1974). Seasonal histological changes in the restes of a hillstream teleost, *Schizothorax richardsonii* (Gray/Hard.). *Acta, Acad.*, 88: 398–410.

Clark, F.N. (1934). Maturity of California sardine (*Sardinella caerulea*) determined by ova diameter measurements. *Fish. Bull. California*, pp. 42–49.

Dhasmana, N. (1990). Fishery biology of *Garra gotyla gotyla* (Gray) from Garhwal Hillstreams. *D.Phil. Thesis*. HNB Garhwal University, Srinagar Garhwal.

Dobriyal, A.K. and H.R. Singh (1987). The reproductive biology of the hillstream carp *Barilius bendelisis* (Ham.) from Garhwal Himalaya. *Vest Cs Spolec Zool.*, 51: 1–10.

Dobriyal, A.K. and H.R. Singh (1989). Ecology of rhithrofauna in the torrential waters of Garhwal Himalaya, India: Fecundity and sex ratio of *Glyptothorax pectinopterus* (*Pisces*). *Vest. Cs. Spolec. Zool.*, 53: 17–25.

Dobriyal, A.K. and H.R. Singh (1993). Reproductive biology of a hillstream catfish *Glyptothorax madraspatanum* (Day) from Garhwal, Central Himalaya, India. *Aquaculture and Fisheries Management, UK*, 24: 699–706.

Hickling, C.F. and E. Rutenberg (1936). The ovary as an indicator of the spawning period of fishes. *J. Marine Biol. Assoc. UK*, 21: 311–318.

Jhingran, V.G. (1982). *Fish and Fisheries of India*. Hindustan Publishing Corporation, New Delhi, India.

Malhotra, Y.R. (1966). Breeding in some fishes of Kashmir valley. *Ichthyologica*, 5: 53–58.

Malhotra, Y.R. (1971). Studies on the seasonal changes in the ovary of *Schizothorax niger* (*Hackel*) from the lake of Kashmir. *Jap. J. Ichthyol.*, 17(3): 110–116.

Misra, M. (1982). Studies on the fishery biology of *Schizothorax richardsonii* (Gray), an economically important food fish of Garhwal Himalaya. *D. Phil. Thesis*. HNB Garhwal University, Srinagar Garhwal.

Nautiyal, P. (1994). *Mahseer the Game Fish: Natural History, Status and Conservation Practices in India and Nepal*, (Compiled and Edited). Rachna, Srinagar Garhwal.

Negi, K.S. and A.K. Dobriyal (1997). Sexual maturity and spawning ecology of a hillstream carp *Crossocheilus latius latius* (ham.) from glacier-fed river Mandakini, Garhwal Himalaya. *J. Inland Fish. Soc. India*, 29: 26–33.

Qadri, M.Y., Mir, S. and A.R. Yousuf (1983). Breeding biology of *S. richardsonii* (*Gray and Hard*). *J. Indian Inst. Sci.*, 64: 73–81.

Raina, H.S. (1976). Seasonal histological changes in the ovary of *Schizothorax esocinus* (Heckel). *Matsya*, 2: 66–71.

Raina, H.S. (1977). Observations on the fecundity and spawning behaviour of *Schizothorax esocinus* (Heckel) from Dal Lake Kashmir. *Indian J. Fish.*, 24(1 and 2): 201–203.

Rautela, K.K. (1999). Ecological studies on the spawning biology of some coldwater fishes from the Khoh stream. *D.Phil. Thesis*. HNB Garhwal University Srinagar Garhwal.

Sehgal, K.L., Shah, K.L. and J.P. Shukla (1971). Observations on the fish and fisheries of Kangra valley and adjacent areas with special reference to the mahseer and other fishes. *J. Inland Fish. Soc. India*, 3: 63–71.

Sunder, S. (1984). Studies on the maturation and spawning of *Schizothorax Curvifrons* Heckel from river Jhelum, Kashmir. *J. Indian Inst. Sci.*, 65: 41–51.

Sunder, S., Bhagat, M.J., Joshi, C.B. and K.V. Ramakrishna (1979). On the fishery and some aspects of biology of *S. niger* (Heckel) from Dal Lake Kashmir. *J. Inland Fish. Soc. India*, 11: 82–87.

Wood, H. (1930). Scottish herring shoals. Pre-spawning and spawning movements. *Scotland Fish Bd. S. Invest.*, 1: 1–71.

Chapter 11
Bioaccumulation of Trace Metals in Marine Algae (Chaetomorpha) at Tharangambadi Coast, South East Coast of India

P. Martin Deva Prasath

Reader, P.G Department of Chemistry, TBML College, Porayar – 609 307, Tamil Nadu, India

ABSTRACT

Growing industrialization and population explosion are the major factors responsible for marine pollution. Disposal of industrial, anthropogenic and municipal wastes into the neighbouring streams and rivers causes many health hazards. This plays a vital role in raising the levels of heavy metals in the marine environment. Hence, monitoring seasonal variations in the concentrations of heavy metals has become an important concern of the environmental researchers. Therefore, this study is aimed to assess the trace metal concentrations (Zn, Cu, Fe and Mn) in water, sediments and marine algae (Chaetomorpha) of Tranquebar coast in Nagapattinam District of Tamil Nadu.

Keywords: *Trace metals; Marine algae, Tharangambadi, Bioaccumulation, South East coast, India.*

Introduction

Growing industrialization, population explosion and migration to urban centers are major factors responsible for environmental pollution. The disposal of anthropogenic wastes, dumping of municipal and industrial wastes in the neighbouring streams and rivers have quite significant role in raising the level of toxic metals in coastal areas. Metals are biologically non-degradable and through food-chain, they may finally pass on to man (Thomas and Jaquet, 1976). Numerous heavy metal studies have been made to understand the effect of heavy metals on animals, but relatively little attention has been paid to study the trace metal accumulation on seaweeds. Hence, along with the marine algae, Chaetomorpha, accumulation of trace metals (Zn, Cu, Fe and Mn) in water and sediment in Tharangambadi coast were also studied.

Materials and Methods

Samples of water, sediments and *Chaetomorpha* were collected at monthly interval in Tharangambadi coast (Lat. 11°2′N; Long. 79°49′E) in Nagapattinam district of Tamil Nadu, India for a period of two years from January 2000 to December 2001. Tharangambadi coast receives domestic and municipal sewages, shrimp farm effluents and agricultural discharges mainly through Uppanar estuary of Tharangambadi. The following methods are adopted for the detection of Zn, Cu, Fe and Mn in Water, Sediments and Marine Algae.

Water (Brooks *et al.*, 1967)

Surface water samples were collected in precleaned and acid washed polypropylene bottles and the samples were filtered in Millipore filter paper (Pore size 0.45 µ). The samples were preconcentrated with APDC-MIBK extraction procedure (Brooks *et al.*, 1967). The resulting solution was aspirated to the Flame Atomic Absorption Spectrophotometer (Perkin-Elmer model 373) for the detection of Zn, Cu, Fe and Mn in water.

Sediment and Algae (Chester and Hughes, 1967)

Sediments samples were collected in precleaned, acid washed PVC corer and the algae *Chaetomorpha* were collected by hand picking in intertidal Tharangambadi coast for a period of two years from January 2000 to December 2001. Both the samples were washed with metal free double distilled water and dried in hot air oven at 110°C for 5–6 hrs and ground to powder in a glass mortar and stored in precleaned polythene bags. 500 mg of sample was taken and digested with a mixture of 1 ml of Con H_2SO_4, 5 ml of Con HNO_3 and 2 ml of $HClO_4$. A few drops of hydrofluoric acid is added to achieve complete digestion and filtered the sample to make up to 25 ml with metal free double distilled water for the estimation of Zn, Cu, Fe and Mn using Flame Atomic Absorption Spectrophotometer (Perkin-Elmer model 373).

Results

Monthly variation in the bioaccumulation of Zn, Cu, Fe and Mn in water, sediments and marine algae recorded at Tharangambadi coast are shown in Figures 11.1–11.12. The annual mean concentration of the metals was also calculated.

Monthly variation in the accumulation of Zn in water varied from 4.23 μgl^{-1} (May 2000) to 102 μgl^{-1} (September 2000); Cu varied from BDL (Below Detectable Level, July and August 2001) to 40.2 μgl^{-1} (April 2000); Fe varied from 84.82 μgl^{-1} (May 2000) to 260.4 μgl^{-1} (February 2000); and Mn varied from 0.82 μgl^{-1} (January 2000) to 47.6 μgl^{-1} (May 2000) during the year January 2000 to December 2001.

Monthly variation in the accumulation of Zn in sediment varied from 10 μgg^{-1} (May 2000) to 192.2 μgg^{-1} (December 2001); Cu varied from BDL (February 2000) to 174 μgg^{-1} (December 2000); Fe varied from 343 μgg^{-1} (January 2001) to 20800 μgg^{-1} (February 2000); and Mn varied from 79 μgg^{-1} (December 2000) to 881 μgg^{-1} (February 2001) during the year January 2000 to December 2001.

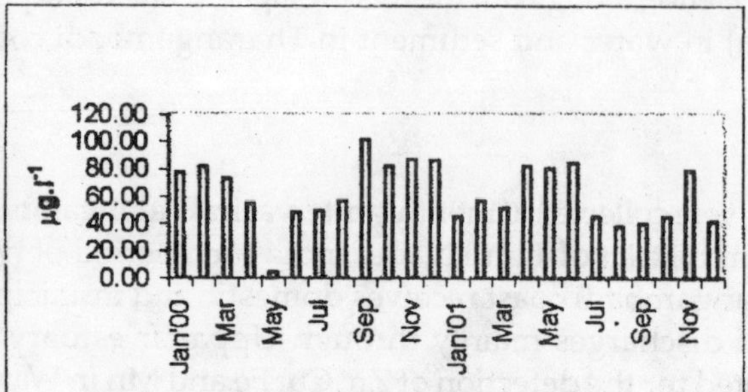

Figure 11.1: Monthly Variations in Zn Concentration (µg.l⁻¹) in Water Recorded at Tharangambadi Coast During January 2000 to December 2001

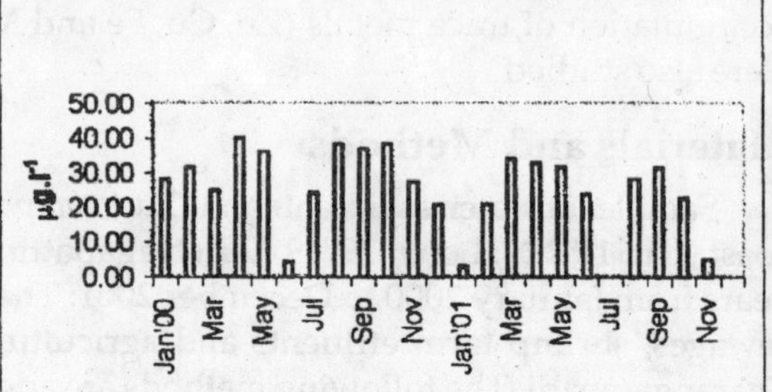

Figure 11.2: Monthly Variations in Cu Concentration (µg.l⁻¹) in Water Recorded at Tharangambadi Coast During January 2000 to December 2001

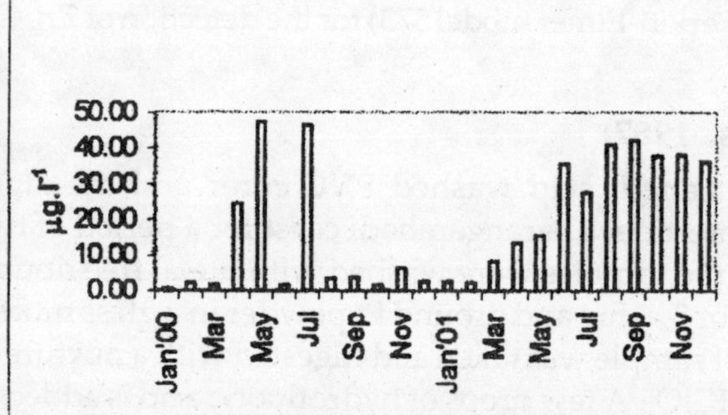

Figure 11.3: Monthly Variations in Fe Concentration (µg.l⁻¹) in Water Recorded at Tharangambadi Coast During January 2000 to December 2001

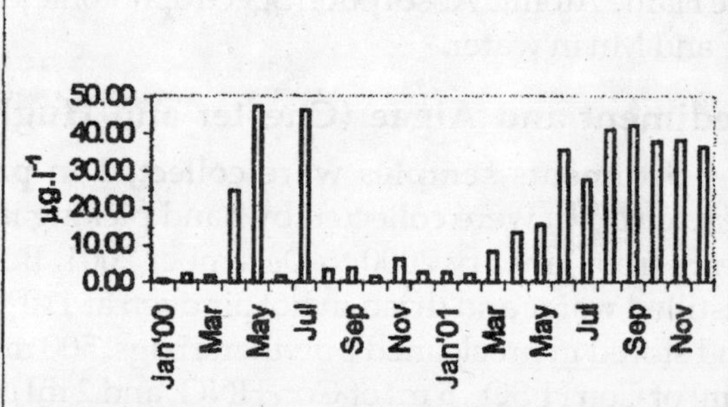

Figure 11.4: Monthly Variations in Mn Concentration (µg.l⁻¹) in Water Recorded at Tharangambadi Coast During January 2000 to December 2001

Monthly variation in the accumulation of Zn in *Chaetomorpha* varied from 7.5 µgg⁻¹ (August 2000) to 68.1 µgg⁻¹ (October 2001); Cu varied from 9.5 µgg⁻¹ (July 2000) to 69 µgg⁻¹ (September 2001); Fe varied from 381.5 µgg⁻¹ (July 2000) to 1274.5 µgg⁻¹ (September 2001); and Mn varied from 43 µgg⁻¹ (May 2000) to 392 µgg⁻¹ (November 2001) during the year January 2000 to December 2001.

Discussion

The concentration of Zn, Cu, Fe and Mn in water was high during monsoon season and low during summer season. The high values of Zn, Cu, Fe and Mn during monsoon season are due to disposal of agricultural, domestic, municipal and shrimp farm discharges. Further, natural sources of heavy metals are through land and river runoff and from mechanical and chemical weathering of rocks (Bryan, 1984). The low values of metals during summer and premonsoon season are due to

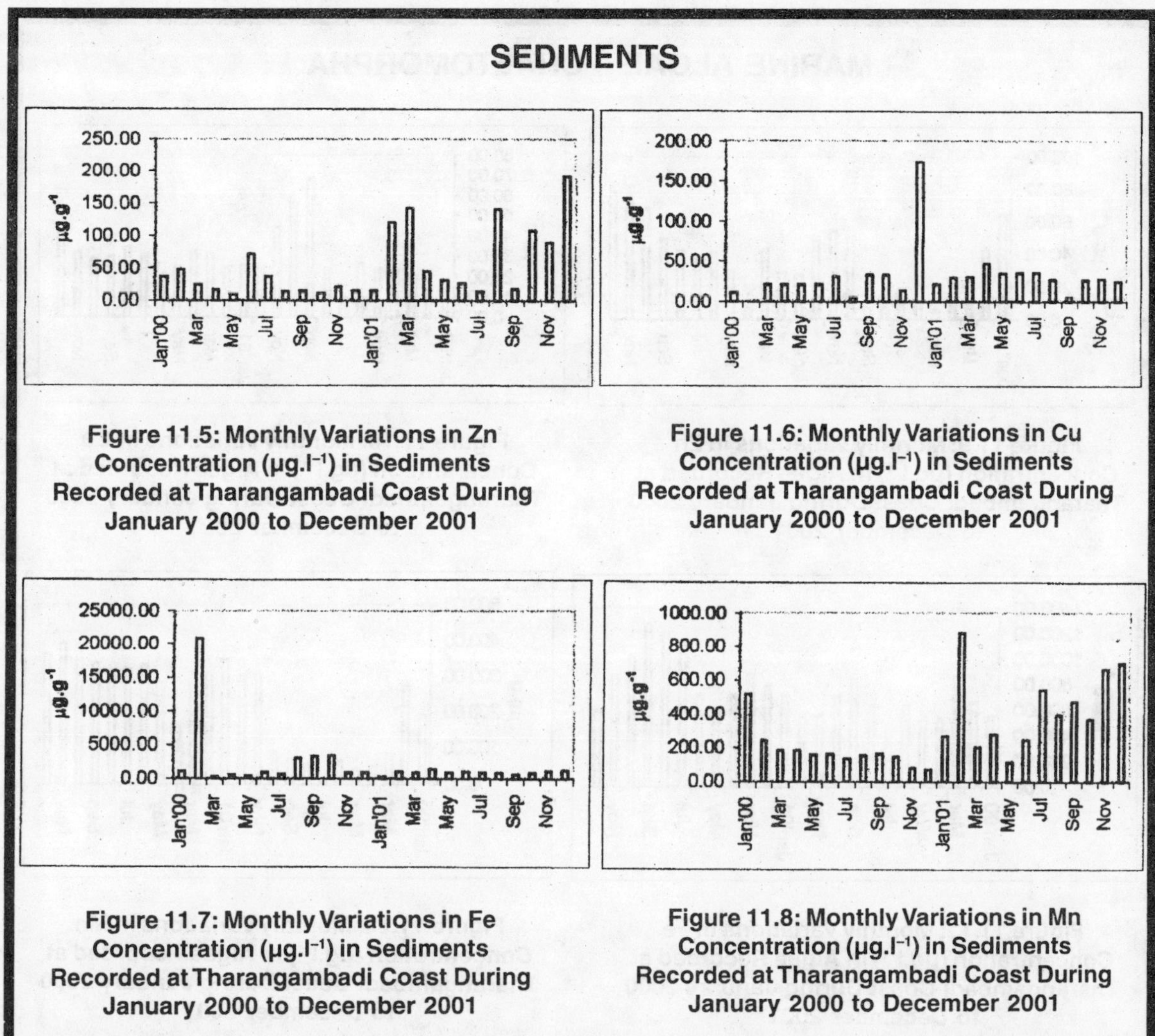

Figure 11.5: Monthly Variations in Zn Concentration (μg.l^{-1}) in Sediments Recorded at Tharangambadi Coast During January 2000 to December 2001

Figure 11.6: Monthly Variations in Cu Concentration (μg.l^{-1}) in Sediments Recorded at Tharangambadi Coast During January 2000 to December 2001

Figure 11.7: Monthly Variations in Fe Concentration (μg.l^{-1}) in Sediments Recorded at Tharangambadi Coast During January 2000 to December 2001

Figure 11.8: Monthly Variations in Mn Concentration (μg.l^{-1}) in Sediments Recorded at Tharangambadi Coast During January 2000 to December 2001

utilization and absorption of metals by primary producers like algae and phytoplankton which showed high density due to stagnant environmental condition (Rajaram, 2002 and Sampathkumar *et al.*, 1992).

Sediment metal concentrations are mainly controlled by river inflow, sediment particle type and size and organic contrast which influences the absorption and accumulation of metals in sediments (Loring, 1978). The higher values of metals in sediment during monsoon seasons are due to land runoff and the influx of the metal rich fresh water increased the particulate matter and suspended sediment load along with shrimp farm effluents. The lower values of trace metals noted during summer season are due to low fresh water inflow, biological utilization, dominance of high saline water, precipitation of particulate matter, decreased land drainage, sediment particle size and type, mud content, sulphide content and microbial activity (Craig and Moreton, 1983 and Rajaram, 2002).

MARINE ALGAE – CHAETOMORPHA

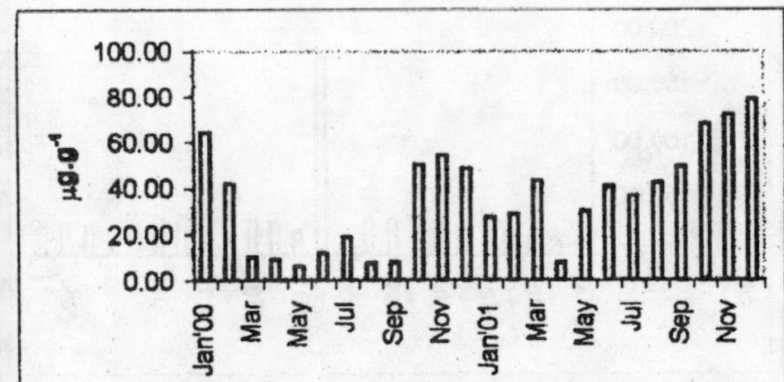

Figure 11.9: Monthly Variations in Zn Concentration (µg.l^{-1}) in Algae Recorded at Tharangambadi Coast During January 2000 to December 2001

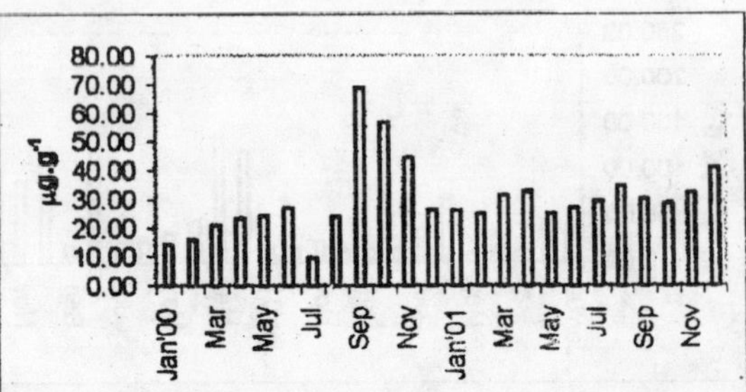

Figure 11.10: Monthly Variations in Cu Concentration (µg.l^{-1}) in Algae Recorded at Tharangambadi Coast During January 2000 to December 2001

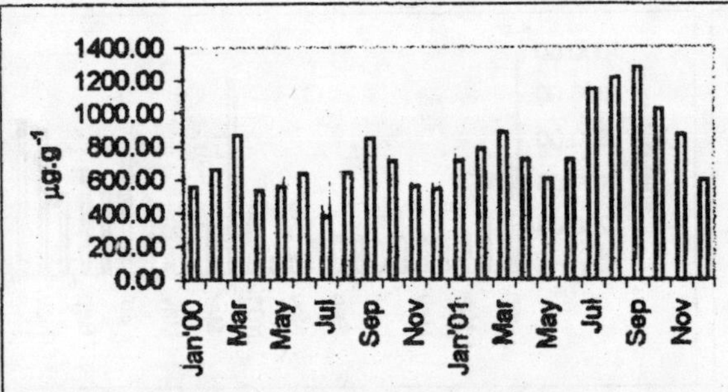

Figure 11.11: Monthly Variations in Fe Concentration (µg.l^{-1}) in Algae Recorded at Tharangambadi Coast During January 2000 to December 2001

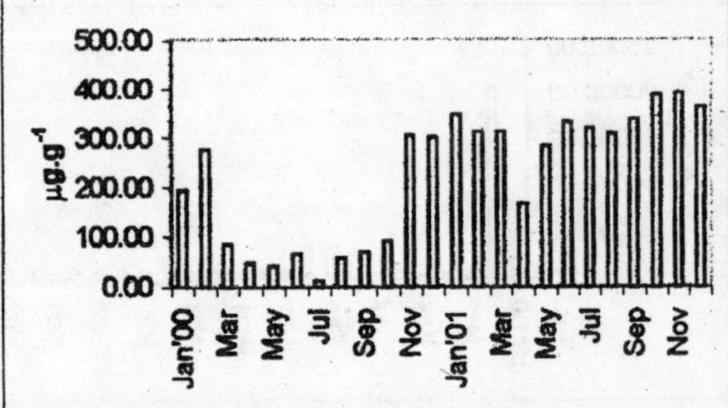

Figure 11.12: Monthly Variations in Mn Concentration (µg.l^{-1}) in Algae Recorded at Tharangambadi Coast During January 2000 to December 2001

Marine algae are good accumulator of trace metals. Uptake and accumulation of trace metals by the algae depends on availability of elements in water, environmental condition and species composition and population size (Kumarguru, 1980). In the present study, the high levels of trace metals Zn, Cu, Fe and Mn in algae during monsoon season are due to the pre existence of high concentration of metals found in water and sediment which are transferred to algae, low salinity and algal density in monsoon season increased the accumulation rate by adsorption and absorption mechanisms (Boyle *et al.*, 1976) and low values during premonsoon season may be due to lack of rainfall, resultant runoff and low algal density.

Hence, from the present study, based on the bio-accumulation of trace metals in Chaetomorpha, it is concluded that algae can be used as monitors of aquatic heavy metal pollution, on a long-term basis, as this species shows greater heavy metal accumulation capacity.

References

Boyle, E.A., Sclater, F.R. and J.H. Edmond (1976). On the marine geochemistry of cadmium. *Nature*, London, 263: 42–44.

Brooks, R.R., Presley, B.J. and I.R. Kalplan (1967). APDC-MIBK extraction system for the determination of trace metals in saline water by atomic absorption spectroscopy. *Talanta*, 14: 809–816.

Chester, R. and M.J. Hughes (1967). A chemical technique for the separation of ferromanganese materials, carbonate minerals and absorbed trace elements from pelagic sediments. *Chem. Geol.*, 2: 249–263.

Craig, P.J and P.A Moreton (1983). Total Hg, methyl Hg and sulphide in River Carron sediments. *Mar. Pollut. Bull.*, 14: 408–411.

Kumarguru, A.K. (1980). Studies on the chemical and biological transport of heavy metal pollutants Copper, Zinc and Mercury in Vellar estuary and the toxicity of these pollutants to some estuarine fish and shell fish. *PhD. Thesis*, Annamalai University.

Loring, D.H. (1978). Environmental biogeochemistry and geomicrobiology: Methods, metals and assessment. In: *Proceedings of 3rd International Symposium on Environmental Biogeochemistry* (Ed.) W.E. Krumbein. Wolfenbuttal, W. Germany, Ann. Arbor. Science Publication, Inc., Ann Arbor Michigam, p. 1025–1010.

Rajaram, R. (2002). Studies on hydrobiology and mercury content in Uppanar estuary. Cuddalore, Southeast coast of India. *Ph.D. Thesis*, A.U., India, pp. 128.

Sampathkumar, P. (1992). Investigation on plankton in relation to hydrobiology and heavy metals in the Tranquebar, Nagapattinam Coast. *Ph.D. Thesis*, A.U., India, pp. 184.

Thomas, R.L. and J.M. Jaquet (1976). *Fish Res. Board*, Canada, 33: 404.

Chapter 12

Agrometeorological Assessment of Soil Moisture Stress of Kharif Crops on a Weekly Basis for Real-time Use

S. Venkataraman

Retired Director, India Meteorological Department and Former WMO/FAO Agromet Expert
59/19, Navsahyadri Society, Pune – 411 052, India
E-mail: fintech@vsnl.com

ABSTRACT

The terms used in studies on Crop-Weather-Soil-Water relationships and their connotations are briefly explained. A methodology requiring (a) weekly values of Evaporative Power of Air and (b) rainfall data on (i) daily basis up to one month of crop growth and (ii) on a weekly basis thereafter for estimating (i) root zone soil moisture accretion and (ii) soil moisture needs for Potential Transpiration is detailed. Procedure for computation of Stress days, SDs, is outlined. Format for entry of basic and derived data to compute SDs is presented. Possible applications for the methodology are indicated.

Keywords: *Potential moisture needs, Actual transpiration, Effective rainfall, Yield response.*

Introduction

In view of the plethora of terms used in the literature on Crop-Weather-Soil-Water relations it was considered prudent to refer to them and define their meaning. The terms Potential Evapotranspiration (PET), Reference Crop Evapotranspiration and Reference Evapotranspiration (ETO), Potential Crop

Transpiration (PCT), Evaporative Power of Air (EPA) and Pan Evaporation (EP) have been suggested as measures of peak water needs of extensive and uniform crops in the active, vegetative stage. However, the entities refer to different surfaces ranging from Turf-grass of PET through tall crops (PCT tall crops) to water filled containers (EP). In PET the ratio of evaporation from the soil surface, E to PET will be the evaporative potential at the soil surface, PEs. Even over bare soil and also from fields with incomplete crop cover the E and/or Transpiration losses, T, from a field on the day following an irrigation will be equal to PET. For agricultural crops the Peak Water Need, PWN, will be much greater than that of a similarly exposed grass cover (Allen *et al.*, 1989) and the PWN of aerodynamically rough crops will be greater than that of a similarly exposed aerodynamically smoother crop (Monteith, 1964). The transpiratory component of PWN will be Potential Transpiration, PT and the ratio of PT/PWN will be the Transpiratory Coefficient Tc. The term Crop-Coefficient Kc (Allen *et al.*, 1992) is the ratio PWN/PET. The increase in quantum of peak water loss from agricultural crops compared to that of a grass cover is due to an increase in the transpiratory component. With crop growth, PEs and Tc decrease and increase respectively in a reciprocal way. PEs reaches a limiting value of 0.20 PET at the ground-cover crop stage. Tc on the other hand will continue to increase beyond the ground-cover stage and up to full canopy stage for agricultural crops and the ratio will be highest for aerodynamically rough crops. Again, in the maturity phase the water needs varies amongst crops as it is controlled by crop physiology (Hattendorf *et al.*,1988; Venkataraman, 1995). Thus while the value of PEs will be crop-age specific the values of Tc will be both crop-stage and crop specific.

In the case of Kharif crops, root zone Soil Moisture is derived from rainfall and is termed as Soil Moisture Recharge (SMR). Thus SMR will be the portion of rainfall that is not lost due to evaporation and percolation beyond the root zone. The former is also referred to as Soil Moisture Accretion (SMA) (Venkataraman, 2001) while the latter is also called Deep Drainage (DD). SMR is sometimes referred to as Effective Rainfall (ER). The actual quantity of water transpired by the crop AT, vis-à-vis the PT relevant to the crop and its age/stage, will depend on the quantum and ease of availability of root zone soil moisture. The former varies amongst crops while the latter is influenced by the level of PET and soil type (Allen *et al.*, 1992). When SMA is greater than PT the excess moisture called Soil Moisture Storage, SMS, can be carried over for subsequent use by the crop. SMS has a limiting value which varies with crop and soil type. SMS is sometimes referred to as Root Constant (RC). When AT is less than PT the crop will be suffering a moisture stress. Over any given time interval the ratio of AT/PT will be the Transpiratory Satiation Index (TSI) while 1–(AT/PT) will be the Moisture Stress Index (MSI). The product of MSI and N, where N is the number of days to which MSI relates will be the Stress Days (SDs). MSI will also be equal to SDs/N. The parameters PET, ET, E, T, AT, PT, SMA, DD, SMR and SMS are expressed in same units as precipitation *viz.*, equivalent depths of water per unit area. The parameters PEs, Tc, MSI and TSI will be ratios. In the present communications some of the above terms have been used with connotations as mentioned above. New term(s) if any used have been clearly spelt out relating to what they refer to.

Materials and Methods

FAO has adopted the ten-day period, called the Dekad, as the time-unit in the FAO methodology for forecasting yields of rainfed crops. The methodology for computation of Stress days on a Dekadal basis has been given by Venkataraman (2001). The "Standard Week" is in use in India as the time-unit for agrometeorological studies. It was decided to examine if any changes in the format and procedure for computation of the various parameters used for assessing SDs on a Dekadal basis will be required for the weekly period. The salient features relating to such an examination with reference to the rainfall budgeting factors are detailed below.

Evaporative Power of Air (EPA)

A new term "Evaporative Power of Air" (EPA) has been used (Venkataraman, 2001) to denote the potential of air to evapotranspire water from agricultural crop surfaces given 100 per cent opportunity. For operational purposes an easily measured value of EPA is necessary. The parameter Pan Evaporation (EP), would appear suitable to provide a measure of EPA. However, the dimensions, manner of mounting and operation and material of construction ofwater-filled pans to measure EP vary. Fortunately in India measurement of EP over a network of stations has been standardized and standard data of EP are fairly widely available. For assessment of soil moisture stress of Kharif crops only EPA during the monsoon season, in which weather conditions are diffuse and uniform in time and space, need to be considered. Keeping the above in view and based on a critical examination of available findings, Venkataraman (2001) had concluded that in Kharif season in India EPA will be 1.1 times that of EP. The above ratio will apply for the weekly period also.

Evaporative Depletion of Rainfall

Evaporation via the soil surface from bare soil and from irrigated and rainfed cropped surfaces has been dealt with at length by Venkataraman (2001). Basic features of the study are recapitulated below. From a soil column raised to field capacity moisture status, evaporation initially proceeds from the surface layers of the soil at rate of PEs till a quantum of 10 mm of water in Fine and Medium soils, 7 mm in Sandy soils and 5 mm in coarse soil is lost. After this stage evaporation occurs from a sub-surface layer, of moisture capacity equal to that of the surface layer, at a rate controlled by the Hydraulic Conductivity of the soil in a time-dependant manner. Over bare soil PEs is 1.0. Kharif crops reach the ground-cover stage in 6 weeks from sowing when PEs is 0.20. The decrease in PEs with crop growth will be slower during germination and nearer the ground-cover stage and can reasonably be taken as 0.9, 0.7, 0.5, 0.4, 0.3 and 0.2 at the end of weeks 1, 2, 3, 4, 5 and 6 respectively. Thus the mean PEs in the 5th and 6th weeks will be 0.35 and 0.25 respectively. As PET in Kharif season is about 4.5 mm per day the potential for evaporation from the soil surface will be 1.6 and 1.1 mm respectively in the 5th and 6th weeks. In view of the foregoing it is easily seen that evaporation from the soil in a growing crop will occur only from the surface layer from the 5th week of crop growth. This leads to the corollary that evaporative depletion of rainfall can be allowed for at a fixed rate of (a) 0.35 and 0.25 times EPA respectively in the 5th and 6th weeks and (b) 0.20 EPA from the 7th week onwards till harvest of the crop.

Over bare soil and in the period up to one month of crop growth the irregular wetting of the soil column with varying amounts of rainfall renders the assessment of evaporative depletion on a weekly basis not possible and one must consider daily falls of rain in this period to assess evaporative depletion. The procedure for computation of evaporative depletion in the first 4 weeks of crop growth will be the same as suggested for the Dekadal period (Venkataraman, 2001) with PEs values as outlined above being used for weeks 1 to 4 and is detailed below under the discussion on entry of derived data in the format for computation of Stress Days (SDs).

Transpiratory Co-efficient (Tc) and Soil Moisture Storage (SMS)

Values of Tc on a Dekadal basis for a number of typical Kharif crops have been given by Venkataraman (2001). These values would need to be interpolated to obtain weekly values. Weekly Tc values for a few typical Kharif crops are given in Table 12.1. There will be no change in values of SMS for the weekly computations, *viz.*, 80 mm in non-sandy soils and 50 mm in sandy soils for normal crops and 100 mm in non-sandy soils and 60 mm in sandy soils for deep rooted crops.

Table 12.1: Values of Tc for Some Kharif Crops of Varying Maturity Durations

Crop	Life	Crop Age in Weeks																						
		1	2	3	4	5	6	7	8	9	10	11	12	13	14	15	16	17	18	19	20	21	22	23
Hybrid Maize																								
	14	0.1	0.3	0.45	0.6	0.7	0.8	0.85	0.9	1.0	1.0	1.0	0.9	0.7	0.5	–	–	–	–	–	–	–	–	–
	17	0.1	0.3	0.45	0.6	0.7	0.8	0.85	0.9	1.0	1.0	1.0	1.0	1.0	1.0	0.85	0.7	0.5	–	–	–	–	–	–
	20	0.1	0.3	0.45	0.6	0.7	0.8	0.85	0.9	1.0	1.0	1.0	1.0	1.0	1.0	1.0	0.95	0.9	0.7	0.6	0.5	–	–	–
Hybrid Millets Sorghum																								
	14	0.1	0.3	0.45	0.6	0.7	0.8	0.8	0.8	0.8	0.8	0.8	0.8	0.8	0.8	–	–	–	–	–	–	–	–	–
	17	0.1	0.3	0.45	0.6	0.7	0.8	0.8	0.8	0.8	0.8	0.8	0.8	0.8	0.8	0.8	0.8	0.8	–	–	–	–	–	–
	20	0.1	0.3	0.45	0.6	0.7	0.8	0.8	0.8	0.8	0.8	0.8	0.8	0.8	0.8	0.8	0.8	0.8	0.8	0.8	0.8	–	–	–
Local Sorghum																								
	17	0.1	0.2	0.3	0.4	0.5	0.6	0.7	0.8	0.8	0.8	0.8	0.7	0.6	0.5	0.5	0.4	0.35	–	–	–	–	–	–
	20	0.1	0.2	0.3	0.4	0.5	0.6	0.7	0.8	0.8	0.8	0.8	0.8	0.75	0.7	0.6	0.55	0.5	–	–	–	–	–	–
	23	0.1	0.2	0.3	0.4	0.5	0.6	0.7	0.8	0.8	0.8	0.8	0.8	0.8	0.8	0.75	0.7	0.7	0.6	0.55	0.5	0.4	0.4	0.35
Peanut																								
	14	0.1	0.2	0.4	0.6	0.7	0.8	0.9	0.9	0.9	0.9	0.9	0.9	0.8	0.75	–	–	–	–	–	–	–	–	–
	17	0.1	0.25	0.4	0.6	0.7	0.8	0.9	0.9	0.9	0.9	0.9	0.9	0.9	0.9	0.9	0.8	0.75	–	–	–	–	–	–
	20	0.1	0.2	0.4	0.6	0.7	0.8	0.9	0.9	0.9	0.9	0.9	0.9	0.9	0.9	0.9	0.9	0.9	0.9	0.8	0.7	0.7	–	–
Safflower																								
	14	0.1	0.15	0.25	0.3	0.5	0.6	0.7	0.8	0.85	0.9	0.7	0.6	0.4	0.2	–	–	–	–	–	–	–	–	–
	17	0.1	0.15	0.25	0.3	0.4	0.6	0.6	0.7	0.75	0.8	0.9	0.8	0.7	0.6	0.5	0.3	0.2	–	–	–	–	–	–
	20	0.1	0.15	0.25	0.3	0.4	0.5	0.5	0.7	0.7	0.7	0.8	0.85	0.9	0.9	0.8	0.7	0.6	0.4	0.3	0.2	–	–	–
	23	0.1	0:2	0.2	0.3	0.4	0.45	0.5	0.6	0.65	0.7	0.8	0.8	0.85	0.9	0.9	0.9	0.9	0.7	0.6	0.5	0.4	0.3	0.2

Results and Discussion

The format, similar to that of Frere and Popov (1986), detailed by Venkataraman (2001) for dekadal period is applicable to the weekly period also. The format for entry of basic and derived data, are set out at Table 12.2. Exemplary data-entries are also given therein. The entry of basic data in the format is straight-forward. The principles and guidelines for computation of the derived parameters detailed by Venkataraman (2001) are briefly given below.

Table 12.2: Format and Sample Computation of Stress Days
Station: A; Year: B; Crop: Maize; Crop Life: 14 weeks; Soil Type: Medium; Max. SMS: 100 mm

1.	Weeks	25	26	27	28	29	30	31	32	33	34	35	36	37	38	39
2.	Rain, mm	26	68	31	25	16	23	62	12	83	7	24	26	16	26	14
3.	EPA, mm	32	35	34	31	31	31	31	31	31	31	31	29	28	27	30
4.	SMA, mm	0	25	3	8	5	12	54	6	77	1	18	20	10	21	8
5.	Crop week	–	1	2	3	4	5	6	7	8	9	10	11	12	13	14
6.	Tc	–	0.10	0.30	0.40	0.60	0.70	0.80	0.85	0.90	1.0	1.0	1.0	0.90	0.90	0.50
7.	TN = 3*6	–	3	10	12	19	22	25	26	28	31	31	29	25	24	15
8.	SMA–TN, mm	–	22	– 7	– 4	– 14	– 10	29	– 20	49	– 30	– 13	– 9	– 15	– 3	– 7
9.	SMS, mm	–	22	15	11	0	0	29	9	58	28.	15	6	0	0	0
10.	AT, mm	–	3	10	12	16	12	25	26	28	31	31	29	16	21	14
11.	Stress Days	–	0	0	0	1	3	0	0	0	0	3	0	3	1	0

Weekly values of EP multiplied by 1.1 will give weekly values of EPA. Following Venkataraman (2001) for computation of evaporative depletion over bare non-sandy soils, go on adding the difference between daily rainfall and EPA. When summation becomes zero or negative take moisture accretion as zero and proceed to next day. If summation is more than 10 mm assign amount in excess of 10 mm as accretion in the sub-surface layer. Proceed adding 10 mm to next day's difference between rainfall and EPA. At the end of the week, sum up the accretions in the sub-surface layer. The amount in excess of 10 mm will be SMA. Enter the value of SMA in row No. 4 against that week. Continue the procedure in the next week. In case of Sandy and Coarse soils use values of 7 and 5 mm respectively instead of 10 mm. During the crop period starting from sowing week use values of 0.95, 0.8, 0.6 and 0.45 EPA respectively to assess SMA, as per procedure for bare soil, in the 1st, 2nd, 3rd and 4th weeks of crop growth. As mentioned earlier evaporative depletion of rainfall can be allowed at (a) 0.35 and 0.25 times EPA respectively in the 5th and 6th weeks and (b) at 0.20 EPA from the 7th week onwards till harvest of the crop. The entity Tc required to estimate PT from EPA can be had from Table 12.1 for a few crops and can also be culled out from the literature on lysimetric studies on ET of other given crops. For computing AT from values SMA, PT and SMS proceed as follows:

1. Algebraically add values of SMA-PT in column 8 and enter positive values as SMS in column 9. When the sum is negative or Zero enter value of SMS as "0" in column 9.
2. Enter non-zero values of column 8 as AT in corresponding column in Row 10.
3. When value is Zero in column 8, enter AT in same column in row 10 as equal to SMS in column of previous week in Row 9 plus SMA in same column in Row 4.

Possible Applications

The moisture stress index, MSI derived from those of SDs as assessed above will, like all agromet indices, be required to be translated into a measurable parameter through use of well validated models requiring use of easily available accessory data. The MSI can be used as a Yield Response Factor (Doorenbos and Kasam, 1979). The MSI would be required to be weighted vis-à-vis the crop growth stage of incidence of moisture Stress. From the weighted Yield Response Factors one can arrive at the likely reduction in crop yield from that of a non-stressed crop. Initially the validation can be done with data of research stations. Later the validated models can be tested with yield data of recent years recorded in farmers, fields through scientific crop-cutting experiments. Since technology is fast changing all yield data must relate to years of the recent past. The model can then be adopted to provide inputs for forecasts of crop yield under the National Crop Weather Watch. Application of the methodology to long-series rainfall data at a given location can also assist in Kharif crop planning. The assessed frequency of incidence of various amounts of moisture stress on a weekly basis can be used to select a suitable cropping practice in tune with local rainfall climatology.

References

Allen, R.G., Jensen, M.E., Wright, J.L. and R.D. Burman (1989). Operational estimates of reference evapotranspiration. *Agronomy J.*, 81: 650–662.

Allen, R.G., Pereira, L.S., Raes, D. and M. Smith (1992). *Crop Evapotranspiration: Guidelines for Computing Crop Water Requirements*. Food and Agriculture Organisation (FAO) Irrigation and Drainage Paper No. 56: 300.

Doorenbos, J. and A.H. Kasam (1979). *Yield Response to Water*. FAO Irrigation and Drainage Paper No. 56: 193.

Frere, M. and G.F. Popov (1986). *Early Agrometeorological Yield Assessment*. FAO Plant Production and Protection Paper No. 73: 154.

Hattendorf, M.J., Redelfs, M.S., Amos, B., Stone, L.R. and R.E. Gwin (1988). Comparative water use characteristics of six row crops. *Agronomy J.* 80: 80–85.

Monteith, J.L. (1964). Evaporation and Environment. *Proceedings 19th Symposium Society of Experimental Biology*, p. 205–234.

Venkataraman, S. (1995). Agrometeorological determination of the optimum distribution of total water requirements of crops. *International J. Ecol. & Env. Sci.*, 21: 251–261.

Venkataraman, S. (2001). Agrometeorological anticipation of yields of rainfed crops. *Ind. J. Env. & Ecoplan.*, 5: 135–144.

Chapter 13

Environmental Impact and Utilization of Fly Ash: A Study of IB-Thermal Power Plant

D.K. Sahoo, A. Behera**, Pramila Mishra*** and N.S. Meher*****

*Officer, R&D, Tata Refractories Ltd., Belpahar, District Jharsuguda, Orissa – 768 218

**Regional Officer, State Pollution Control Board, Modipara, Sambalpur – 768 001

***Reader in Chemistry, P.G. Department, Sambalpur University, Jyoti Vihar, Burla, Sambalpur

****Chemist, SPAA Papers Titlagarh, District Bolangir, Orissa

ABSTRACT

The present piece of research deals various environmental problem related to generation of Fly Ash as negative impact and some of the important Fly Ash utilization of management as positive impact techniques with proper remedial measures.

Introduction of Fly Ash

The nation to face 21st Century needs electricity to fulfil its needs and accelerate its growth and one of the most abundant source of fuel for generation of electricity is Coal. Where coal is fire power plan compress 65 per cent of total energy generating capacity and 70 per cent of electricity output.

The inorganic residue which is left after the burning of pulverized coal is known as Coal.

Combustion By Product (CCBP)

There are two main types of Fly Ash.

Furnace Bottom Ash (FBA)

This is the fraction that is collected at the bottom of the Incinerator and consist of Coarse and heavy particles.

Pulverized Fuel Ash (PFA) of Fly Ash

This is the lighter particles that has been carried away by fuel gases.

In powder fixes Coal Fired Electricity cants 75 to 80 per cent of total ash consist of fly ash. Fly ash readily accumulate and causes enormous problems for its disposal. The indiscriminate disposal of fly ash causes large volume of land, water and energy resources world over and also causes great environmental concern.

In India three-fourth of total electricity is generated is thermal out of which almost 90 per cent is coal base. At present 90 million tons of fly ash is produced annually in India. And most of it is dumped in ash pond consuming nearly 65000 acres of land though most of it is recoverable and can be utilized if proper techniques are employed.

IB-Thermal Power Plant

IB-thermal power plant started as a capacity of 960MW electricity plant near Banaharpali 25 KM from Jharsuguda is situated in the coal mining area of Mahanadi Coal Field, as subsidiary of Coal India Limited. Coal are mainly brought from BOCM open cast project and Lakhanpur open cast project are semi-bituminus type. The typical analysis of coal and fly ash is given in Table 13.1. Water is used from the Hirakud Dam reservoir. Before the establishment of the plant the surrounding area at a core distance of 10 km. radius was environmentally clean area. The sulphur dioxide was in average 65 microgram per cubic meter in summer 62 in winter 38 in rainy season, and SPM values are 298, 288, 178 microgram per cubic meter in summer, winter and rainy season respectively. Now it is found the sulphur dioxide values are 91, 82, 52 and SPM values are 310, 304, 211 microgram per cubic meter respectively in summer, winter and rainy season with a large amount of fly ash and bottom ash are produced in the power plant.

Table 13.1: Proximate Analysis of Coal: Average

Area	*V.M.*	*F.C.*	*Ash*
Talcher Area	30.74	49.08	12.18
BOCM of IB	36.56	28.76	34.68
Lakhanpur OCP of IB	34.11	27.79	38.10

In general fly ash is a fine Alumino-Silicate material which consist of solid and spherical particle. It mainly consist of oxides of which Al, Fe, Si are the main constituent. Apart from this many trace elements some of which may be radioactive and toxic are found. The relative percentage of these oxide determine the character and use of the fly ash.

Mineralogically it consist of small sphere of glass of complex chemical composition and crystalline constituent mainly quartz, mullite, hematite and unburnt carbon. Fly ash from different source varies in 3 basic properties like lime reactivity, specific surface density and unburnt carbon content. The fly ash dry hopper content lesser unburnt carbon content and higher lime reactivity surface area as compared to pond ash from the same source.

Table 13.2: Chemical Analysis of Fly Ash of BOCM Coal (Sub-bituminus type)

13.2(a): Analysis of Fly Ash from BOCM Coal		13.2(b): Composition of Fly Ash	
Moisture	0.02	LOI	0.3
LOI	0.26	SiO_2	40–60
SiO_2	59.20	Al_2O_3	20–30
Al_2O_3	29.87	Fe_2O_3	4–10
Fe_2O_3	7.06	CaO	1–5
CaO	0.46	MgO	0.2
MgO	0.17	Na_2O	0.5
Na_2O	0.46	K_2O	0.5
K_2O	0.60	SO_2	0.2
SO_2	0.12		

Challenges Faced by 21st Century

Air Pollution

The world wide increase in burning of coal has increased the percentage of CO_2 in the atmosphere thus causing global warming due to the greenhouse effect (since 1995). It is the main cause of ozone depletion. Fine particle of fly ash in the form of suspended particle matter (S and M), produced as residue after burning of coal causes several lung and other related health problems in humans and animals. Fly ash is reported to cause ailments like allergic bronchitis, silicosis, and asthma. The low particle density and definite selecting gradient.

Water Pollution

Fly ash contaminate water bodies in its contact and also indirectly. Water pollution can adversely affect human and environmental health. Fly ash contains heavy metals and such inorganic materials (nitrate, cadmium etc.) are especially known to pollute the water bodies. The combination of toxic substances and heavy deposits of materials at the bottom cause serious threat to marine environment (Pervez and Pandey, 1994). Fly ash pollutes both ground and surface water. This can be checked by contouring and lining fly ash files, diverting drainage and selecting proper soil sites, which are not thermal plant affects the aquatic life also, it increases the temperature of water bed and decreases oxygen content of water causing extinction of rare species. The effluents also result inland fish productivity of water bodies (Singh and Singh, 1994).

Pollution Caused by Radioactivity and Other Toxic Materials

Coal contains a number of trace elements in the form of impurities. These trace elements have long durability as and when it react will heat in thermal power plants. Sometimes it may be converted into other elements but most of the time it is in the present form, which will be hazardous when contaminated with water of soil. Fly ash contains radioactive materials and other toxins thus causing threat to the environment and giving rise to high risk of radiation related health problems.

Land and Soil

Huge land is required for disposal of fly ash. Fly ash which get incorporated in soil above a certain threshold limit can change the physical, chemical and biological characteristics of soil but below that level it is useful for crop yield. Fly ash contaminates land resources to a very large extent. Most of the fly ash produced in the power plants is dumped into landfills and as ponds. This converts large areas of land into useable and other ash tips. In this way large areas of agricultural and other land is wasted. Fly ash dumped in this way makes the land barren and pollutes the topsoil (Sing *et al.*, 2000).

Biodiversity

The disposal of fly ash in fly ash ponds renders the redundant and devoid of nutrients thus making conditions unfavourable for some plants to grow. Also fly ash contains many heavy and toxic materials unfit for plant and animal consumption. Fly ash also polluted ground and surface water endangering the marine life. All these factors accumulate to reduce the biodiversity of nature and thus posing a severe threat to the environment.

Effect of Human

The finer particles of fly ash of 0.1 mm, are normally harmful to the human being residing close to the power plant as these get deposited in the pulmonary tissues of respiratory tracks. The fine particles inhaled by human being normally in early morning and evening hour due to temperature variation. These small fractions are impact found to the most mutagenic and carcinogenic.

Table 13.3 shows the advantages and disadvantages of fly ash disposal.

Table 13.3: Advantages and Disadvantages of Wet and Dry Disposal of Fly Ash

Disposal Method	*Advantage*	*Disadvantage*
Wet (100 per cent Recycling)	1. Ash handling is simple	1. No scope for ash utilization.
	2. Less costly	2. Land requirement is high (1.5 to 2.0 Ac/MW to last for 20 years)
		3. There is a chance of groundwater pollution.
		4. Often complete recycling of water budgeting.
Dry	1. Land requirement is less (0.8 to 1.00 Ac/MW) to last for 20 years	1. Require a good management practice failure can cause nuisance in the locality.
	2. Water conserved saving from cess.	2. Cost of ash handling is high which may increase the operating cost of production.
	3. Segmented use of land helps better rehabilitation.	3. Environmental problem during transportation.

The quality and huge amount of coal deposited in Mahanadi Coal Fields area some new thermal plants are under construction rapidly. Table 13.4 shows the future plants of Orissa and Table 13.5 shows the ash disposal status of present plants.

Disposal

Most of the fly ash is still disposed of in landfills and storage causing serious ecological problems. But much of this fly ash can be recovered and utilized as a resource (Adriano *et al.*, 1980).

Ash Utilization

1. Dense packing of minefills.
2. Roller compacted concrete technology for hydraulic structure.

3. Reclamation/Structural filling of low laying areas.
4. Building components.
5. Dyke raising.
6. High value added applications like extraction of alumina, cenospheres etc.
7. Long term studies for use of fly ash in agriculture and forestry.

Table 13.4: Future Power Plants of Orissa

Plant	Place	Capacity (MW)
NTPC (Expansion)	Kaniha, Angul	2000
IB-Thermal (OPGC) Expansion	Banharpalli, Jharsuguda	420
AET Trans Power	Banharpalli, Jharsuguda	1000
OEP Asia	Banharpalli, Jharsuguda	1320
CEP Asia	Hirma, Jharsuguda	3950
Samli Power	Lapanga, Sambalpur	500
Samli Power	Rengali, Sambalpur	500

Table 13.5: Ash Disposal Status in the Existing Power Plants

Power Plant	Installed Capacity (MW)	Coal Consumption* (Tonnes/Day)	Ash Generations (Tonnes/Day)	Disposal System**	Ash Pond Area	Discharge of Overflows
NTPC, Kaniha	1000	15,800	7300	Wet	800	100% recycle
TTPS, Talcher	490	5000	2200	Wet	38	Nandira
INDAL HPCL, Hirakud	72	1530	640	Dry	99	–
IB-Thermal, Banharpalli	420	4500	1090	Wet	246	100% recycle***
RSP, Rourkela	260	2500	1000	Wet	64	Brahmani

* The coal consumption and ash generation figures for 1995 is based on actual generation of power which may be considered as approximate figures.

** Present overflow is into Nandira. There is a proposal for complete recycling.

*** Although the industry has originally proposed 100 per cent recycling of ash pond overflow, it now proposes to discharge a part of its Ash Pond overflow to Hirakud reservoir along with the industrial effluent.

Use in Agriculture

Fly ash can be used as additive or amendment material in agricultural application. So far fly ash has been used in the manufacture of manure and even added to the soil directly to improve productivity. These are various benefits of using fly ash in agriculture applications. The addition of fly ash alters the structure of clayey and sandy soil to loamy. Its addition is known to increase the water holding capacity of soil phenomenally. As fly ash is alkaline in nature its use increases the pH of the soil thus making it useful for neutralizing acidic soil. Also it adds essential plant nutrients and enhances the crop yield. The use of fly ash has greatly improved the physical health of the soil by improving the water holding capacity and porosity of the soil. It has also been holding capacity and porosity of the soil. It has also been observed that the resistance levels of crop towards diseases and pest infections have also increased due to the use of fly ash (Kumar, 1996).

Though not much evidence has been found regarding the radioactivity levels in fly ash but still precaution should be taken while using fly ash in agriculture applications.

Bricks Making

Different fly ash from the production of fly ash, sand lime bricks indicate that their compressive strength increases with the increase in moulding and autoclaving pressure. The strength of brick cures at atmospheric pressure with steam for six hours attend the value of 50 to 60 kg/cm. sq., which an autoclaving at 14 kg/cm. sq. for six hours increase to 100 to 150 kg/cm. sq. depending upon the characteristic of fly ash. Fly ash contain more unburnt carbon and lesser lime reactivity and fineness yield bricks of lesser strength under same production condition.

Recommendations

The land requirement for dry disposal is less and besides the dry fly ash can find alternate use. Alternate use of fly ash particularly cement making, should be encouraged. Government may invite private entrepreneur for establishment of a large cement mill in Jharsuguda District where a large number of thermal power plants are coming up as Ambuja Cement, and ACC cement has started production cement from fly ash in Gujarat and Madhya Pradesh.

Thermal power plants of very low plant load factor should be allowed to operate. Transmission and distribution loss can be minimized if small power plants are established near lead cements.

In areas where large number of power plants are coming up stricter standards for SPM emission and World Bank standards for sulphur dioxide and no emissions are required to be imposed.

Acknowledgement

The authors are thankful to Prof. (Retd.) B.K. Sinha, Environment Science, Dr. Ashutosh Naik, Reader, Earth Science of Sambalpur University and Mr. Sibananda Bhoi, Computer Faculty, B.R. High School, Belpahar for their encouragement and time to time help for the preparation of the present study.

References

Ash Utilisation in National Thermal Power Plants.

Mathur, A.K. (2000). NTPC, Noida, Research Article, Seminar Collection.

Sahsame, Mohini and F. Ashoka (2000). *Cow Ash Utilization in Building and Agricultural*, RRL, Bhopal, Seminar Collection.

The Hetordorn Work Consultants, 1998, Elesvier Science Ins., 16(6): 1–15.

Chapter 14
A Study of Lead Contamination in Groundwater, Soil, Paddy Straw, and Milk and Blood of Buffaloes

K. Ayyadurai, P. Arockia Sahayaraj and M. Govindarajulu

Institute of Food and Dairy Technology, Tamil Nadu Veterinary and Animal Sciences University, Koduvalli, Alamathi Post, Chennai – 600 052

ABSTRACT

A pilot study was conducted to investigate the contamination of lead in buffaloes reared near the Cooum river flowing through Chennai city. Samples of water, soil, paddy straw from different zones adjoining the Cooum river and milk and blood of buffaloes of the respective zones were collected, and analysed using AAS (Perkin Elmer–2380). Among the samples water and milk prevailed a lead concentration above the permissible limits with a significant positive (r= 0.94) correlation.

Keywords: *Groundwater, Soil, Paddy straw, Milk, Blood and Lead.*

Introduction

The Cooum river flowing through the city of Chennai was once a fresh water stream used for bathing and washing clothes. In due course, the stream water was polluted due to the mixing of untreated sewage, effluents of the industries located on the banks of the river. The perennial disposal of domestic and industrial waste into the river quite-often caused stagnation of water and a free flow established only during monsoon rains. Effluents from metal industries, cow dung and urine from cattle sheds and untreated sewage from slums contribute much to the increase heavy metal concentration. Heavy metal concentration in the water and sediment of the Cooum were reported by many scientific workers (Rao, 1986; Bernice *et al.*, 1987). However, a detailed study on bioaccumulation of heavy metals in terrestrial animals is not available.

Lead is a typical toxic metal and enters the animal through ingestion of water, soil, fodder, feed and inhalation of polluted air and retained in the bones, organs and blood (Coulston and Korte, 1974, Chandra 1980). It is excreted in milk, hair, urine and dung (Elinder *et al.*, 1994) of animals. Milk being an essential food especially for children (IDF, 1992), it is necessary to check the concentration of lead in the milk of cows and buffaloes.

The concentration of heavy metals (Pb, Cd, Se, As) in milk of cows and buffaloes of various zones in the city of Chennai has been reported by (Vivekanandan *et al.*, 1993). The milk of buffaloes had a significantly higher concentration of lead than the cow (Ayyadurai *et al.*, 1998). The milk of buffaloes reared near the three water courses of the city showed higher concentration of lead than from other areas of the city. Hence, an investigation on the source of pollution of lead and their related contamination in the water and milk and blood of buffaloes was conducted.

Materials and Methods

Well water fed to buffaloes reared near the Cooum and soil samples were collected in six zones (Koyambedu, Shenoy Nagar, Aminjikarai, Choolaimedu, Chetpet and Chindatripet) of the city of Chennai. The contamination of lead in the buffaloes was studied by grouping the zones as follows:

Table 14.1: Source of Pollution in the Study Area

Industrial	*Residential*	*Estuarine*
Koyambedu (1) Shenoy nagar (2)	Aminjikarai (3) Choolaimedu (4) Chetpet (5)	Chindatripet (6)
Lorry washings and metal industries	Cattle rearing near the Cooum	Effluent from metal workshops

Samples of water and soil were collected and pretreated (EPA, 1974). The milk of buffaloes was also simultaneously collected and processed (Brooks *et al.*, 1970).

Samples of blood from the buffaloes were collected in haparinated vials and digested (Sandell, 1959). Feed and paddy straw were also collected and digested using acid mixture ($HClO_4$ and HNO_3) of Heckman (1967). The digested/pretreated samples were analysed for the presence of lead using AAS (Perkin Elmer–2380) at a wave length of 283.3nm.

Results and Discussion

The well water, soil, paddy straw, blood and milk of the buffaloes showed the presence of lead. The concentration of lead in each of the samples collected in the six zones are tabulated in Table 14.2.

Well water

The mean concentration of lead in groundwater ranged from 0.07 to 0.31 µg/ml. The highest mean concentration was registered at Zone 1, 2 and 6 being the industrialized. The lead concentration in the groundwater was gradually lowered when the Cooum passed through the residential zone of 3, 4 and 5. The lowest mean concentration of lead in groundwater was recorded at zone 4 (0.07 ± 0.0004 µg/ml) with an increase of 0.30 µg/ml at zone 6. Such an increase may partly be attributed to the intrusion of highly polluted sea water in to the rivulet and partly to the activities of the adjoining metal workshops. The data indicated that the concentration of lead in the well water differed significantly ($P < 0.1$) at industrial/estuarine and residential zones. The overall mean concentration of lead in groundwater exceeded the standards of BIS and ICMR the highest desirable limit being (0.1 μg/ml). It is observed that the dwellers who use the well and bore well water to bathing avoiding drinking. However, the buffaloes have an access to the well water through oil cakes, bran and well water mix. The stipulation of NRC (1989) limit the concentration of lead to 0.1 μg/ml in the water fed to cattle. On the contrary, the mean concentration of lead (0.21 μg/ml) in the well water of the present study exceeded the permissible limit. The potable water supported by the corporation was also analysed for lead and compared with the water from bore wells and wells situated near the banks of Cooum (Table 14.3).

Table 14.2: Mean ± SE Concentration of Lead in Well Water, Soil, Paddy Straw and Blood and Milk of Buffaloes

Zone	*Well Water (μg/ml)*	*Soil (μg/g) Dry Weight*	*Paddy Straw (μg/g) Dry Weight*	*Blood (μg/dl)*	*Milk (ng/ml)*
	Mean ± SE	*Mean ± SE*	*Mean ± SE*	*Mean ± SE*	*Mean ± SE*
Koyambedu (1)	0.31 ± 0.0023	3.4 ± 0.73	0.64 ± 0.002	3.90 ± 0.74	68 ± 1.6
Shenoy Nagar (2)	0.30 ± 0.0021	3.2 ± 0.78	ND	2.32 ± 0.43	64 ± 1.5
Aminjikarai (3)	0.16 ± 0.0006	2.2 ± 0.38	0.89 ± 0.003	2.82 ± 0.71	65 ± 1.6
Choolaimedu (4)	0.07 ± 0.0004	1.0 ± 0.01	ND	2.50 ± 0.63	57 ± 1.3
Chetpet (5)	0.14 ± 0.0005	0.6 ± 0.00	ND	4.90 ± 0.92	60 ± 1.2
Chindatripet (6)	0.30 ± 0.0010	3.9 ± 0.80	0.15 ± 0.001	5.11 ± 1.29	68 ± 1.6
Overall Mean	**0.21 ± 0.0011**	**2.38 ± 0.45**	**0.28 ± 0.001**	**3.60 ± 0.79**	**63.70 ± 1.4**

Table 14.3: Range and Mean Concentration of Lead in Well Water, Bore Well Water and Corporation Water (μg/ml)

Sources	*Range*	*Mean ± SE*	*Standards*
Well water (n = 86)	N–1.73	0.360 ± 0.0019	0.1 (ICMR/BIS/NRC) 0.5 (WHO)
Bore well (n = 110)	ND–1.62	0.270 ± 0.0015	–
Corporation water (n = 7)	ND–0.04	0.008 ± 0.0000	–
Overall mean	–	**0.212 ± 0.0011**	–

The mean concentration of lead in the corporation water was found to be 0.008 μg/ml which was far less than the well water (0.36 μg/ml) and bore well water (0.27 μg/ml). Pandya *et al.* (1983) reported the median concentration of lead in wells, city water supply, tanks and house hold taps of the Kolkatta city as 0.09 μg/ml which was nearly twice the permissible level of 0. 05 μg/ml WHO (1984). While the concentration of lead in corporation water of the Chennai city was almost negligible. The well water near the Cooum area exceeded the permissible limit. It is interesting to note that the seepage of the Cooum to the adjoining wells contribute to high concentration of lead.

The concentration of lead in the well water (0.36 μg/ml) and in the borewell water was found to be 34 times higher than that of the corporation. Similarly 3.6 and 2.7 times 17.2 and 5.4 times respectively of the ICMR and WHO.

Soil

The concentration of lead in the samples of soil collected near the banks of the Cooum rivulet ranged from 0.6 to 3.9 μg/g. A Highest concentration of lead in soil 3.9 μg/g was registered in the estuarine zone followed by industrial and residential zone. Pongsakul *et al.* (1999) reported that a lead concentration of 17.5 μg/g in the soil near a river in Thailand was not a potential hazard to animals through food chain. The soluble toxic metals if added as a pollutant to soil could serve as a pool of labile toxicant available to the agricultural food chain (Brams *et al.*,1989). As the lead concentration in the soil was well within the permissible limit (10–100 μg/g), the buffaloes presumably be unaffected through food chain. On contrary, Bernice *et al.* (1987) are of the opinion that the lead may enter buffaloes by adsorption (wallowing) and food chain of consuming grass and fodder grown on the silt of the Cooum in addition to the oral intake of Cooum water.

Paddy Straw

Paddy straw fed to buffaloes was analyzed for the concentration of lead to ascertain feed as a source of plumbism. The mean concentration of lead in paddy straw ranged from ND to 0.89 μg/g (dry weight). The samples of paddy straw collected from the residential zone (Aminjikarai) showed the highest concentration of lead (0.89 μg/g). Since the paddy straw purchased in the Chennai city market by the owners of the animals was from different Agro climatic zones of Tamil Nadu, and near by States, the bioaccumulation of lead in the milk through the paddy straw cannot be assessed directly. It is welcoming feature that the lead concentration in paddy straw is for less than permissible limit (30 μg/g) as stipulated by NRC (1989). It appears that a limited increase of lead in animal feed could be tolerated without causing concern over adverse effects on animal health (Sharma *et al.*, 1982).

Blood

The mean lead concentration in the blood of buffaloes ranged from 2.32 to 5.11 μg/g. Higher mean concentrations of 4.90 and 5.11 μg/dl were found in the blood of buffaloes of Chetpet and Chindatripet reared on the banks of the Cooum.

They drink water from the well on the banks and wallow in the river fed with untreated effluent, sewage and domestic wastes. Licking of posters by the animals also contribute to the increased lead concentration in blood. The lead in the blood bind to erythrocytes and plasma proteins and pass on to muscle, milk or the nervous system and remains in the body for limited time and excreted in milk, dung and urine and hair (Humphreys, 1991). Lead content of blood of normal farm animals of different species ranged from 12.9 to 14.0 μg/dl (Allerofft, 1950). The mean range of lead in blood reported in the present investigation is below the permissible limit ranged from 5 to 25 μg/dl (Buck, 1975).

Milk

The mean lead concentration in milk observed in six zones ranged from 57 to 68 ng/ml. There is no significant difference in the concentration of lead in the milk of buffaloes between the zones. However, the overall lead concentration in the buffaloes was above the permissible limit of 40 ng/ml. (Webb and Johnson, 1978). The buffaloes reared near the banks of Cooum are exposed to automobile exhaust. Srividya (2000) reported that the vehicles plying in the Chennai city is 8 per cent of the total number of registered vehicles in India. Mackay (1991) indicated that the urban population of animal is exposed to lead concentration primarily through air. A recent study on lead content of cattle milk samples from three sites in Varanasi, along the road with different automobile traffic densities showed that milk collected from an area of heavy traffic contained 46–72 ng/ml of lead which is much higher than the FDA permissible level of 30 ng/ml (Bhatia and Choudhri, 1996).

A positive significant correlation was observed between the concentration of lead in water and the concentration of lead in milk of buffaloes (r = 0.94).

Acknowledgement

The authors are thankful to University Grants Commission, New Delhi, for the financial assistance.

References

Alleroft, R. (1950). Lead as a nutritional hazard to farm livestock: Distribution of lead in the tissues of bovine after ingestion of various lead compounds. *J. Comp. Path.*, 60: 190–218.

Ayyadurai, K., Eswara Moorthy, M., John Jeberathinam, N., Swaminathan, C.S and V. Krishnasamy (1998). Studies on the concentration of lead and cadmium in milk of cow and buffalo. *Indian J. Environ. Hlth.*, 40 (4): 367–371.

Bernice, A., Baghyalakshmi V. and R. Lakshmi (1987). Liminology of river Cooum with special reference to sewage and heavy metal pollution. *Proc. Indian Acad. Sci. (Anim. Sci.)*, 96: 141–149.

Bhatia, I. and G.N. Choudhri (1996). Lead poisoning of milk the basic need for the foundation of human civilization. *Indian J. Public Hlth.*, 40(1): 24–26.

Brams, E., Antony, W. and L. Weatherspoon (1989). Biological monitoring of an agricultural food chain: Soil cadmium and lead in ruminant tissues. *J. Environ. Qual.*, 18: 317–323.

Brooks, B., Luster, G.A. and D.G. Easterly (1970). Spectrophotometric determination of trace elements in milk. *Absorp. Newsl.*, 9: 93.

Buck, W.B. (1975). J. Am. Vet. Med. Assoc., 166: 222. In: *Veterinary Medicine*, (Eds). O.M Rodostits, D.C. Blood and C.C. Gay. ELBS, London.

Chandra, S.V. (1980). Toxic Metal in Environment: A study report of R&D work done in Indian Industrial Toxicology Research Center, pp. 65.

Coulston, F. and F. Kortem (Eds.) (1974). *Environmental Quality and Safety Supplement Vol. II: Lead.* Suttgart, New York.

Elinder, C.G., Fruberg, L., Kjellstrom, T., Nordberg, G.F. and V. Vouk (1994). Biological monitoring of metals. *World Health Organization*, Geneva.

EPA (1974). *Methods for Chemical Analysis*. EPA 625/6–74–003a.

Heckman, M.J. (1967). Mineral in feeds by atomic absorption spectrophotometry. *AOAC*, 50: 45.

Humphreys, D.J. (1991). Effects of exposure to excessive quantities of lead in animals. *British Vet. J.*, 147: 18–30.

IDF (1992). Bulletin of the IDF 278/92 Chapter IV lead.

Mackay, D. (1991). *Multimedia Environmental Models: The Fugacity Approach*. Lewis Publishing Chelsea. MI.

National Research Council (1989). *Nutritional Requirement of Dairy Cattle*, 6th Revised edn. National Academy of Sciences, Washington D.C.

Pandya, C.B., Patel, J.S., Parikh, D.J., Chatterjee, S.K. and N.L. Ramanathan (1983). Environment lead Exposure as health problem in India: An overview. *J. Environ. Biol.*, 4(3): 127–148.

Pongsakul, P., Zarcinas, B.A, Cozens, A., and M.J. Mclaugblin (1999). Assessment of heavy metal Pollution of soils and crops in Thailand; Second international conference on contaminants in the soil environment in the Australasia-Pacific Region. 12–11 December 1999, New Delhi, p.333.

Rao, K.V. (1986). Health hazards of the Cooum river. Seminar on River Cooum–Let it be resource held during June, 1986 in Chennai, p. 8–18.

Sandell, E.B. (1959). *Colorimetric Determination of Trace Metals*. Inter Science Publishers Inc., New York, p. 468 and 964.

Sharma, RP., Street, J.C., Shupe, J.L and D.R. Bourcier (1982). Accumulation and depletion of Cadmium and lead in tissues and milk of lactating cows. *J. Dairy Sci.*, 65(6): 972–979.

Vivekanandan, O., Krishnasamy, V., Ayyadurai, K. and C.S. Swaminathan (1993). Studies on the concentration of selenium and arsenic in cow and buffalo milk. *J. Environ. Biol.*, 14(4): 261–266.

Webb, B.H. and A.H. Johnson (1978). *Fundamentals of Dairy Chemistry*, 2nd edn. AVI Publishing Company, Inc. Westport Connecticut, p. 25.

Chapter 15
Present Pollution Level in Kolkata and its Abatement

Debojyoti Mitra

Lecturer, Department of Mechanical Engineering, Jadavpur University, Kolkata – 700 032

ABSTRACT

The present article senses the urgeness in evaluating the present situation of air pollution. The causes of air pollution in industrial and metropolitan areas are ventured with special attention to the grave situation at Kolkata and its suburbs. The various preventive measures, already taken or to be taken, to abate air pollution in Kolkata, are put forward for the general environmental awareness.

Keywords: *Air pollution, Kolkata, SPM, RPM,* SO_2, NO_2, *Abatement.*

Introduction

The environment is the totality of the physical conditions on earth or part of it, especially as affected by human activities. An environmentalist is concerned with the biosphere which extends from the deep sea to flying heights in the sky. The man is the best and the most intelligent creature in the biosphere. He continually changes his environment to meet his biological and social needs. He extracts the materials from the environment to sustain his life or for his ease and comfort. After utilization of the materials, the end products are wastes of all description some of which are harmful for the biosphere. The wastes that are harmful are called pollutants. They contaminate or defile the environment.

The byproducts of the industrial activities of human beings are mostly harmful which cause pollution of air, water and the land. These are known accordingly as atmospheric, hydrospheric and lithospheric pollution. Amongst them, the most detrimental is the air or atmospheric pollution. The human being inhales about 22000 times a day and takes about 16 kg of air per day. Air inhaled is far greater than the amount of food or water taken by a human being per day. Polluted air seriously affects the health and depending on the degree of pollution it can become life taking! Therefore, air pollution is the greatest environmental evil towards mankind.

The man cannot prosper and be comfortable without industrial activity, consider the situation if we stop power producing industries to reduce pollution. So, its growth must continue. But the industrial activities must be regulated and controlled in a manner such that the pollution caused by them is within statutory limits.

Another potential source for air pollution are the automobiles. The numbers of automobiles are larger in industrial areas. Therefore the pollution loads of the automobiles are superimposed on the existing pollution load of the industries, which make the situation grave. Congested cities like Kolkata, Delhi and Mumbai are most adversely affected by vehicular pollution.

The situation is becoming alarming day by day in metropolitan and industrial areas. Hence, proper measures must be taken to abate pollution to protect the environment from harmful effects arising out of all kinds of socio-economic activities and impact of growing population. The abatement must be in a planned way–it must be effective as well as economic. The present article concentrates on the present situation of pollution in and around the industrial belt of Kolkata and discusses various measures to abate such a menace.

Air Pollution in and around Kolkata

By most accounts, Kolkata is a prime candidate for any award for the most polluted city in the world. Long on culture and political consciousness, the city is, paradoxically, an environmentalist's nightmare because of the poor quality of life it offers. The near-poisonous air, deafening noise levels and degraded water reserves have contributed to low quality environment. In conventional wisdom, Kolkata's woes arise mostly from an unsuitable geographical location and other hydro-geological factors. But the city's environment appears to have worsened over the years, primarily because of haphazard land use, inadequate civic facilities and poor public response to the need for environmental protection.

Figure 15.1 sums up the condition of air pollution in Kolkata as found out by the West Bengal Pollution Control Board during the last one year. It indicates planning conditions so far as SPM and RPM concentrations are concerned, especially during the winter season.

Abatement of Air Pollution in Kolkata

Abatement of air pollution calls for proper monitoring and control. Monitoring of pollutants in the environment may either be made continuously or intermittently. West Bengal Pollution Control Board has been working quite efficiently in monitoring the pollution level in Kolkata and its suburbs.

Pollution control is made either by (*i*) diluting and dispersing or (*ii*) by concentration and containing. A third method may be recycling. While speaking of dilution and dispersion, one must also plan to disperse the industries. The places, which are already very polluted *e.g.* Kolkata or Durgapur should not be made available for further growth of industries. If by dilution and dispersion the pollution density can be kept below the statutory limit it will not be very harmful for the biosphere. For this reason, the chimney heights of the boilers are increased so that the exhaust gases and the SPM may get dispersed.

The problem of pollution control has become difficult to solve due to indiscriminate growth of industries in and around some cities like Kolkata, Mumbai etc. and also because most part of such industries had their birth in an age when people were not environmentally conscious and hence

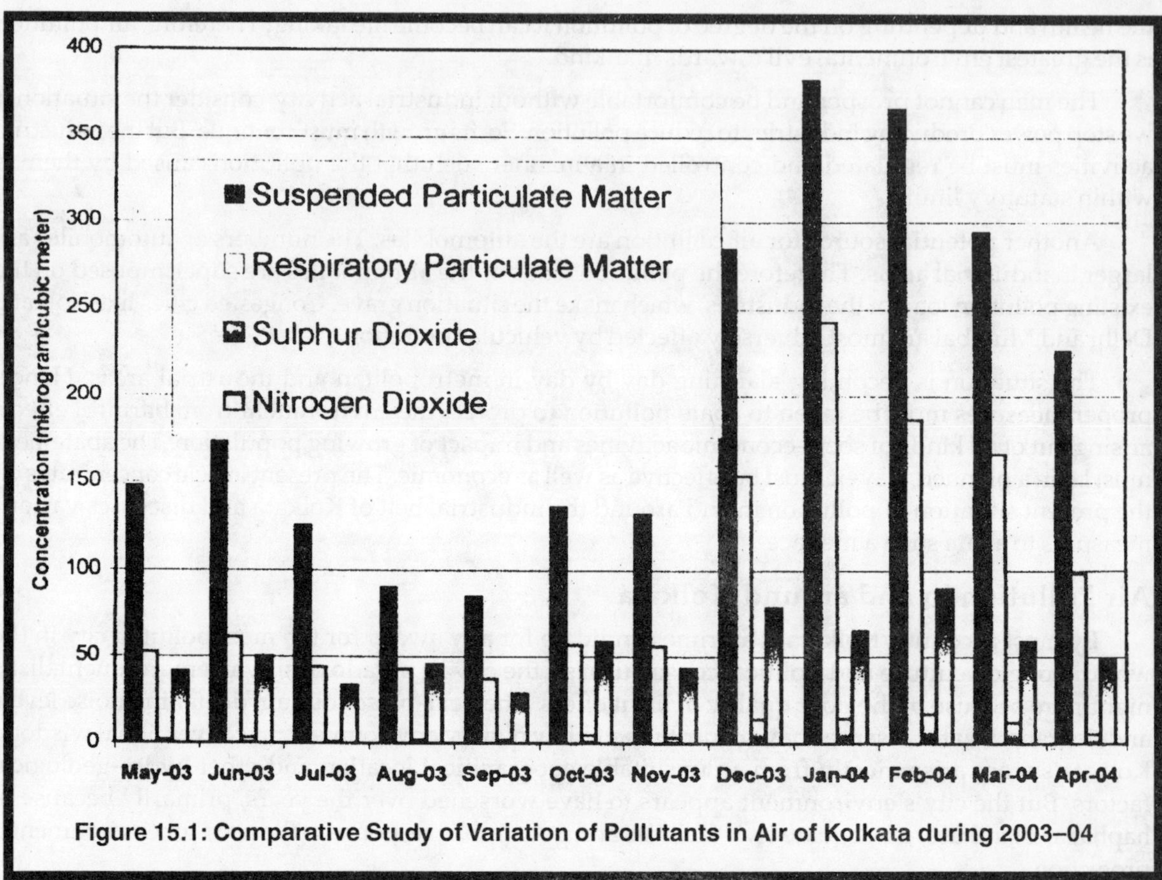

Figure 15.1: Comparative Study of Variation of Pollutants in Air of Kolkata during 2003–04

proper protective measures were not taken during their birth. But fortunately now the people have become aware of the environment and they have been able to appreciate that a corporate ethics must be developed in order to protect it.

The control of pollution caused by the industries should be given top priority. Standards of air pollution control have been prescribed for different areas and are given in Table 15.1.

Table 15.1

Area Category	*Concentration in Micro-gm/m³*			
	SPM	*SO_2*	*CO*	*NO_2*
Industrial and mixed area	500	120	5000	120
Residential and rural area	200	80	2000	80
Sensitive area	100	30	1000	30

The automobile exhausts contain many harmful gases and the pollution caused by them is significant in the ground level. Pollutants are exhausted from three points in an automobile–the exhaust point, crankcase blow by and the evaporative losses. Only hydrocarbons are emitted from the last two sources as a result of fuel evaporation.

The Government of West Bengal has now become quite aware of the devastation the menace of air pollution can produce in the near future. Some strong measures have already being planned by the Government to abate pollution in Kolkata and its suburbs. The then Hon'ble Minister-in-Charge Environment, Tourism and Youth Affairs. West Bengal, Sri Manabendra Mukhopadhyay in a Press Conference held at Paribesh Bhawan on 27th January, 2001 announced the actions which are under consideration of West Bengal Pollution Control Board and Department of Environment, Government of West Bengal for improving air quality of the city. The proposed steps include conversion of all the coal fired boilers operating within the city limit to oil fired ones and exploring the option of closing the operation of New Cossipore Thermal Power Station of CESC during the winter months when the air pollution level of the city is at its worst.

In terms of resolutions adopted by the Board in its 117th meeting held on 19.12.2000 and in exercise of the powers conferred upon the Board under section 15 of the Air (Prevention and Control of Pollution) Act, 1981 and sub-section (3B) of section 12 of the Water (Prevention and Control of Pollution) Act, 1974, the West Bengal Pollution Control Board is pleased to empower in addition to the Member Secretary of the State Board under section 25 of the Water (Prevention and Control of Pollution) Act, 1974 and under sections 21, 24, 25, 26 of the Air {Prevention and Control of Pollution) Act, 1981 as amended from time to time the in-charge of Regional Offices of the State Board within their respective region to grant or to refuse 'Consent to Establish' for the Small Scale Special Red category (excepting those attracting EIA notification of Ministry of Environment and Forests, Government of India) and other Small Scale ordinary Red/Orange/Green category of industries which are not looked after by the General Manager, District Industries Centres.

West Bengal Pollution Control Board and the Department of Environment, Government of West Bengal in exercise of the powers conferred under section 17(1)(g) and section 20 of the Air (Prevention and Control of Pollution) Act, 1981 issued the following order: "All four wheeled vehicles (other than passenger vehicles) with GVW equal to or less than 3500 Kg and vehicles with GVW exceeding 3500 Kg brought for registration before Registering Authorities in Kolkata Metropolitan Area shall on and from 23rd October, 2001 conform to the respective norms specified for these categories of vehicles in the Central Motor Vehicles (2nd Amendment) Rules, 2001 published vide Government of India Notification No. G.S.R. 286(E) dated 24.04.2001."

Conclusion

All the socio-economic activities aiming at betterment of life causes environmental degradation - the degradation of air in particular. So, time is ripe to take strong measures for the abatement of pollution all over the world. New laws are being introduced and standards are being set on pollution concentration. Industries have been enforced to accept the legislation and to keep their discharges within the statutory limits. The environmental standards have been fixed up with a look to the trend lines of industrial growth, population growth and increasing need for food and the changes in social patterns and economic activities. If by proper regulatory measures the pollution is kept within the specified standards, it will not be very much harmful for the biosphere.

References

West Bengal Pollution Control Board Website

The Telegraph, various issues

Chapter 16

The Effect of Dairy Effluent on Neighbouring Ecosystem

Jyoti Sharma and Anil Kumar Yadav

Department of Chemistry, Government Raj Rishi P.G. College, Alwar – 301 001, Rajasthan
Email: drjyotisharma@coolgoose.com, anil131177@yahoo.com

ABSTRACT

A study has been carried out to determine the quality of effluent of milk dairy of Alwar (Rajasthan). The study was conducted during the month of January to December 2002. Nearly sixteen physico-chemical and biological parameters were studied and result have been presented in Table 16.1, which shows that the water is totally unfit for irrigation, fish culture, cloth washing and drinking purposes for animals because this wastewater contains high level of inorganic salt and coliform.

Keywords: *Wastewater, Water quality, Total solid and Coliform.*

Introduction

Man is damaging the earth in various ways; one of which is through environmental pollution. Product of erosion, sewages, industrial wastage, deforestation, agricultural practices, uses of pesticides, insecticides and chemical fertilizers causes water pollution (Cunningham and Saigo, 1994).

Even water a precious natural resource is not exempted from this damage. Sadly this wonderful gift of nature is now under threat, from none other than man himself in its totality. But the product of human activities such as sewage, industrialization, deforestation, construction, agricultural practices may sum up to the pollution caused over long period (Kamble and Tayade, 2001).

The present study was conducted keeping this fact in mind to determine the quality of wastewater of milk dairy of Alwar district which is used for irrigation, washing and drinking purposes for animals. Milk dairy is a major industry of the Alwar city. It supplies the milk for three districts *i.e.*

Alwar, Bharatpur and Dausa. This industry is situated on the Alwar–Jaipur by pass. It discharges the effluents into a Nallaha, which reaches the field.

This contaminated water is absorbed by the soil and thus the groundwater is also polluted.

The farmers use this water for irrigation purposes and grow the vegetables by this water and this contaminated vegetables are sold in the market. These vegetables causes various diseases and this wastewater has very foul smell which causes problem to the residents nearby.

Taking all these things into consideration the author has investigated some physico-chemical and biological parameters of the effluent coming out such as pH, EC, chloride, fluoride, nitrate, total hardness, Ca-hardness, Mg-hardness, alkalinity, dissolved oxygen, total dissolved solids, total suspended solid, total solid, iron, arsenic and total coliform by standard method.

Experimental

Sample Collection

Samples were collected in polythene bottles of good quality of half litre capacity and brought to the laboratory without adding any preservative. To avoid contamination the sample was collected directly in rinsed bottles. Monitoring was done during the first week of every month for a period of one year.

Analytical

Only highly pure (Analar grade) chemicals were used for analysis. pH, dissolved oxygen and electrical conductance were determined within six hours of collection. The other physico-chemical and biological parameters were estimated by the standard procedures. The estimated parameters and methods used are listed in Table 16.1.

Table 16.1: Estimation of Various Physico-chemical and Biological Parameters by Different Methods

Sl.No.	*Parameter*	*Method Used*
1.	pH	pH-Metry
2.	Electrical Conductance	Conductometry
3.	Chloride	Argentometric method
4.	Nitrate	Nitrate Ion Meter
5.	Fluoride	Fluoride Ion Meter
6.	Total hardness	EDTA Titration
7.	Ca-Hardness	EDTA Titration
8.	Mg-Hardness	EDTA Titration
9.	Alkalinity	Titrimetry
10.	Sulphate	Titration Method
11.	Total dissolved solid	Conductometry
12.	Total suspended solid	Filteration method
13.	Iron	Double layer method
14.	Arsenic	Double layer method
15.	Coliform (MPN/100ml)	Multiple tube dilution method
16.	Dissolved Oxygen	Titration method
17.	Biological oxygen demand	Titration method

Results and Discussion

The physico-chemical and biological parameters of the milk dairy wastewater sample collected, are given in Table 16.2. These parameters clearly indicate that the water is highly polluted due to high coliform and solids.

Table 16.2: Physico-chemical and Biological Analysis

Sl.No.	Name of Parameters	Jan.	Feb.	Mar.	Apr.	May	Jun.	Jul.	Aug.	Sept.	Oct.	Nov.	Dec.
1.	pH	7.62	7.34	7.64	7.88	8.04	8.10	8.01	7.89	7.44	7.59	7.80	8.30
2.	Chloride	170	180	190	250	200	210	240	240	160	320	370	390
3.	Fluoride	0.4	0.8	1.4	1.2	0.2	0.4	0.2	0.4	0.8	1.2	1.8	1.6
4.	Total Hardness	180	170	200	220	200	230	270	200	170	200	240	260
5.	Ca-Hardness	100	80	130	140	110	120	90	70	60	90	110	190
6.	Mg-Hardness	80	90	70	80	90	110	180	130	110	110	130	70
7.	Nitrate	5	15	25	5	40	20	15	20	30	40	50	35
8.	Sulphate	21.4	20.2	24.8	22.9	29.6	35.8	14.8	14.7	29	30.7	31.2	31.6
9.	TDS	1400	1176	1456	2240	1848	2016	1960	1736	1456	1240	2352	2464
10.	TSS	70	66	78	92	87	98	72	70	75	68	72	85
11.	Alkalinity	1210	1200	1280	1210	1240	1230	1220	1200	1220	900	920	940
12.	Dissolved Oxygen	6.8	6.9	6.8	6.7	6.5	6.3	6.1	6.7	7.0	6.8	6.6	6.9
13.	BOD	2.8	2.9	2.7	2.9	2.6	2.4	2.0	3.0	2.4	2.9	2.8	3.0
14.	EC	2000	1680	2080	3200	2640	2880	2800	2480	2080	3200	3360	3520
15.	Coliform (MPN/100ML)	2400	2400	2400	2400	2400	2400	2400	2400	2400	2400	2400	2400
16.	Iron	ND	ND	ND	ND	ND	ND	ND	ND	ND	ND	ND	ND
17..	Arsenic	ND	ND	ND	ND	ND	ND	ND	ND	ND	ND	ND	ND

TDS: Total Dissolved Solid; TSS: Total Suspended Solid; BOD: Biological oxygen demand; EC: Electrical Conductance.

All Values are in mg/l except pH, EC and coliform.

It is totally unfit for irrigation, fish culture, washing and drinking purposes for animals.

pH values were always greater than 7.0 implying that the water remains in the alkalinity range throughout the year. Chloride values varied from 170 mg/l to 390 mg/l. Chloride values are higher than the Indian standards. Similarly fluoride values varied from 0.2 PPM to 1.8 PPM and these values are also higher than the Indian standard. Total hardness values varies from 170 mg/l to 270 mg/l. Nitrate values were also high in the month of the November *i.e.* 50 mg/l and these values varies from 5 mg/l to 50 mg/l. Sulphate values varies from 15 mg/l to 36 mg/l. Total dissolved solid and total suspended solid values were very high and varied from 1176 mg/l to 2464 mg/l and 66 mg/l to 98 mg/l. Total solids values changes according to TSS and TDS values. Dissolved oxygen values varied from 6.3 mg/l to 7.0 mg/l and biological oxygen demand values varied 2.4 mg/l to 3.0 mg/l. Both values shows that the water is highly polluted and electrical conductance values varied from 1680 micromhos/cm to 3520 micromhos/cm and this wastewater contains high coliform above 2400 MPN/100 ml from January to December. Iron and Arsenic are totally absent in the wastewater.

When the values of milk dairy wastewater's parameters are compared with the recommended standard for drinking purpose for animals, irrigation, fish culture and washing, we find that the most of the parameters are above the recommended limit. This water is highly polluted because it contains high level of inorganic salts and coliform. Many diseases are caused due to polluted water because farmers use this water for irrigation purpose and grow vegetables and this contaminated vegetables are sold in the market.

So it may be concluded that the wastewater of the milk dairy is highly polluted and totally unfit for any use as is used with out further treatment (mainly disinfection).

References

APHA (1995). *Standard Methods for the Examination of Water and Wastewater*, 19th Ed. American Public Health Association, New York.

Baruah, B.K., D. Baruah and M. Das (1996). Study on the effect of paper mill effluent on the water quality of receiving wetland. *Pollution Res. J.*, 15(4): 389–393.

Basu, A.K. (1996). Studies on the effluents from pulp and paper mills and its role in bringing the physico-chemical changes around the several discharge points in the Hoogly river estuary. *J Irstn. Engrs. India*, 46(10): 108–116.

Cunningham and Saigo (1994). *Environmental Science.* Mcgraw-Hill Publishing Company, p. 435–442.

Kamble, G.C. and D.T. Tayade (2001). Studies of some physico-chemical parameters and hydrobiological algal pollutants of water sources in Revasa. *Oriental J. Chem.*, 17(3): 493–496.

Major ion correlation in groundwater of Kancheepuram Region, South India. (2003). *Indian J. Environ. Health*, 45(1): 5–10.

Manivasakam, N. (1996). *Physico-chemical Examination of Water Sewage and Industrial Effluent*, Pragati Prakashan, Meerut.

Mishra, K.D. Impact of sewage and industrial pollution on physico-chemical characteristics of water in river Betwa at Vidisha, (M.P.). In: *Assessment of Water Pollution*, (Ed.) S.R. Mishra, pp. 453–463.

Rajurkar, N.S. *et al.* (2003). Physico-chemical and biological investigations of river Umshyrpi at Shillong, (Meghalaya). 45(1): 83–92.

Sharma, Jyoti and Anil Kumar Yadav (2003). Evaluation of groundwater quality in the village of tehsil Ramgarh (Alwar). *Chemistry: In Indian Journal*, 1(2): 137–143.

Sharma, Jyoti and Anil Kumar Yadav (2003). Monthly valiation of a few parameters of water of two lakes (Jaisamand lake and Silisade lake) of Alwar (Rajasthan). *J. Applied Science Periodical*, 5(3): 125–130.

Singh, Anil Pratap and Jaswant Singh. Impact of paper mill effluents on fish fauna of Amit river at Sant Kabir Nagar (U.P.) India, *J Liv. World.*

Trivedi, R.K and P.K. Goel (1984). *Chemical and Biological Methods for Water Pollution Studies.* Environmental Publications, Karad.

Chapter 17

Energy Content of the Agro-based Industrial Solid Waste

B.G. Pachpande, V.S. Patel, S.R. Kulkarni, S.B. Attarde and S.T. Ingle

School of Environmental and Earth Sciences,
North Maharashtra University, Jalgaon – 425 001

ABSTRACT

The energy demand is increasing with increase in population all over the world. To meet this demand the use of fossil fuels has increased and is responsible for the increased concentration of gases pollutants in the atmosphere. Utilization of biomass as a alternative source of energy to the fossil fuel will reduce the load of gases pollutants to the atmosphere. The biomass is generated in the form of agricultura1 residue or waste generated in the fruit and food processing industry is a appreciable source of energy. In the present investigations the calorific value moisture content and ash content of various food and fruit processing industries was evaluated. The manorial characteristics of this waste were also evaluated. The results of the study shows that the energy content of the water vary as per the waste characteristics.

Introduction

Biomass contributes as the world's fourth largest energy source today up to 14 per cent of the world's primary energy demand. In developing countries it can be as high as 35 per cent of the primary energy supply. Biomass is a versatile source of energy in that it can he readily stored and transformed into electricity and heat. It has also the potential that it is used as a raw material for production of fuel and chemical feedstock.

India has an abundant supply of biomass resources, which could be a potentially significant source of energy. Some resources are already being exploited for energy, but considerable amounts are still treated as waste, and remain untrapped. Biomass products serve a number of needs, including fuel, feed and fodder. It has long been identified as a sustainable source of renewable energy. The agricultural and food processing industry considered to be the largest contributor to the total annual production of solid waste (Parikh, 1984; Katyal and Satake, 2001). Now biomass is considered as one of the most prospective energy sources in the future for reducing carbon dioxide emissions and conserving fossil fuel resources (Yamamoto, 2001, Jingjing *et al.*, 2001). However, power generation using a solid fuel had significant limitations with respect to materials handling and efficient energy conversion. Converting biomass fuel into a liquid eliminates many of these problems and makes possible the use of higher efficiency combined cycle systems for power generation (Kumasaki, 2000). The fast depletion of fossil fuel and large-scale deforestation has created serious energy crises in the developing nations (Kulshrestha *et al.*, 1996). The characteristics of agro and agro industrial waste vary as per the source of generation. Due to diverse nature the utility of this material for energy becomes limited (Grover, 1996).

The biomass generated from the agro-based industries is putricible, if placed open it gives foul smelling. Open dumping and uncontrolled burning of agro residue, agro industrial waste add pollutant in the environment (Katyal and Satake, 2001). The proper way to dispose the agro residue and agro industrial waste is its conversion in to organic manure or energy fuel.

In recent years the numbers of agro based industries are emerged in the country. Amongst the agro based industries food and fruit processing industries are well developed throughout the country. In most of the food and fruit processing industries the waste is generated in the solid form. The pulp and juice industries generates fruit peels and covers *e.g.*, Mango, Onion, Mosambi, Chiku, Banana, Pineapple, Dalimb peels and fruit covers etc. as a waste.

The agro based industries like sugar mill generates bagasse as a waste in the form of fibers, press mud in the form of solid particles. Oil industries generate groundnut peels, cottonseed covers and other seed covers, which are used for extraction of oil. The waste generated in the industries is organic in nature carrying appreciable amount of energy. The industries should encourage exploring the energy from such waste (Katyal and Satake, 2001). Since bagasse possesses a good calorific value majority of industries use it as a substitute for coal in the boiler house for steam generation. It is estimated that 2000 MW energy can be explored only from bagasse generated by sugar industries (Kulshrestha *et al.*, 1996). Food processing, brewery, sugar mills, distilleries, paper mills are the major industries of solid waste generation.

The present work was carried out to investigate the energy content of the solid wastes and its suitability for use as fuel. Moisture content, density and calorific values are examined from estimating the energy content of solid wastes. The solid waste was also analysed for the parameters *viz.* organic matter, % nitrogen, % phosphorous and % potassium to know its suitability for composting.

Materials and Methods

Energy Content

Samples were collected from the food and fruit processing industries. The waste used to anlayze were piles of dalimb, mosambi, chiku, banana, ground nut, potato, onion, arhar, cottonseed cover, baggase and press mud. The samples were dried for five days in sunlight. These were crushed and ground to fine powder. Quarterization method was used to obtain the representative sample. The dried sample was prepared by grinding. Bomb calorimeter (Petroleum Instrument Corporation) was used to determine the calorific values of the samples containing little water at the bottom of the bomb. The bomb was assembled and then charged with the oxygen at the pressure 3.0×10^6 N/m^2 (30 atm.) without displacing the air from it. The temperature changes were noted with the high accuracy digital thermometer. The time when the rate of change of temperature becomes constant was noted as a chief period and after it was considered as the after period. The bomb solution was titrated for sulphuric and nitric acid correction.

Manurial Characteristics

The manurial characteristics of the water *viz.*, nitrogen, phosphorous, potassium, moisture and density were estimated by standard methods.

Total nitrogen was estimated by Kjeldahl nitrogen estimation method, potassium by flame photometry and phosphorous by ammonium molybdate-stannous chloride method.

The calorific values were calculated using the Regnault-Pfaundler correction as follows:

Firing Wire Correction

1. Weight of firing wire between the pole pieces of the electrodes in grams.
2. Heat released from the wire in cal. (Calorific value of the Nichrome wire = 335 cal/gm).

Cotton Weak Correction

1. Weight of cotton weak in grams.
2. Heat released from the ignition of the cotton weak in cal. (Calorific value of cellulose = 4180 cal/gm)

Sulphuric Acid Correction [3.60 (a + b – 20)] in Cal.

A = Titer for the bomb solution with standard 0.1 N HCl

B = Titer for bomb solution with standard 0.1N $Ba(OH)_2$

Cooling Correction for the Bomb Calorimeter

n = No. of minutes in chief period

v' = Rate of fall of temperature per minute in preliminary period in °C

v'' = Rate of fall of temperature per minute in after period in °C

t' = Average temperature during preliminary period in °C

t'' = Average temperature during after period in °C

t_0 = Initial temperature at the moment of firing of bomb in °C

t_a = Maximum temperature attained by calorimeter in °C

$$k = \frac{v'' - v'}{t'' - t'}$$

$$\Sigma(t) = (t_1 + t_2 + t_3 + t_4 + \ldots\ldots\ldots\ldots\ldots\ldots\ldots\ldots t_{n-1})$$

$$S = \{\Sigma(t) + 0.5(t_0 + t_a) - nt^7\}$$

Temperature decreases due to calorimeter cooling *i.e.*, the cooling correction was given by

$$nv^7 + kS \text{ in } ^\circ C$$

Calorific Value of the Sample

1. Uncorrected temperature rise ($t_n + t_0$) in °C
2. Corrected temperature rise (Uncorrected temperature rise + cooling correction) in °C
3. Uncorrected heat liberated (Effective mean heat capacity of calorimeter × corrected temperature rise) in cal.
4. Corrected heat liberated [Uncorrected heat – (Firing wire correction + cotton weak correction + sulphuric acid correction + nitric acid correction] in cal.
5. Calorific value of sample in cal/gm

Standard Deviation

$$\overline{X}\,(\text{Avg}) = \frac{X_1 + X_2 + X_3 + \ldots\ldots\ldots\ldots\ldots\ldots\ldots X_n}{n}$$

$$S\,(\text{Standard Deviation}) = \sqrt{(X_1 - X)^2 + (X_2 - X)^2 \ldots\ldots\ldots\ldots (X_n - X)^2/n-1}$$

$$n - 1 = n\ldots\ldots\ldots\ldots \text{ when n is smaller}$$

Results and Discussion

All organic material produced by plants or any conversion process involving life is called biomass. Biomass produced by dedicated cultivation and also the by-products from agricultural business are being used for energy purposes. The main product of agricultural business is food and biomass for energy production and should in no way interfere negatively with food production. The biomass streams contain more or less organic material, which has ever been produced by plants. A growth of the use of biomass as primary energy resource is feasible, but break through is necessary to bring this estimated potential within reach.

Present study highlights on the decomposition abilities and calorific values of the selected food and feed industry wastes such as dalimb, chiku, groundnut peels, and baggase etc. shows that all the samples vary in organic matter content. Baggase contains 98.3 per cent of the organic matter. Nitrogen estimation indicates that the Banana peels and Potato peels are deficient in nitrogen with Arhar (Tur) was rich in nitrogen content with value of 0.42 per cent and rest of them have moderate amount of nitrogen in the samples. Tur and dalimb peels have maximum of 3.5 and 4.0 per cent phosphorous content respectively. Onion peel has least phosphorous content (0.015 per cent) while other has on an average phosphorous content. As far as the potassium concerns onion peels and press mud has least concentration while the others are rich in the potassium content.

From the data obtained we can conclude that though it has less amount of N, P, K values but due to high amount of putricible content the waste can be used as organic manure after composing. During stabilization of the waste the nutrients are recycled. Too low moisture content of the waste is a critical parameter as it affects on the metabolic activities of the organisms (Bhide and Sundersan, 1983).

Table 17.2 shows the results of energy content of the solid wastes generated in food and feed industry. In the study some samples have shown considerable calorific values such as dalimb (3792.38 ± 100 Kcal/kg), groundnut (3898.81 ± 44.0 Kcal/kg), cottonseed (4448.11 ± 44.0 Kcal/kg), press mud (3661.30 ± 36.0 Kcal/kg), baggase (4920.79 ± 20.0 Kcal/Kg) and onion (4200.57 ± 60.0 Kcal/kg).

Properly dried waste can be used as an alternative cheap source for the industry that requires low energy and also for domestic purposes.

Table 17.1: Manurial Characteristics (on Dry Basis) of the Solid Wastes Generated in Some Food and Fruit Processing Industries

Sl.No.	Type of Peels	Organic Matter %	Nitrogen %	Phosphorous %	Potassium %
1.	Dalimb	97.7 ± 1.2	0.28 ± 1.1	1.0 ± 1.1	1.62 ± 0.4
2.	Mosambi	95.40 ± 1.8	0.21 ± 0.14	0.35 ± 0.28	1.04 ± 0.5
3.	Chiku	95.4 ± 2.1	0.28 ± 0.18	2.3 ± 1.1	1.02 ± 3.0
4.	Banana	89.2 ± 4.3	N.D.	0.69 ± 0.54	2.01 ± 0.2
5.	Groundnut	54.0 ± 6.3	Trace	0.815 ± 0.21	2.20 ± 0.4
6.	Cottonseed	84.0 ± 2.8	Trace	0.84 ± 0.41	2.40 ± 0.3
7.	Potato	66.0 ± 9.5	Trace	0.835 ± 0.61	1.60 ± 0.6
8.	Press mud	82.9 ± 5.3	0.15 ± 0.11	0.205 ± 0.25	0.64 ± 0.25
9.	Baggase	98.3 ± 1.2	0.24 ± 0.12	0.65 ± 0.35	1.72 ± 0.3
10.	Onion	93.0 ± 2.5	0.14 ± 0.14	0.015 ± 0.05	0.74 ± 2.5
11.	Arhar	97.6 ± 1.8	0.42 ± 0.21	3.5 ± 1.1	1.92 ± 2.1

Values are mean ± S.D. of five estimations.

Table 17.2: Energy Content of the Solid Wastes Generated in some Food and Fruit Processing Industries

Sl.No.	Type of Peels	Moisture %	Density ug/cm³	Calorific Value kcal/kg
1.	Dalimb	19.88 ± 5.1	20.0 ± 5.0	3792.38 ± 100.0
2.	Mosambi	74.84 ± 4.8	38.0 ± 6.0	3152.13 ± 50.0
3.	Chiku	42.04 ± 5.4	28.0 ± 4.0	2987.16 ± 41.0
4.	Banana	84.44 ± 4.0	116.0 ± 4.40	3133.40 ± 80.0
5.	Groundnut	31.2 ± 4.5	220.0 ± 6.0	3898.81 ± 44.0
6.	Cottonseed	42.5 ± 2.3	234.0 ± 5.0	4448.11 ± 44.0
7.	Potato	82.0 ± 5.3	22.0 ± 4.0	3103.70 ± 33.0
8.	Press mud	41.12 ± 6.8	14.0 ± 6.0	3661.30 ± 36.0
9.	Baggase	70.00 ± 6.3	67.0 ± 5.0	4920.74 ± 20.0
10.	Onion	51.60 ± 7.2	67.0 ± 5.0	4200.57 ± 60.0
11.	Arhar (Tur)	25.6 ± 3.8	30.0 ± 4.0	3002.56 ± 30.0

Values are mean ± S.D. of five estimations.

In India a large proportion of electricity is generated from coal. The coal-fired power plants in our country are extremely modern and have fuel gas cleaning facilities with state of the art technology. These power plants are suitable for co-combination of biomass or organic waste. Meanwhile, most of the sustainably generated energy is produced by co-combination of firing of biomass with coal. Co-firing is the simplest form of biomass used in a coal fired power plants. In this process the biomass is carried with the pulverized coal to the boiler. In this way a part of coal used is replaced by biomass and the proportionate part of the calorific value of the biomass used can be considered as renewable energy.

References

Bhide, A.D. and B.B. Sundersan (1993). In: *Solid Waste Management in Developing Countries.*

Grover, P.D. (1996). Biomass feed processing for energy conservation. *International Conference on Biomass Energy System*, TERI Publication, India.

Jingjing, Li., Zhuang Xing, Pat DeLaquil and Eric D. Larson (2001). *Biomass Energy in China and its Potential, Energy for Sustainable Development*, p. 4.

Katya, Timmy and M. Satake (2001). In: *Environmental Pollution.*

Kulshrestha, M., M.V.S. Iyerand T. Thakur (1996). *Energy from Biomass: Prospects in India.* TERI Publication, India.

Kumasaki, M. (2000). Role of bioenergy in a society based on sustainable resource use. *Japan Tappi J.*, 154: 69–74.

Parikh, P.P. (1984). *State of the Art Report on Gasification of Biomass.* Report submitted to DNES, Government of India.

Yamamoto Koji (2001). Biomass Power Generation by CFB Boiler. *NKK Technical Review*, 85: 29–34.

Chapter 18

Effects of Nutrients and Oxidising Agent on Degradation of Crude Oil Hydrocarbons in Soil Under Natural Environment

Krishna G. Bhattacharyya, Satyendra K. Choudhury** and Prahash C. Sarma***

**Department of Chemistry, Gauhati University, Guwahati – 781 014*
***Department of Chemistry, Cotton College, Guwahati – 781 001*

ABSTRACT

Soil pollution due to crude oil and its different fractions is of known occurrence. Soil microorganisms use such hydrocarbons as the source of carbon necessary for their energy. Microbial degradation is a process of oxidation, limited by the availability of moisture, air, nutrients etc. In the present experiment, extent of degradation of crude oil hydrocarbons in a pre-characterized soil sample is determined in three sets under natural environment in a period of 120 days. Soil, soil mixed with NPK nutrients and soil mixed with hydrogen peroxide are taken in three different sets and the experiment is done by polluting these with measured quantities of crude oil. It has been found that nutrient supplementation can greatly enhance the extent of degradation. The effect of hydrogen peroxide is moderate. The organic carbon content of the samples is found to be increased.

Keywords: *Crude oil, Degradation, Microorganisms, Texture, Soil pH, Electrical conductivity, Bulk density, Porosity, Hydraulic conductivity, Water holding capacity, Total organic carbon.*

Introduction

Plants are the best supporter of animal life and soil is the basis of rooted plants. Therefore prevention of land degradation and rehabilitation of degraded soil needs attention. Leaks, spillages and accidental fallout of crude oil and its different fractions have greatly affected land resources (Atlas, 1973). Experiments showed that initial rate of degradation of some crude oil hydrocarbons in soil are linear (Fu, 1992) but overall decrease of hydrocarbon from soil may be logarithmic with time. The saturated fraction, normal and branched alkanes are totally eliminated within a short period. Cycloalkanes and aromatic compounds take time for degradation. The resin fraction and some polycyclic aromatic compounds are persistent to degradation (Oudot, 1995). Hopanes are common constituents of crude oil and they are very resistant to biodegradation (Prince, 1994). Biodegradation involves the process of oxidation (Jone, 1968).

Types and quantity of mineral oil and the nature of contaminated soil and climate conditions have a major influence on the course of recovery of a polluted soil. The joint action of micro-organisms such as bacteria, fungi etc. result in an increase in the rate and amount of oil biodegradation (Davies, 1979). There are some common measures by which the process of rehabilitation can be expedited. Liming, fertilization, tiling, aeration are some other measures, which are normally adopted. Also decrease in oil concentration shows a definite correlation with temperature that has strong effects on biodegradation and evaporation (Dibble, 1979). Crude petroleum is converted to soil organic matter by bacteria and fungi. During the conversion the organisms which are free livers, fix fairly large amounts atmospheric nitrogen in their soil. Later, this nitrogen becomes available for plant growth and the organic matter improves soil physical conditions (Plice, 1948).

Materials and Methods

The soil sample collected from spots having no background of oil pollution within Guwahati city (Latitude 26° 11′ N, Longitude 91° 45′E) are characterized with respect to texture by Bouyoucos Hydrometer method (Sing, 2000), soil organic carbon by chromic acid titration method (Walky, 1934), pH by digital pH meter using glass electrode, electrical conductivity by digital conductivity meter, bulk density, porosity, hydraulic conductivity based on Darcey's law and water holding capacity (Trivedy, 1987). Known quantities of soil samples are taken in a total of 18 (6 × 3) polythene bags and then polluted by crude oil addition with the help of an emulsifier. NPK nutrients and hydrogen peroxide are then mixed in the second and third sets respectively as in Table 18.1. These are then placed indoor (Plate 18.1).

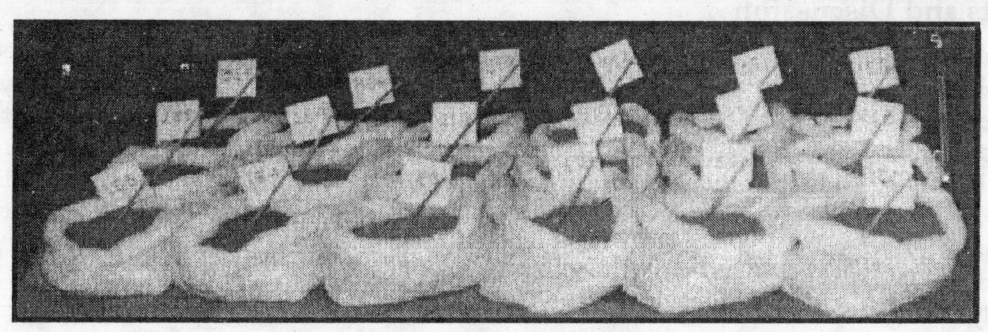

Plate 18.1: Indoor Placement of Samples

Table 18.1: Samples and Their Contents

Set	Sample Number	Soil Taken	NPK Added	H_2O_2 Added	Emulsifer Added (Labolene)	Crude Oil Added
		g	g	mL	mL	g
1	1A0	3000	00	00	2	00
	1A1	2997	00	00	2	03
	1A2	2985	00	00	2	15
	1A3	2970	00	00	2	30
	1A4	2955	00	00	2	45
	1A5	2940	00	00	2	60
2	2A0	2990	10	00	2	00
	2A1	2987	10	00	2	03
	2A2	2975	10	00	2	15
	2A3	2960	10	00	2	30
	2A4	2945	10	00	2	45
	2A5	2930	10	00	2	60
3	3A0	3000	00	10	2	00
	3A1	2997	00	10	2	03
	3A2	2985	00	10	2	15
	3A3	2970	00	10	2	30
	3A4	2955	00	10	2	45
	3A5	2940	00	10	2	60

At an interval of 25–30 days around 20g of accurately measured soil from each samples are withdrawn and Soxhlet extraction (Clesceri, 1989) with the help of petroleum benzine of 60–80°C boiling range is done and the amount of undegraded material is determined by subtracting the same obtained in the control samples. Number of components having bp ≤ 220°C is then determined by GC instrument under a fixed set of conditions in some of the samples extracted with HPLC grade n-hexane. At the end of the experiment the total organic carbon of the samples are determined.

Results and Discussion

The background of the soil sample is given in Table 18.2. The amount of undegraded petroleum ether soluble part recovered from the soil samples are presented in Table 18.3. The extent of degradation in percentage of the applied amount is given in Figures 18.1–18.3. It has been found that the NPK supplementation is an effective measure of rehabilitation of oil inundated soil. Supply of nutrients to the soil bags helps the micro-organisms to grow abundantly, consequently the degradation becomes faster. Hydrogen peroxide decomposes some of the organic matter present in the polluted soil and it also contributes towards the degradation to some extent. It has been seen that almost 50 per cent degradation in all the three sets have occurred during the first two months of placement. The city biweekly average ambient temperature and relative humidity data during the period of the experiment are given in Table 18.4. The experiment was done during summer. There was mild increasing trend of temperature during the study period. G.C. analysis of three samples *viz.*, 1A5, 2A5 and 3A5 from three sets in the third month since placement have shown that number of volatile components (bp ≤ 220°C) becomes highest in the NPK applied samples, and least in H_2O_2 applied samples. The number of peaks including solvent becomes 13, 21 and 7 respectively. Since extent of degradation in the 2nd set is highest, number of degraded component are highest. Since hydrogen peroxide removes organic compounds from soil, a good number of organic compounds have been removed from soil, and number of components remained in soil becomes least.

Table 18.2: Physico-chemical Properties of the Collected Soil Sample

Parameter	*Value*	*Unit*	*Parameter*	*Value*	*Unit*
Sand	06.0000	Per cent	Electrical Conductivity	00.358	$mScm^{-1}$
Silt	20.0000	– do –	Bulk Density	01.120	gcc^{-1}
Clay	74.0000	– do –	Porosity	57.740	Per cent
Organic Carbon	00.3946	– do –	Hydraulic Conductivity	00.169	$cmmin^{-1}$
pH	06.0000	–	WHC	26.980	w/w %

Table 18.3: Amount of Undegraded Petroleum Benzine Soluble Part of the Oil Obtained after Subtracting the Value Obtained in Control Bags

Sample Number	*Time in Months*				*Mean*	*Corr. Coeff.*
	1 m	*2 m*	*3 m*	*4 m*		
1A1	01.930	01.532	00.569	00.223	01.063	– 1
1A2	08.114	07.391	0.5.702	02.133	05.835	– 1
1A3	19.191	15.312	10.832	05.151	12.621	– 1
1A4	24.132	22.494	13.874	09.322	17.455	– 1
1A5	33.932	30.631	19.532	10.527	32.658	– 1
2A1	01.413	01.200	00.454	00.000	00.772	– 1
2A2	07.700	06.575	04.321	01.897	05.123	– 1
2A3	14.913	12.030	08.432	04.023	09.849	– 1
2A4	20.312	17.757	12.391	08.413	14.718	– 1
2A5	29.432	24.603	12.757	09.657	19.112	– 1
3A1	01.613	01.400	00.465	00.110	00.897	– 1
3A2	09.413	07.342	05.003	02.003	05.940	– 1
3A3	18.973	14.362	09.319	06.312	12.241	– 1
3A4	23.727	21.632	13.737	08.499	16.899	– 1
3A5	33.419	25.319	14.783	10.322	20.961	– 1

The values of organic carbon determined on completion of 120 days in all the samples are given in Table 18.5. It is seen that there is an increase of organic carbon in the entire sets compared to the original soil. The average increase is 0.2219, 0.3599 and 0.2647 per cent in the three sets respectively. The set with highest degradation has the highest organic carbon content. A plot of land unsuitable for agricultural purpose due to lack of organic carbon perhaps be made suitable by degrading hydrocarbon wastes effectively.

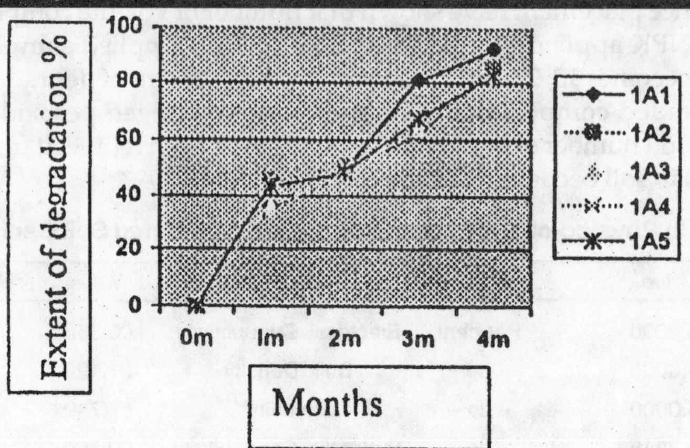

Figure 18.1: Extent of Degradation of Hydrocarbons with Respect to Time in Months for Set 1

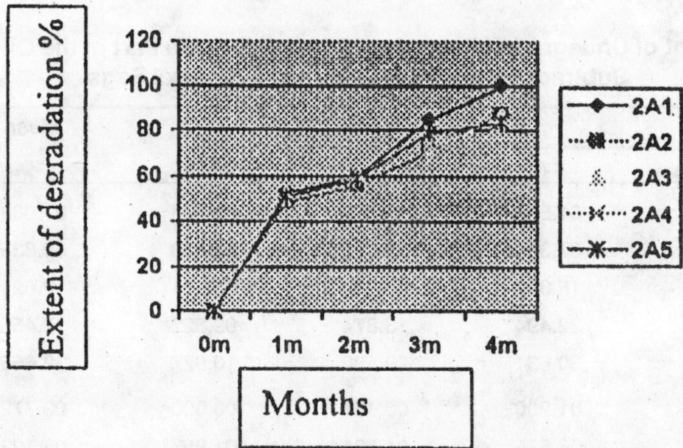

Figure 18.2: Extent of Degradation of Hydrocarbons with Respect to Time in Months for Set 2

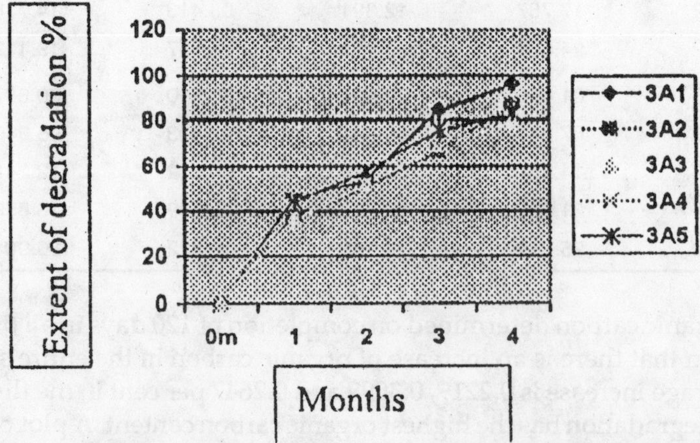

Figure 18.3: Extent of Degradation of Hydrocarbons with Respect to Time in Months for Set 3

Table 18.4: Biweekly Average Ambient Temperature and Relative Humidity of the City During the Period of the Experiment

Parameter	*1 m*		*2 m*		*3 m*		*4 m*		*Mean*	*Corr. Coeff.*
	1st Half	*2nd Half*	*1st Half*	*2nd Half*	*1st Half*	*2nd Half*	*1st Half*	*2nd Half*		
Temperature	28.03	28.45	28.60	30.07	30.36	29.55	29.37	28.74	29.1	+ 0.5
R Humidity	81.82	83.53	84.70	78.27	76.68	83.53	82.90	83.46	81.86	0

Table 18.5: Organic Carbon Content of the Samples at the End of the Experiment

Sample No.	*Organic Carbon %*	*Sample No.*	*Organic Carbon %*	*Sample No.*	*Organic Carbon %*
1A0	0.3946	2A0	0.4215	3A0	0.4315
IA1	0.4215	2A1	0.6087	3A1	0.6207
1A2	0.4294	2A2	0.6939	3A2	0.7057
1A3	0.9051	2A3	0.8563	3A3	0.9612
1A4	0.6928	2A4	0.9076	3A4	0.5980
1A5	0.6339	2A5	0.8405	3A5	0.5954
Mean (last five)	0.6165	Mean (last five)	0.7814	Mean (last five)	0.6962
Average increase (Mean-control data)	+ 0.2219	Average increase (Mean-control data)	+ 0.3599	Average increase (Mean-control data)	+ 0.2647

Acknowledgement

Authors are thankful to UGC, NER for financial assistance.

References

Atlas, R.M. and R. Bartha (1973). Abundance distribution and oil biodegradation potential of micro-organisms in Raritan Bay. *Environ. Pollut.*, 4: 291–300.

Clesceri, L.S., A.E. Greenberg and R.R. Trussell (1989). *Standard Methods for the Examination of Water and Wastewater–APHA*, 17th Edn. Washington DC.

Davies, J.S. and D.W.S. Wastelaki (1979). Crude oil utilisation by fungi. *Can. J. Microbiol.*, 25: 146–156.

Dibble, J.T. and R. Bartha (1979). Rehabilitation of oil inundated agricultural land: A case history. *Soil Sci.*, 128: 56–60.

Fu, M.H., and M. Alexander (1992). Biodegradation of styrene in sample of natural environment. *Environ. Sci. Technol.*, 26: 1540–1544.

Jones, D.F. and R. Howe (1968). Microbial oxidation of long chain aliphatic compounds. Part–I Alkanes and Alk-l-enes. *J. Chem. Soc.* (C) Org., p. 2801–2808.

Oudot, J., H. Chainean and L. Morel (1995). Microbial degradation in soil microcosms of fuel oil hydrocarbons from drilling cuttings. *Environ. Sci. Technol.*, 29: 1615–1621.

Plice, M.J. (1948) Same effects of crude petroleum on soil fertility. *Soil Science Soc. Amer. Proc.*, 13: 413–416.

Prince, R.C. *et al.* (1994). 17 alpha H, 21 beta H hopanes as a conserve internal marker for estimating the biodegradation of crude oil. *Environ. Sci. Technol.*, 28: 142–145.

Singh, D., P.K. Chonkar and R.N. Pandey (2000). *Soil Plant Water Analysis: A Methods Manual.* ICAR, p. 9–11.

Trivedi, R.K., P.K. Goel and C.L. Trisal (1987). *Practical Methods in Ecology and Environmental Science.* Environmental Publication, Karad, p. 120–121.

Walky, A.J. and I.A. Black (1934). Estimation of soil organic carbon by the chromic acid titration method. *Soil Sci.*, 37: 29–38.

Chapter 19
Ecological Management to Control Sal Tree Mortality

S.G. Ahmad

Department of Agricultural Structures and Environmental Engineering, Jawaharlal Nehru Krishi Vishwa Vidyalaya, Jabalpur – 482 004 (M.P.)

ABSTRACT

Sal (*Shorea robusta*) a fairly larger deciduous tree with majestic shining foliage is one of the most important trees of South Asia, found extensively in India and also in parts of Bangladesh and Myanmar.

Sal would stand a better chance of survival and development under "Ecological System of Management" rather than commercial management. Climatic parameters in relation to the management practices were studied as drought at shorter intervals are the main cause of Sal mortality.

Introduction

One of the worst uncertainties on earth is the climate. The geographic location and the physical features of land which are fast changing in the globe now pronounce the abnormal meteorological behaviour in environment. Meteorology and forestry which have corelative influences play a great deal in determining an effective model to minimize the decreasing trend and regional imbalance in Sal forest production specially in rain-dependent areas. India is a land of varied climate affording scope for much diversity in forestry.

The Himalayas run along its entire length in the north, to south of this barrier are the alluvial plains watered by mighty rivers. In the further south, there lies the plateau of peninsular India skirted by narrow coastal strips, the Arabian Sea to the west, the Bay of Bengal to the east and the Indian Ocean to the South. The vast enclosure certainly creates variation in the pattern and distribution of rainfall and other meteorological events. The Indian forestry mainly depends on the monsoon rains. Delay in the onset of rains and breaks in the rainy spells during the flowering, seeding and fruiting season are the problems which the Indian forester to confront with.

Weather elements in addition to precipition temperature, sunshine and wind influence the growth of Sal forest. As there is a variation from year to year in the advent of the monsoon over different divisions the Sal forest growth is fluctuating. Now the ability to conserve moistures of uncertainties in weather–events holds the key to sustainable rainfed forest growth.

As per 1981 census nearly 62 per cent of the tribal population in the country live in and around Sal forest. The problems of conservation, protection development and management of Sal forest can not be separated from the lives of the tribals. Sal seed is a valuable source of income, which has been yielding about Rs. 100 million per year since nationalization of the product. As the collection rates rise this income will increase. Sal seed collection have been identified a potential of 200,000 tones per year.

Gregarious Sal forests occurs in two almost continuous belts on either side of the Gangetic Plain in India. The northern belt stretches along the Sub-Himalayan tract, ascending, occasionally to 1200 to 1500 m-elevation in outer Himalayan tract, valleys *i.e.* from Kangra district in Himachal Pradesh in north west to Daraang district in Assam in the east. The second belt, in the south of Gangetic Plain tracts near Ganges in Santhal Paraganas in Bihar extending to Gujarat and Orissa and Shrikakulam in Andhra Pradesh in the south and upto Jabalpur district in M.P. in the west.

Madhya Pradesh and Chattisgarh is the largest forest and tribal states of the country. Almost 70 per cent of forest cover is dense forest. The combine total forest area of the State is 155,414 km^2 which is 35.09 per cent of the total geographical area. Teak forests account for 27,783 km^2 (18 per cent), Sal forests, 25,703 km^2 (16.5 per cent) and mixed miscellaneous species are found over an area of 101,927 km^2 (65.6 per cent).

The State of Madhya Pradesh and Chattisgarh forest covering an area of 22,152 sq km. Sal forests are found on old granites, metamorphics or on Gondwanas tracts. Sal also occurs on lateritic parts in Surguja and Jashpur, on Cuddapahs in parts of Bastar and even on basalts in Mandla and Jabalpur. The soils under Sal forests here are mostly red and yellow soils these are largely found in the Mahanadi basin and Chattisgarh region.

Sitandi has ideal growing conditions for Sal, because the rock is largely granite and organic, with soft sandy loam overlying the rock. The soil reaction is acidic, making it ideal for Sal. The area is free of forest, which makes Sal regeneration from seed that much the easier. Sal is a great colonizer after teak. It is a tree with a very long cycle of growth, ranging from between 100 and 160 years, of course Sal does not grow a trap, which makes the 80 degree inner radius the dividing line between teak and Sal zone. The Sal region of the country provides home to a considerable husk of the country tribal population of India on account of its social economic significance. The Sal has occupied a soft corner in the heart of the tribal.

Materials and Methods

The Sal Expert Committee in their report (1984) have observed that between 1969 to 1981, M.P. State lost about 11,000 km^2 Sal area, diverted for various developmental works, such as irrigation projects, industries, roads townships etc. However, there is no established research evidence to show that decline in forest area was on account of any ecological disturbances. Large scale mortality of Sal trees in sixties caused erosion of growing stock but was not found to have any impact on the total extent of Sal forests in the State.

Large scale mortality of Sal has occurred in various forest divisions, more pronouncedly in Kondagaon (Makdai Reserve), south Raipur (Risgaon, Sitanadi, Nagri and Birgudi Ranges), North Bilaspur Division (Achanak Mar Range), Surguja and east Mandla (Motinala Range). Forest Divisions engaged the attention of State Government, working plan was suspended in south Raipur division to salvage the dead trees. However, no such systematic works were done in other affected areas (Bilaspur and Surguja) on account of the fact that these areas had much less trees to be cut than south Raipur division. Severe drought in 1967 resulted in the death of 47,715 trees spread over 75 compartments in south Raipur division (Birgudi, Risgaon, Sitanandi and Nagri Ranges).

Drought was again experienced in 1973 and 1975. Reconnaissance survey revealed 233 compartments affected by Sal mortality. Yield from the cutting of dead trees was assessed to be about 28,000 cu m of timber. As a result of recurring droughts in 1976–77 and again in 1979–80, in all 309 compartments spread over an area of 94,898 hectares were found to be affected by the problem of large-scale mortality. The enumeration of drought affected trees recorded a figure of 9,01,737damaged trees for Sal. It was estimated that on an average 22.8 per cent trees per hectare (Sal area) have died due to drought during the period from 1975–76 to 1979–80.

For the management strategy for maintaining healthy Sal ecosystem the causative factors could be specified as followed:

1. The Sal crop is mostly of coppice origin and therefore trees dry up after a certain age die to repeated coppicing and consequent loss of vigour.
2. As against 5.2m spacing the prevailing spacing is 3.5m. Higher plant population would creates more hospitable micro-climatic regime and can be expected to reduce aridity and tree mortality.
3. Deficient natural regeneration is attributed to excessive biotic pressure.
4. Construction of large reservoirs, dames have disturbed the hydrological cycle.
5. Extensive heart rot (Khairi) also conjectured as a prominent causative factor.
6. Biotic factors such as (a) uncontrolled heavy grazing (b) frequent fires and (c) uncontrolled and unregulated removal of firewood by Nistaris; and headloaders.

The protection against these biotic pressure, may be useful in maintaining the Sal ecosystem. Many of these pressures cannot be eliminated but can be regulated to a great extent.

Results and Discussion

After touring various Sal divisions of the State, discussing the problem, with the local forest officials and scientists of SFRI/RFRC, Jabalpur and scanned available literature on Sal research, among the various recommendation, the emphasis is placed on regulatory measures to control, grazing and fuel headloads, stoppage of Sal seed collection after rains, soil and moisture conservation measures etc. These recommendations are yet to be implemented in the field.

Recurring drought years were attributed to be responsible for this malady especially following years of low rainfall. Lower limit of rainfall was calculated as 1271mm and higher limit as 1581.15mm. The problem of mortality was seen magnified in years receiving rainfall below the lower limit. In many years the annual rainfall was lower than 1000mm. The period following such a scanty rainfall witnessed maximum Sal mortality.

Tree species also depend upon secondary sources of moisture (atmospheric moisture through dew, fog etc.). Recurring fires and grazing, and the fuel head removal were stated to have eliminated the ground flora which otherwise acted as surface mulch and regulated evaporation from soil surface. Concentrated collection of Sal seeds, Mahuwa flowers and Tendu leaves, cutting of small poles and saplings for fuel wood for sale (even in small towns big villages) have eliminated the protective soil cover. During summer months when trees trigger-off moisture less (evapo-transpiration) through leaf shedding it is infact nullified by the surface evaporation on account of loss of plant cover. In last 10–15 years, the fires have assumed alarming proportion because to facilitate collection of different produce the forests are subjected to 3–4 deliberate fires caused by forest produce collection. Earlier, the magnitude of fire was not so pronounced as many produce were not collected for trade. General rural poverty is also said to have magnified the problem. In brief the research findings clearly indicated the need for soil and moisture conservation emphasized the urgency for biotic pressures.

Although no conclusive proofs are available about the causative factors of Sal mortality, recurring droughts at shorter Intervals are definitely related to this phenomenon. Studies need to be carried out of overcome the, problems created by recurring drought conditions. Man-made droughts also, need to be looked into as this aggravates the problem of moisture stress.

References

Anonymous (2000–2001). *Annual Reports of Meteorological Observatory of India of Indian Meteorological Department of Adhartal Farm*, JNKVV, Jabalpur.

Ahmad, S.G. (1996). Ecological and economic importance of Sal forest. *Udyanki Jeewan*, Bhopal.

Buch, M.N. *The Forest of Madhya Pradesh.*

Dube, S.B. and V.K. Sen. *Report of the Sal Expert Committee.*

Hewetspm, C.E. *Discussion on Ecological Position of Sal in Central India.*

Kulkarni, D.N. (1956). Comparative distributional characteristics of Sal and Teak in Madhya Pradesh, 9th *Silva Conference.*

Parsad, R. *Status of Sal (Shorea robusta) Forest of Madhya Pradesh: A Report.*

Tewari, D.N. (1995). *A Monograph on Sal* (*Shorea robusta* Ganertin f).

Chapter 20

Estimation of Efficacy and Economics of Chemical and Bio-insecticides for Management of Shoot and Fruit Borer of Okra

Rabindra Prasad, Devendra Prasad and Sanjay Kumar Sathi

Department of Entomology, Birsa Agricultural University, Kanke, Ranchi – 6

ABSTRACT

The over all results of two years field experiments indicated that avoidable loss in marketable fruits yield, caused by *Earia vitella* Fab., ranging from 34.01 to 52.85 per cent could be prevented by certain chemical and bio-insecticides. As such, the highest increase in the yield, over control, was recorded with the protection provided by alphamethrin (112.11 per cent) followed by lambdacyhalothrin (105.29 per cent), BTK, Delfin @ 1.0 kg/ha (104.12 per cent), BTK, Delfin @ 0.75 kg/ha (94.44 per cent) diflubenzuron (68.95 per cent), thiodicarb (87.62 per cent), dimethoate (77.62 per cent) and endosulfan (63.05 per cent) on account of substantial reduction in shoot damage (*i.e.* 5.10 to 7.06 per cent) and fruit damage (*i.e.* 9.12 to 20.06 per cent) caused by *E. vitella* in okra in the field condition. Five foliar spraying given at 15 days intervals, commencing first spray at 35th days after sowing, with alphamethrin (0.006 per cent) resulted to minimum shoot and fruit damage (2.99. and 3.35 per cent) and maximum yield of okra fruit (97.70q/ha) with highest net return (Rs. 18,186/ha) and BCR (7.36). BTK, Delfin WG@ 1.0 kg/ha remained at par with alphamethrin (0.006 per cent) giving rise to lower shoot and fruit damage (4.45 per cent and 3.69 per cent) and higher fruit yield (94.02 q/ha) and net return (Rs. 16,124/ha) with greater BCR (5.26) Although, the maximum damage of shoot (20.26 per cent) and fruit (23.41 per cent) caused by *E. vitella* with the lowest yield of marketable okra fruits (46.06 q/ha) were obtained when the crop was left unprotected for natural pest situation.

Keywords: *Bio-insecticides, Okra, Earias vitella, Avoidable loss, Net return.*

Introduction

Fruit and shoot borer (*Earias vitella* Fab.) is one of the most important insect pest of okra. The insect may cause enormous loss in the fruit yield varying from 35 to 53 per cent (Prasad, 2001). With the purpose to minimize the avoidable yield loss, the present experiment was undertaken to explore the information pertaining to the relative bioefficacy and economics of some chemical and bio-insecticide for control of *E. vitella* in the okra ecosystem for searching out alternative insecticides one as of the compatible component of IPM.

Materials and Methods

Field experiment was conducted during spring summer season in Madanpur village nearby main campus of Birsa Agricultural University Ranchi with okra (var. Pusa Sawani) sown in the 2nd week of February in during 1996 and 1997 in a randomized block design with 11 treatments and three replications in the plot size of 4.00 × 3.0 m^2. Eight insecticides belonging to three groups of chemical and two groups of bio-insecticides were taken for estimation of their relative bio-efficacy and economics for the control of *Earias vitella* Fab. infesting okra (Table 20.1.) Recommended package of practices were followed for raising the crop. Foliar spraying were applied, at fortnightly intervals given four times, commencing first spray at 35 days after sowing. Number of shoots (healthy and damaged) were recorded on all the plants on per plot basis to compute percentage of shoot damage. Weight of fruits (healthy and damaged) were recorded treatment-wise during each plucking and percentage of fruit damage was calculated. As such, yield of fruits recorded on per plot basis (kg/ha) were calculated in terms of q/ha after pooling yield of all the plucking in the form of year-wise and pooled mean of two years. Based on pooled data of two years result avoidable loss in yield and gain (increase) in yield, over control, was calculated by following formula:

$$\text{Avoidable loss in yield (per cent)} = \frac{T - C}{T} \times 100$$

$$\text{Increase (gain) in yield (per cent)} = \frac{T - C}{C} \times 100$$

where,

T = Yield of treated plot in the respective, treatment;

C = yield of obtained from untreated plot.

Finally, comparative economics in terms of net return and benefit cost ratio (BCR) of each treatment were worked out, as compared to unprotected crop based on mean data of two years result.

Results and Discussion

Results of relative bio-efficacy of the insecticides in terms of incidence of shoot and fruit borer and yield of fruits are presented in Table 20.1 and that of relative economics treatment-wise are shown in Table 20.2.

Table 20.1: Impact of Some Chemical and Bio-inseticides on the Incidence of Fruit Borer (*Earias vitella* Fab.) and Yield of Okra

Treatment	*Dose (% a.i).*	*Shoot Damage (%)*			*Fruit Damage (%)*			*Yield of Marketable Fruits of Okra (q/ha)***			*Overall Gain in Yield Over Control (%)*	*Overall Loss in Yield Over Treatment (%)*
		1996	*1997*	*Pooled Mean*	*1996*	*1997*	*Pooled Mean*	*1996*	*1997*	*Pooled Mean*		
Thiodicarb	0.15	6.55 (14.78)	7.80 (16.23)	7.17 (15.50)	7.46 (15.82)	9.03 (17.57)	8.25 (16.69)	88.46	84.38	86.42	87.62	46.7
Dimethoate	0.050	8.22 (16.66)	9.88 (18.24)	9.05 (17.45)	9.23 (17.67)	11.06 (19.39)	10.14 (18.53)	83.38	80.24	81.81	77.62	43.69
Endosulfan	0.080	10.24 (18.64)	12.76 (20.89)	11.50 (19.76)	12.43 (20.64)	13.06 (21.63)	12.74 (20.88)	76.40	73.80	75.10	63.05	38.66
Lambdacyhalothrin	0.007	2.34 (10.48)	4.06 (11.58)	3.20 (11.03)	3.85 (11.26)	5.08 (12.96)	4.46 (12.11)	96.43	92.70	94.56	105.29	51.29
Alphamethrin	0.006	2.14 (8.14)	3.84 (11.25)	2.99 (9.69)	2.96 (9.82)	3.75 (11.09)	3.35 (10.45)	99.60	95.80	97.70	112.11	52.85
Diflubenzuron	0.040	9.26 (17.68)	11.44 (19.75)	10.35 (18.71)	10.63 (19.03)	10.99 (19.29)	10.81 (19.16)	79.20	76.45	77.82	68.95	40.01
Neem oil	4.000	14.33 (22.24)	16.00 (23.59)	15.16 (22.91)	15.86 (23.46)	18.03 (25.12)	16.94 (24.29)	70.40	69.20	69.8	51.54	34.01
Vanguard (Azadirachtin 0.03%)	0.500	12.89 (20.98)	13.46 (21.49)	13.17 (21.23)	13.45 (24.48)	15.14 (22.78)	14.29 (22.13)	74.28	72.48	73.38	59.31	37.23
BTK (Delfin WG)	1.000 kg-ha*	3.86 (11.26)	5.05 (12.94)	4.45 (12.10)	3.06 (9.98)	4.33 (11.98)	3.69 (10.98)	95.00	93.05	94.02	104.12	51.01
BTK (Delfin WG)	0.750 kg-ha*	4.75 (12.55)	6.43 (14.67)	5.59 (13.61)	3.99 (11.42)	5.87 (13.95)	4.93 (12.68)	90.48	88.65	89.56	94.44	48.57
Untreated check		19.50 (26.14)	21.02 (27.25)	20.26 (26.58)	22.45 (28.26)	24.38 (29.58)	23.41 (28.92)	49.63	42.50	46.06	–	–
SEm (±)		1.08	1.06	1.09	1.05	1.07	1.06	2.18	3.08	3.24	–	–
CD (P = 0.05)		3.26	3.19	3.28	3.17	3.23	3.18	6.55	9.89	9.69	–	–
CV (%)		7.82	9.22	6.40	8.23	8.74	8.8	7.88	8.16	6.88	–	–

Figure under the parentheses are angular transformed values.

*Dose of the formulated products.

**Mean 20 observations in terms of number of plucking of fruits.

Table 20.2: Relative Economics of Some Chemical and Bio-insecticides in Protecting Okra Against *Earias vitella* Fab., Based on Pooled Results of 1996 and 1997

Insecticidal Treatments: Insecticides	Dose (% a.i.)	Reduction in Shoot Damage Over Control (%)	Reduction in Fruit Damage Over Control (%)	Yield Increase Over Unprotected Crop (q/ha)	Money Value of Increased Yield @ Rs. 400/q	Cost of Insecticidal Control Comprising of Cost of Labour and Insecticides (Rs./ha)	Net Return Over Control (Rs./ha)	Benefit Cost Ratio (BCR)
Thiodicarb	0.15	13.09	15.16	40.36	16144.00	2690.00	13454.00	5.00
Dimethoate	0.050	11.21	13.27	35.75	14300.00	2275.00	12025.00	5.28
Endosulfan	0.080	8.76	10.67	29.04	11616.00	2150.00	9466.00	4.40
Lambdacyhalothrin	0.007	17.06	18.95	40.5	16200.00	2380.00	13820.00	5.81
Alphamethrin	0.006	17.27	20.06	51.64	20656.00	2470.00	18186.00	7.36
Diflubenzuron	0.040	9.91	12.60	31.76	12704.00	2540.00	10164.00	4.00
Neem oil	4.000	5.10	6.47	23.74	9496.00	1980.00	7516.00	3.79
Vanguard (Azadirachtin 0.03%)	0.500	7.09	9.12	27.32	10928.00	2250.00	13478.00	5.99
BTK (Delfin WG)	1.000 kg/ha*	15.81	19.72	47.96	19184.00	3060.00	16124.00	5.26
BTK (Delfin WG)	0.75 kg/ha*	14.67	18.48	43.5	17400.00	2170.00	15230.00	7.02
Untreated check	–	–	–	–	–	–	–	–

*Dose of the formulated commercial formulation of BTK in the name of Delfin.

Shoot and Fruit Borer Infestation

The lowest shoot and fruit damage were obtained when the crop was treated with alphamethrin, during both the years, which remained at par with lamdacyhalothrin followed by BTK, and thiodicarb, (Table 20.1). However, all the treatments were found to be significantly superior over control. A perusal of results (Table 20.1) revealed that the test insecticides in terms of their efficacy in descending sequence were found in order of : alphamethrin > lamdacyhalothrin > BTK > thiodicarb > dimethoate > endosulfan > diflubenzuron > Amrutguard > neem oil : in reducing both shoot and fruit damage caused by the insect pest, *Earias vitella* Fab. during both the years.

Pooled results of the two years investigation followed almost similar trend (Table 20.1). Accordingly, the lowest shoot damage (2.99 per cent) was recorded with alphamethrin which remained at par with lambdacyhalothrin (3.20 per cent), followed by BTK (4.45–5.59 per cent), thiodicarb (7.17 per cent), dimethoate (9.55 per cent), diflubenzuron (10.35 per cent), endosulfan (11.5 per cent) and neem based insecticides (13. 17–15. 16 per cent) based on pooled results of two years experimentation.

In general, similar results were found in terms of suppression of fruit infestation (Table 20.1). However, slightly higher reduction in fruit damage were obtained by the insecticidal treatments almost in the similar fashion as that of minimization of shoot damage during both the years.

Earlier, Sardana and Tiwari (1987) and Srinivasan *et al.* (2000) found that diflubenzuron proved to be highly effective against *E. vitella* Mishra (1989) reported that synthetic pyrethroid provided primsising protection to okra plants against *E. vitella.* Sreelatha and Divakar (1998) and Prasad (2001) recorded higher reduction in the incidence of *E. vitella* on okra with foliar application of alphamethrin. Dhamdhere *et al.* (1988) observed that thiodicarb (0.15 per cent) remained highly effective in reducing okra fruit damage. Raja *el al.* (1997–98) obtained promising control of *E. vitella* of okra with neem oil (4 per cent). Thus results of all these earlier workers were found to be almost in agreement with findings of the present experiment in reducing the incidence of *E. vitella* on okra.

Yield of Healthy and Marketable Fruits of Okra

A perusal of results (Tables 20.1 and 20.2) indicated that the highest yield of 99.60 and 95.80 q/ha were obtained by protecting the crop with lambdacyhalothrin, during 1996 and 1997 respectively, which remained at par with that of alphamethrin and BTK. Earlier, Mishra (1989) obtained higher yield of okra with deltamethrin and fenvalerate and Sreelatha and Divakar (1998) and Prasad (2001) obtained promising yield with alphamethrin as compared to the conventional insecticides. In the present studies, superiority of insecticides were found to be in order of : Synthetic pyrethroids (lambdacyhalothrin > alphamethrin) > BTK (Delfin) > Carbamate product (thiodicarb) > OP product (dimethoate) > IGR (diflubenzuron) > endosulfan > neem products (Amruat guard > Neem oil) in terms of realization of yield of healthy fruits of okra.

The results of the present investigation indicated that avoidable loss in fruit yield caused by shoot and fruit borer (*E. vitella*) was found to be tune of 34.01 to 52.85 per cent which could be prevented by various insecticidal treatments. Highest increase in yield (112.11 per cent), over control, was noticed with alphamethrin followed by lambdacyhalothrin (105.29 per cent); BTK, Delfin WG applied @ 1.0 kg/ha (104.12 per cent); BTK, Delfin @ 0.75 kg/ha (94.44 per cent), thiodicarb (87.62 per cent), dimethoate (77.62 per cent) and endosulfan (63.05 per cent). Neem insecticides could be able to increase the yield of healthy fruits up to the level of 51.54 per cent in case of neem oil, 4 per cent and 59.31 per cent in case of Amrutguard (1.0 per cent). The IGR (diflubenzuron) increased yield upto 68.00 per cent of marketable fruits of okra.

Relative Economics of Insecticides

Economics calculated on the basis of mean results of the two years revealed that the maximum net return of Rs. 18,186/ha was obtained with alphamethrin (0.006 per cent) which was followed by BTK, Delfin WG@ 1.0 kg/ha (Rs. 16,124/ha) Delfm @ 0.75 kg/ha (Rs. 15,230/ha), lambdacyhalothrin (Rs. 13,820/ha), Amrutguard (Rs.13,478/ha), dimethoate (Rs. 12,025/ha) and the chitin inhibitor, diflubenzuron (Rs. 10,164/ha). Considering on the basis of both bio-efficacy and economics, alphamethrin took the lead in suppressing the incidence of shoot and fruit borer upto minimum level of 2.99 and 3.35 per cent which in turn resulted to the highest yield (97.70 q/ha) and maximum net return of Rs. 18,186/ha and benefit cost ratio (7.36) followed by BTK, Delfin @ 1.0 kg ha which gave rise to appreciable fruit yield (94.02 q/ha) and net profit of Rs. 16,124/ha with BCR of 5.26. Earlier Sreelatha and Divakar (1998) obtained higher suppression of *E. vitella* and substantial yield enhancement of yield of okra fruits under protection provided by alphamethrin. Recently, Prasad and Prasad (2000) recorded substantial enhancement in fruit yield with BTK (Delfin) applied @ 0.750 kg/ha used against *E. vitella* infesting okra.

Based on two years results, it is concluded that okra can be prevented from attack of *E. vitella* through foliar spraying of anyone of alphamethrin (0.006 per cent), BTK (Delfin WG OK @ 0.75 to 1.0 kg/ha), thiodicarb (0.15 per cent), lambdacyhalothrin (0.007 per cent) and diflubenzuron (0.04 per cent) for realizing higher marketable fruit yield, net profit and benefit cost ratio.

Acknowledgement

The authors are grateful to the Dean (Agric.), D.R. and the Vice Chancellor, BAU, Ranchi for providing the facilities for conducting this experiment.

References

Dhamdhere, S.V., J. Bahadur and U.J. Mishra (1988). Effect of carbamate insecticides against okra fruit borer, *Earias vitella* Fabricious. *Indian J. Agril. Res.*, 22(1): 19–22.

Mishra. P.N. (1989). Bio-efficacy of some newer insecticides against the pest complex of okra (in India). *Bangladesh Hort.*, V17(1): 1–4.

Prasad, R. (2001). Management of okra fruit borer (*Earias vitella* Fab.) through the interaction of intercropping and. insecticide. *Indian J. Env. & Ecaplan.*, 5(1): 125–129.

Prasad, R. and D. Prasad (2000). Eco-friendly management of fruit borer (*Earias vitella* Fab.) through application of bio–pesticides for sustainable production of okra. In: *Abstracted Research Papers.* An International Meet of Entomo. Congress, 2000: Perspective for the new millennium, held at University of Kerala, Thiruvananthapuram, India, pp. 52.

Raja, J., B. Rajendran, C.M. Pappiah, P.P. Reddy, N.K.K. Kumar and A. Verghese (1997–98). Advances in IPM for horticultural crops. In: *Proce. 1st Nat. Symp. on Pest Management in Horticultural Crops: Environmental Implications and Thrust*, held at IIHR, Bangalore, p 118–120.

Sardana, H.R. and G.C. Tiwari (1987). Effects of diflubenzuron on the pupa of okra shoot and fruit borer (*Earias vitella* Fab.). *Int. J. Tropical Agric.*, V5(2): 150–153.

Sreelatha and B.J. Divakar (1989). Impact of pesticides on okra fruit borer, *Earias vitella* Fab. (Lepidoptera. Noctuidae). *Insect Env.*, 4(2): 40–41.

Srinivasan, R., S. Uthasamy and G. Ravi (2000). Biochemical changes induced by diflubenzuron in the integuments of larvae of cotton bollworms. In: *Abstracted Research Papers.* An International Meet of Entomo. Congress, 2000: Perspective for the new millennium, held at University of Kerala, Thiruvananthapuram, India, pp. 95.

Chapter 21
Modelling the Impact of Developmental Proposals in Mixing Zone Context

Aabha Sargaonkar

National Environmental Engineering Research Institute, Nagpur, Maharashtra, India
E-mail: sohum_ngp@sancharnet.in

ABSTRACT

Wide and perennial rivers often exhibit pockets of land in the riverbed area which are used by local community for different activities in non-monsoon months. Proper planning of these areas can prevent unauthorized encroachments as well as help in generating returns in many ways. Also it will help in protecting various ecosystems in the river bed area and phenomenon in the land-water interface; provide better aesthetic look and life at the river banks; help in generating employment and so on. However, it is desired that any activity takes into consideration an amicable solution for wastewater collection, treatment and disposal. In this context, present paper describes a two-dimensional model simulation study for a hypothetical developmental plan along the river banks. Simulation results describing extent of mixing zone in a wide river are useful as a guideline for developmental planning, in order to achieve desired water quality control.

Keywords: *Two-dimensional model, Mixing zone, Developmental plan, Water quality.*

Introduction

The scenario of unauthorized encroachments and anthropogenic activities in the riverbed area brings our attention towards fast deterioration of water quality in rivers and subsequent damage to the ecosystem thereof. More precisely, number of drains discharging wastewater into the rivers and solid waste dumping sites along the river banks pose threat to the aquatic community, recreational activities, quality of surface and groundwater, and also result into problems such as sedimentation and erosion.

In view of the need for conservation of natural resources, developmental plans in the riverbed areas are being proposed. These consider land allocation, land use planning and landscape development; and protection requirement with risk assessment due to river flooding. However, one primary requirement of any developmental plan in the riverbed area is to maintain the water quality in river within stipulated standards/criteria for the designated best use (DBU). In this paper, a case study is presented for a hypothetical developmental plan along the river banks. The plan considers development for public and semipublic activities and commercial use. The impact of waste-load generated from these activities is simulated using a two-dimensional (2-D) water quality model for wide rivers.

Two-dimensional Model for Wide Rivers

The coupled 2-D model of DO-BOD system is based on stream-tube concept and orthogonal. curvilinear coordinate system (Yotsukura and Cobb, 1972; Yotsukura and Sayre, 1976). The transport equation describing advection and transverse dispersion is coupled with the first order reaction kinetics for BOD decay (*i.e.* sink for DO) and reaeration of the stream through air-water interface (*i.e.* source for DO) as,

$$\frac{\partial L}{\partial x} = D_1 \frac{\partial^2 L}{\partial q^2} - \frac{m_x k_d L}{u} \quad (1)$$

$$\frac{\partial C}{\partial x} = D_x \frac{\partial^2 C}{\partial q^2} - \frac{m_x k_d L}{u} + \frac{m_x k_a (C_5 - C)}{u} \quad (2)$$

where,

L = BOD (mg/l); x = longitudinal distance (m); D_y = transverse diffusion factor (m^5s^{-2}); q = transverse cumulative discharge (m^3s^{-1}); m_x = metric coefficient for x-axis; k_d = BOD decay rate coefficient (s^{-1}); u = Stream velocity (ms^{-1}); C_s = DO saturation concentration (mg/l); C = DO concentration (mg/l); k_a = reaeration rate coefficient (S^{-1}).

In the formulation, the transverse diffusion factor D_y is a function of local depth h_y (m); local velocity U_y (ms^{-1}); and transverse dispersion coefficient e_y (m^2s^{-1}) and the transverse cumulative discharge q, measured from one bank to another is an independent variable. The mathematical details of transformation from variable y (*i.e.* distance across the river width) to q are given by Yotsukura and Sayre (1976). Model also considers reaeration proportional to the DO deficit (C_s–C), where, C_s at stream temperature T° C is estimated as (Covar, 1975)

$$C_s = 14.652 - (0.41022\ T) + (0.007991\ T^2) - (7.7774 * 10^{-5}\ T^3) \quad (3)$$

Salient Features of the Model

1. The 2-D model equations are solved by implicit finite difference scheme wherein, forward difference approximation is used in advection term and Crank-Nicolson scheme in the dispersion term.

2. For the coupled system of DO–BOD, the solution of Eq. (1) becomes the additional input condition for DO simulation.
3. The incomplete mixing of wastewater with river flow and hence mixing across the width due to concentration gradient is the characteristic feature of 2-D modelling.
4. Model supports multiple-reach system simulation, by considering variability in channel characteristics in various stretches of river.
5. Software provides literature values of kinetic constants, which are internally subjected to temperature correction for site-specific conditions.

A Case Study of Developmental Planning

In order to model the impact of two major activities *viz.* commercial and public and semi-public activities on river water quality, a hypothetical situation is considered. Figure 21.1 depicts the schematic diagram of the study stretch and the hypothetical developmental plans along the river banks. The basis for computation of waste load from different plans is the activity, population involved in that activity and BOD generated at the rate of 0.06 5kg/head-day and hydraulic loading at the rate 210 lit/head-day (Rendell, 1999). The public-semipublic activity considers population of 18,425 with total BOD loading of 1110 kg/day; and the commercial area development with activities such as poultry packing, dairy, light industry etc. consider effluent flow of 446 m^3/day. Domestic wastewaters W1 and W2 respectively consider population of 11,200 and 6,750 with 728 and 438 kg/day BOD. Various locations at which these waste loads generated from proposed activities meet the river are also shown in Figure 21.1. The expected rise in the wastewater load due to various activities is given in Table 21.1.

Table 21.1: Expected Rise in Wastewater Load Due to Proposed Activities in Riverbed Area

Wastewater Discharge	*Type of Waste/ Proposed Development*	*Effluent Flow (m^3/s)*	*BOD (mg/l)*
WI	Domestic waste	0.027	309.5
W2	Commercial area	0.016	1900.0
W3	Public-semipublic	0.005	286.8
W4	Domestic waste	0.044	309.1

On the basis of velocity of flow, the river in the study stretch has been characterized as slow moving and at some places sluggish, and hence transverse mixing becomes an important phenomenon when wastewater discharge is located at the bank. Baseline river water quality is considered as 3 mg/l BOD and 7 mg/l DO. Model simulation exercise considers six reaches of 0.5 km each and four wastewater drains coming from four different activities. Based on the river geometry and hydraulic characteristics of the model river, transverse mixing coefficient e_y was estimated as outlined by Lau and Krishnappan (1981). For a given width to depth ratio in a reach, estimated e_y ranged from 0.01–0.013 m^2/s. BOD decay and reaeration rate coefficient were assigned values as 0.5 day^{-1} and 2.3 day^{-1}, respectively.

Results and Discussion

Mixing zone simulations and analysis of results as schematic diagrams are shown in Figures 21.2 and 21.3, These indicate that the mixing zone with high BOD (BOD > 5mg/l) remains attached to the right bank for several kilometres in case of Developmental Plan–1; whereas in case of Plan–2 both

River Stretch Under Consideration = 3 km

Area available for development: Along right bank as well as left bank

Proposed Development

1. Commercial area
2. Public-semi-public activity

W1 and W4 are typical domestic wastewaters generated from catchments/supply zone and are assumed to meet the river without treatment at 0.0 and 2.0 km, respectively.

Plan–1: Developmental Plan–1 assumes proposed development along the right bank

1. Commercial area in the 0.0–0.5 km stretch with wastewater W2 meeting the river at 0.5 km
2. Public and semi-public activity in the 0.5–1.0 km stretch with wastewater generated meeting the river at 1.0 km

Left bank

River flow

Right bank

W1 1 W2 2 W3 W4

0.0km 1.0 km 2.0km 3.0 km

Plan–2: Developmental Plan–2 assumes proposed development along the left bank

1. Commercial area in the 0.0–1.0 km stretch with wastewater W2 meeting the river at 1.0 km
2. Public and semi-public activity in the 1.0–1.5 km stretch with wastewater generated meeting the river at 1.5 km

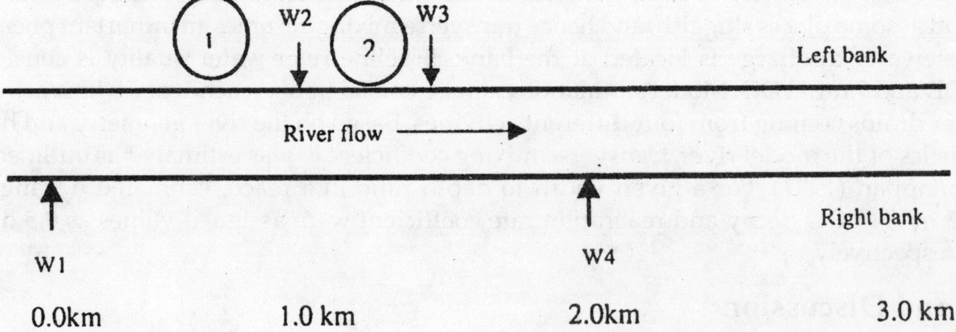

Figure 21.1: Developmental Plans in the Riverbed Area

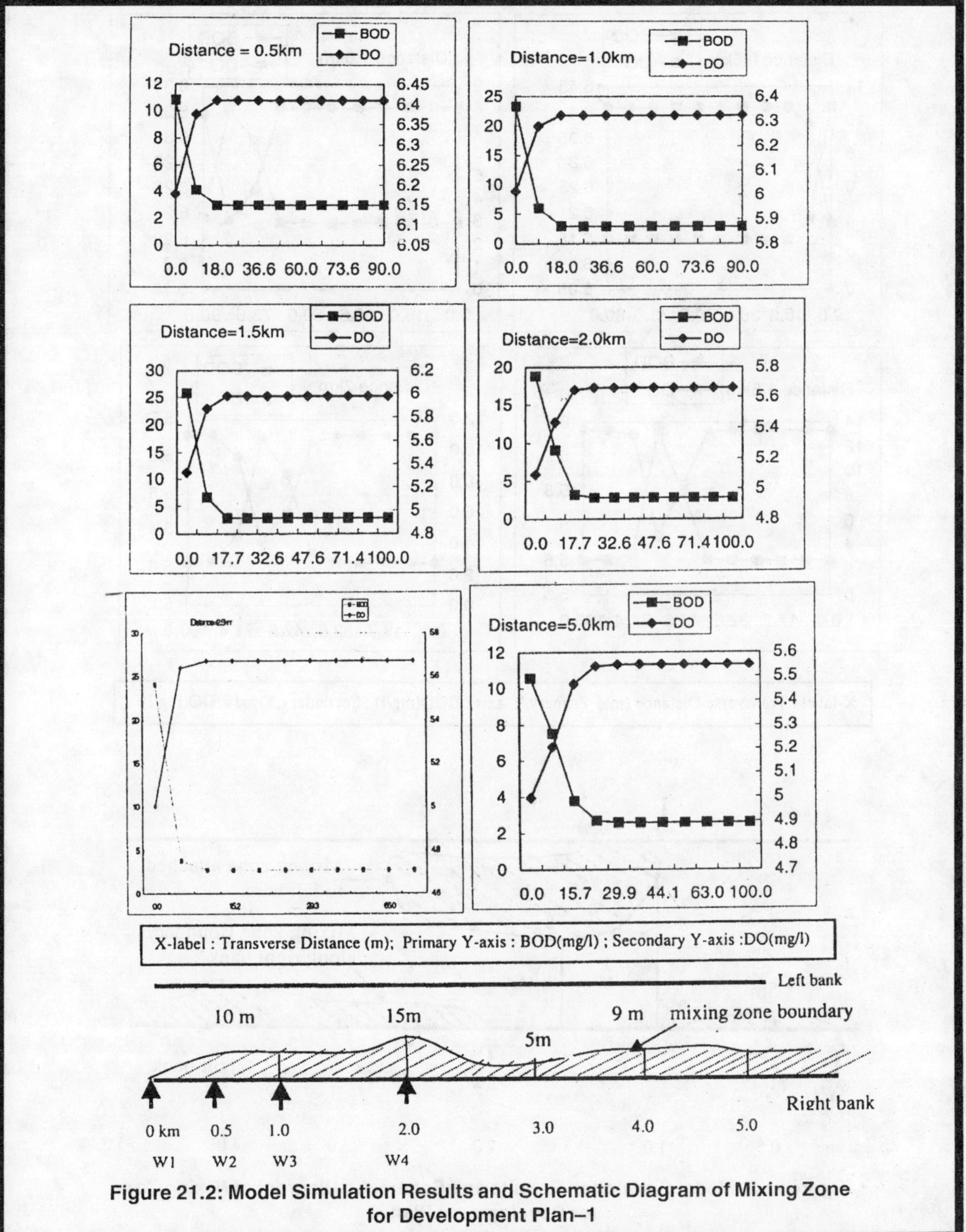

Figure 21.2: Model Simulation Results and Schematic Diagram of Mixing Zone for Development Plan–1

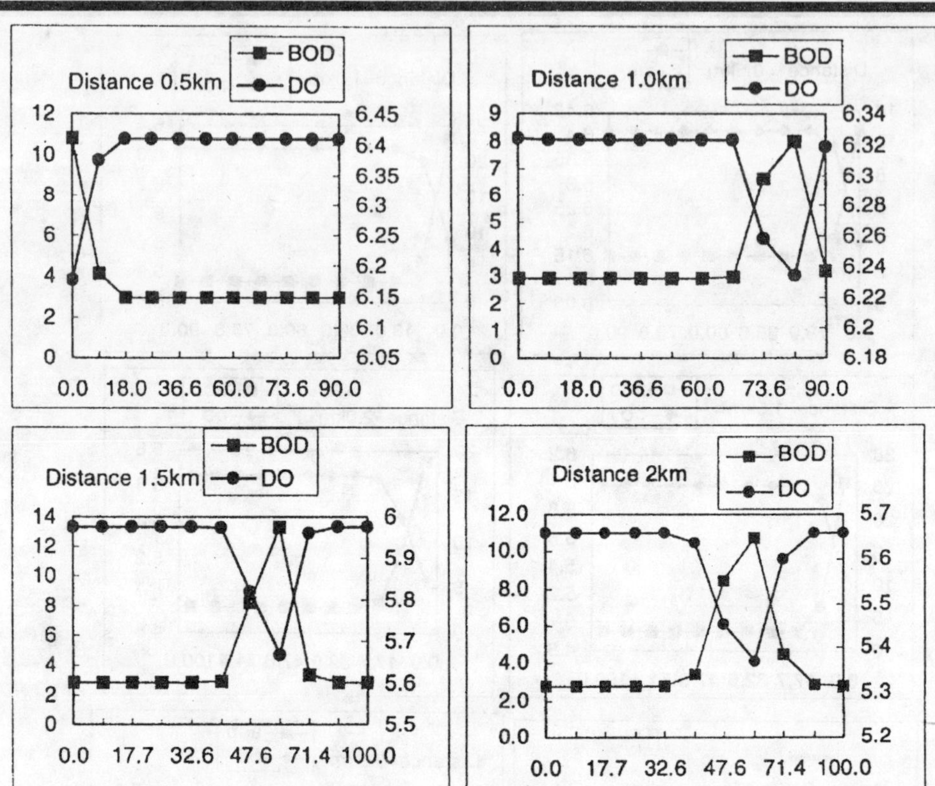

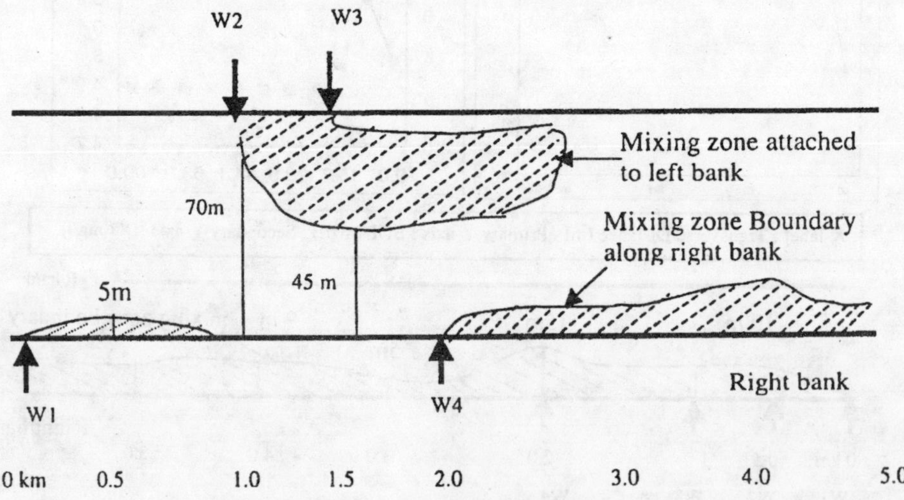

Figure 21.3: Model Simulation Results and Schematic Diagram of Mixing Zone for Development Plan–2

the banks exhibit transverse mixing zone along the area of development. The transverse spread of mixing zone in Plan–1, is from 5 m to 15 m along the right bank with maximum BOD 25mg/l. In Plan–2, maximum BOD in the mixing zone is around 14mg/1 and both the banks exhibit mixing zone for several kilometres downstream of the wastewater release at 1.0 and 2.0 kms.

Conclusion

Model simulation study for two case scenarios indicate that Plan–2 is the preferred scenario if there is no proposed plan of development in the downstream of 2.0 km. This would enhance the mixing and dilution of waste, and therefore help in fast recovery of river water quality. However, the preferred location of wastewater discharge and the permissible effluent load must be estimated from multiple simulations for various loading conditions in order to provide in-stream protection to aquatic biota.

Predictions form 2-D model can provide realistic view of the transverse mixing zone and thus help in deciding the appropriate location of wastewater discharge point in developmental planning. These results can be used as guideline for environmental examination, impact analysis and further detailed planning.

Acknowledgement

The author is grateful to Director, NEERI for encouragement and support to publish this article.

References

Covar, A.P. (1975). Selecting the proper reaeration coefficient for use in water quality models. *Proceedings of the Conference on Environmental Modelling and Simulation*, p. 340–343.

Lau, Y.L. and B.G. Krishnappan (1981). Modelling transverse mixing in natural streams. *J. Hydraulics Engg. Div.*, ASCE, 107, No. HY2: 209–225.

Rendell, F. (1999). *Water and Wastewater Project Development*. Thomas Telford Publishing, London.

Yotsukura, N. and E.B. Cobb (1972). Transverse diffusion of solutes in natural streams. *United States Geological Survey Professional Paper*, 582–C.

Yotsukura, N. and W.W. Sayre (1976). Transverse mixing in natural channels. *Water Resources Res.*, 12: 695–704.

Chapter 22
Detoxification Efficiency of Four Fungal sp. on Dye Effluent

K.T.K. Anandapandian, S. Chandrasekarenthiran, S. Kirupaa and G. Ram Kumar

P.G. Unit of Microbiology, Thiagarajar College, Madurai – 625 009, Tamil Nadu

ABSTRACT

Dye effluent sample was collected from colour yarn limited and tested for parameters such as colour, pH, total dissolved solids, total suspended solids, total solids, chlorinity, Hydrogen sulphide, dissolved oxygen absorbed oxygen, Biological oxygen demand, chemical oxygen demand, phenol, oil and grease and toxicity determined by survival efficiency of fishes. Among seven fungal species isolated from dye amended soil, four isolates were used for detoxification process (Two different *Aspergillus* sp., *Rhizopus* sp. and *Geotrichium* sp.) and were named CRII, CRIII, CRIV and CRV. CRII shown best one for detoxification activity.

Keywords: *Dye effluent, Detoxification, Aspergillus sp., Rhizopus sp., Geotrichium sp.*

Introduction

Toxic chemicals discharged into the environment through various industrial wastes constitute one of the major causes of environment problem (Sudhabai *et al.*, 1998). Textile dye industries are one which contribute to this pollution than other industries. More than eight thousand chemically different types of dyes are being manufactured and they require volumes of water and generate equally volumes of wastes which is highly coloured and complex. For example Thirupur an industrial town, houses about five hundred textile dyeing units located on river bank of the Noyyal releasing effluents untreated into river (Namasivayam *et al.*, 1992). Microbiological cleanup of this types of pollution can be advantageous when compared to other chemical remediation techniques. *Phanerochaete chrysosporium* and other fungus caused 80 per cent detoxification in broth containing 2.5 per cent of effluent. There was reduction in BOD and COD values. *Fusarium* sp. causes 35 to 85 per cent of detoxification, fungi are good in the accumulation of heavy metal such a cadmium, copper, mercury, lead and zinc (Sullia, 2002). Out of eighteen commercially used textile dyes, eight were degraded by white rot fungus, *Phanerochaete chrysosporium* based on decrease of colour (Capalash *et al.*, 1992). *Aspergillus* are efficient in the detoxification of dye effluent due to the production of amylases, proteases and other enzymes. By carefully selecting the fungal isolate and cultural conditions used, degradation of dyes can be obtained. Detoxification of orange colour dye was highest in *Phanerochaete chrysosporium* and it was pre-cultured in phosphate buffered medium. (Chao *et al.*, 1994). In the present study an attempt has been made to detoxify the textile dye industry effluents by using four fungal species, and tested for physico-chemical analysis (pH, TDS, TSS), inorganic analysis (Chlorinity, Hydrogen Sulphide, Salinity), organic analysis (Dissolved oxygen, BOD, COD, Oil and Grease, Adsorbed oxygen, Phenol) and survival efficiency of fishes.

Materials and Methods

The effluent samples were collected from Color Yarn Limited, Allampatty, Madurai at the work time of industry. The samples were taken to the laboratory immediately after collection. Four fungal species were used for treatment and were characterized and confirmed as per Alexopoulos classification of fungi. The freshly collected raw effluent was mixed with the broth culture of fungal species in the ratio of 90 : 10 and incubated for 8 days at 37°C. The parameters such as colour, pH, TDS, TSS, TS, chlorinity, hydrogen sulphide, dissolved oxygen, absorbed oxygen, BOD, COD, oil, grease and phenol were measured in the effluent before and after treatment (APHA, AWWA and WPCF, 1975).

The fresh raw effluent and fungal inoculated effluent were tested for its toxicity as survival efficiency of fishes. 30 fishes (*Lepidocephalicthys thermalis*) which were acclimatized to the lab condition, were exposed to five different types of sample by putting 10 in each type of sample.

Results and Discussion

Tables 22.1 and 2.22 showed that the physico-chemical analysis, inorganic analysis, organic analysis and survival efficiency of fishes of the yellow colour dye effluent collected from Color Yarn Limited.

Table 22.1: Physico-chemical Characteristics of Yellow Coloured Dye Effluent

Parameters	*Raw Effluent*	*CR II*	*CR III*	*CR IV*	*CR V*
Color	Yellow	Light brown	Dark brown	Dark brown	Yellow
pH	8.6	6.2	6.4	6.5	6.7
TDS (mg/l)	2480	720	830	790	817
TS (mg/l)	2240	520	610	530	583
TSS (mg/l)	4720	1240	1440	1320	1400
Chlorinity	596.4	178.9	187.4	193.1	195.9
Salinity	32.2	9.68	10.01	10.4	10.6
Hydrogen sulphide	595	219	243	237	249
Dissolved oxygen	–	1.8	1.89	1.87	1.96
Absorbed oxygen	121	36	39	38.8	41
BOD	87.8	12.3	16.9	14.3	15.3
COD	3907	8608	839	827	889
Oil and grease	178×10^3	54×10^3	61×10^3	55 × 103	60×10^3
Phenol	868	198	213.2	207.3	219.5

Table 22.2: Survival Efficiency of Fishes on Yellow Coloured Dye Effluent

Conditions Exposed	*Raw Effluent*	*CR II*	*CR III*	*CR IV*	*CR V*
100%	–	–	–	–	–
75%	–	1	–	1	–
50%	–	2	1	3	1
25%	–	5	2	4	2

Time exposed = 72 hrs.

No. of fishes (*Lepidocephalicthys thermalis*) taken in each tank = 10

Table 22.3 and 22.4 showed that the physico-chemical analysis, inorganic analysis, organic analysis and survival efficiency of fishes of the blue colour dye effluent collected from Color Yarn Limited. The result also shows that the fungus species CR II was the best decolorizing activity of the yellow and blue colour dye effluent.

Table 22.3: Physico-chemical Characteristics of Blue Coloured Dye Effluent

Parameters	*Raw Effluent*	*CR II*	*CR III*	*CR IV*	*CR V*
Color	Yellow	Light brown	Dark brown	Dark brown	Yellow
pH	8.3	6.2	6.3	6.3	6.4
TDS (mg/l)	2160	751	768	767	788
TS (mg/l)	1890	427	467	438	489
TSS (mg/l)	4050	1178	1235	1205	1277
Chlorinity	724.2	213	218.6	221	231.6
Salinity	39.2	11.5	11.8	11.9	12.5
Hydrogen sulphide	695	318	327	323	336
Dissolved oxygen	–	2.3	2.1	1.9	1.6
Absorbed oxygen	174.6	41	43.6	44.5	45.5
BOD	119.8	36.8	39.7	38.6	39.5
COD	3827	789	813	818	838.5
Oil and grease	188×10^3	39×10^3	46×10^3	47×10^3	49×10^3
Phenol	778	188.3	197.6	191.3	201.5

Table 22.4: Survival Efficiency of Fishes on Blue Coloured Dye Effluent

Conditions Exposed	*Raw Effluent*	*CR II*	*CR III*	*CR IV*	*CR V*
100%	–	–	–	–	–
75%	–	2	–	1	–
50%	–	3	1	2	1
25%	–	4	2	3	2

Time exposed = 72 hrs.

No. of fishes (*Lepidocephalicthys thermalis*) taken in each tank = 10.

The table indicated the difference between the raw effluent and treated effluent. It is noted that BOD and COD values were reduced and dissolved oxygen content was increased. The reduction in BOD and COD values shows that fungal can detoxify the textile dye effluent. Organic nutrients are often presented to fungi as large, insoluble macromolecule complex. These complex must be degraded to smaller substitution which is used as a source of nutrients. Elevated oxygen levels increase the rate of lignin degradation through the production of enzymes and induction of ligninolytic activity by *Phanerochaete chrysosporium* (Das *et al.*, 1999). The dissolved oxygen level is normally at least 5mg/l. for the survival of fish and other aquatic life. (Amudha *et al.*, 1997). Not upto 100 per cent, degradation and detoxification efficiency of fungi on dye effluent, but when compare with the other chemical remediation technique, fungi showed the best detoxifying efficiency of dye effluent. These type of treatment may be used for the development of detoxification of the textile dye effluents.

References

Alexopoulus, C.J. (1996). *Introductory Mycology*. p. 130, 228–229, 311–313.

Amudha, P., R. Nagendran and S. Mahalingam (1997). Studies on the effects of dairy effluents on the behaviour of *Cyprin carpio*. *J. Env. Biol.*, 18(4); 415–418.

APHA, AWWA and WPCF (1975). *Standard Methods for Examination of Water and Wastewater*, 14th ed. APHA, New York, USA.

Capalash, N. and Prince Sharma (1992). Biodegradation of textile azodyes by *Phanerochaete chrysosporium*. *World J. Microb. & Biotech.*, 8: 309–312.

Chao, W.L. and S.L. Lee (1994). Decolouration of azodyes by three white-rot fungi: Influence of carbon source. *World J. Microb. & Biotech.*, 10: 556–559.

Das, M. and S.K. Masud Hussain (1999). Biopulping of non-conventional lignocellulotic agro-waste material (Jute stick) by *Phanerochaete chrysosporium:* A pollution free pulping process. *Indian J. Env. Prot.*, 20: 284 –286.

Namãsivayam, C. and R.T. Yamuna (1992). Removal of congo red from aqueous solutions by biogas waste slurry. *J. Chem. Tech. & Biotech.*, 53: 153–157.

Sudhabai, R. and T.K. Emilia Abraham (1998). Studies on biosorption of chromium (VI) by dead fungal, biomass. *J. Sci. Indus. Res.*, 57: 821–824.

Sullia, S.B. (2002). *Fungal Diversity and Bioremediation*. www.google.com.

Chapter 23
Biodegradation of Tannery Effluent by Using Tannery Effluent Isolate

A. Arun, P. Uma Maheswari and K. Thillai

P.G. Unit of Microbiology, Thiagarajar College, Madurai– 9

ABSTRACT

Tannery is one of the emerging sectors in Industrialization. The tannery effluent has high pollution potential. It creates serious threat to human, and entire ecosystem. Although there are many detoxification procedures, Biological detoxiflcation turns to be simpler, cheaper and significant. In this work *Pseudomonas* sp. was isolated from the tannery effluent and its degradation efficiency was determined by comparing changes in the physico-chemical and organic analysis of the tannery effluent before and after bacterial treatment. The results revealed that the toxic effects where mostly reduced after treatment, making the treated effluent suitable to be discharged in to the environment.

Keywords: *Tannery effluent, Detoxification, Pseudomonas sp.*

Introduction

Leather industries contributes to be one of the major industrial pollution facing the country (Sarkar, 1991) and there are about 2161 Tannery Industries in India and in this 812 Industries were situated in Tamil Nadu. Dindigul is the main area with 61 tanneries located in it (Arora *et al.*, 1981). Tannery effluent is the collection of water which was formed during various stages of processing of leather and collectively called as composite effiuent. The final composite effluent produced is 3500 1/ 100 kg of skin (Rani Perumal and Singaram, 1996).

Total annual skin processed were 58.31 million skins and these wastewaters were discharged in to the environment (Kamini *et al.*, 1999). It induces health hazards to human and other aquatic organisms, it also makes the land and soil unfertile, the ground and surface water turns to be unfit for irrigation and drinking. There are several physical and chemical treatment methods such as electro-chemical method employed to remove heavy metals in the effluent. The primary and pretreatment methods were used for the removal of suspected organic and inorganic solids. But the microbial detoxification procedures were found to be important in complete detoxification of the effluent's toxicity (Chaturvedi, 1992). In the Present work the isolate from the effluent was used for Biodegradation study and was found to be significantly effective in the removal of toxic effects of tannery effluent.

Materials and Methods

The tannery effluent sample was collected from the tannery industries in Dindugul. The effluent was serially diluted and spread plated on Nutrient Agar plates and the isolate was identified to be *Pseudomonas* sp. according to Bergey's manual of Bacteriology (Holt, 1997). The isolated culture was inoculated in to 80 per cent of Tannery effluent which contains 20 per cent of mineral salts medium and incubated for 72 hrs at room temperature.

Physico-chemical Analysis

The colour, pH, odour total dissolved solids (TDS), total solids (TS), total suspended solids (TSS) in the sample before and after treatment was determined.

Organic Analysis

Dissolved oxygen, absorbed oxygen, phenolic content, chemical oxygen demand (COD), biological oxygen demand (BOD) (Winkler's azide method) were also determined in the effluent before and after bacterial treatment.

Results and Discussion

Table 23.1 shows, The colour and odour of the sample was changed after treatment indicating the consumption of phenolic content by *Pseudomonas* sp.

Table 23.1: Physico-chemical Analysis of the Effluent Before and After Treatment

Parameters	*Before Treatment*	*After Treatment*
Colour	Dark bluish green	Pale green
Odour	Phenolic	–
pH	5.7	8.8
TDS (mg/l)	7000	5040
TS (mg/l)	8240	5575
TSS (mg/l)	1240	535

TDS: Total dissolved solids; TS: Total solids; TSS: Total suspended solids (in mg/l milligram/litre).

The pH was changed from acidic to alkaline resulting in the reduction of acidity of the effluent.

TDS, TS and TSS values were also reduced after treatment. Signifying the degradation of toxic solid components in the effluent by the isolate.

Table 23.2 shows, the effluent mainly lacks in the dissolved oxygen content but due to the bacterial treatment it has gained dissolved oxygen content due to microbial metabolic reactions. Absorbed oxygen content was found to be 50 per cent decreased after treatment. Biological oxygen demand (BOD), the indicator of pollution level have decreased significantly indicating that the pollutants level have been reduced. The phenolic content was found to be decreased, it indicates the utilization of phenol as a carbon source by the isolate. The decrease in level of chemical oxygen demand (COD) indicates the reduction of biologically oxidisable and inert organic materials as a result of the degradation by the *Pseudomonas* sp.

Table 23.2: Organic Analysis of the Effluent Before and After Treatment

Parameters	*Before Treatment*	*After Treatment*
Dissolved O_2 (cc/l)	0.106	1.02
Absorbed O_2 (mg/l)	297.5	145.2
BOD (mg/l)	1250	392
Phenol (mg/l)	550	305
COD (mg/l)	4967	1237

BOD: Biological oxygen demand; COD: Chemical oxygen demand (in mg/l milligram/litre).

The results revealed that the *Pseudomonas* sp. isolated from the effluent is efficient enough to degrade the tannic components and it is useful to make the effluent non-toxic after treatment, and these wastewaters can be reused and certainly this biodegradation study will be helpful to some extent for making a pollution free environment.

References

Arora, H.C., Chattapadhya, S.N. and R. Ruth (1975). Treatment of vegetable tannery effiuent by the anaerobic contact filter process. *National Seminar on Assessment and Management of Pollution*, New Delhi.

Chaturvedi, M.K. (1992). Biodegradation of Tannery effluent, isolation and characterization of microbial consortium. *Indian. J. Env. Prot.*, 12: 335–340.

Kamini, N.R., Hemachander, J. Geraldine Sandana Mala and R. Puvana Krishnan (1999). Microbial enzyme technology as an alternative to conventional chemicals in leather industry. *Current Science*, 77: 80–96.

Raniperumal, P. Singaram (1996). Chemical of soil and irrigation water by tannery effluent Tamil Nadu experience, *Indian J. Agri. Chem.*, 1&2: (1–8).

Sarkar, N. (1991). Tannery waste disposal. *Jour. Amer. Leather Chem. Assoc.*, 35: 463.

Chapter 24
Effect of Organic Manures of *Panchagavya* Spray on Nutrient Composition on Raw Rice (*Oryza sativa* L.)

Birendra Kuamar Yadav and A. Christopher Lourduraj

Department of Environmental Sciences, Tamil Nadu Agricultural University, Coimbatore – 641 003

ABSTRACT

A field experiment was conducted at Tamil Nadu Agricultural University, Coimbatore during January to May of 2004 to evaluate the effect of organic sources of manures (farm yard manure, composted poultry manure, composted coir pith and green leaf manure) and *Panchagavya* spray on the nutritive value of rice. The results showed that organic sources of nutrients had significant effect on the nutrient composition of rice as compared to use of chemical fertilizers as nutrient source. *Panchagavya* spray also significantly increased the P, K, Na, Ca, Mg, Fe, Cu, Zn, Mn, carbohydrate, protein, fat and ash content in raw rice as compared to without *Panchagavya* spray. Application of recommended NPK through fertilizers recorded significantly higher P, K, Mn, moisture content, carbohydrate and fat content, followed by application of 50 per cent N through composted poultry manure + 50 per cent N through green leaf manure, which did not differ significantly with each other. However, higher Na, Ca, Mg, Fe, Cu, Zn, protein and ash content was recorded in 50 per cent N through composted poultry manure + 50 per cent N through green leaf manure, which was significantly superior to other treatments.

Keywords: *Panchagavya, Organic manures, Nutritive value, Raw rice.*

Introduction

Rice is the most important food crop accounting for 29 per cent of total calorie intake of the people of developing countries (FAO, 2001). Warman and Howard (1998) compared the organically and conventionally grown potatoes and sweet corn and found that four elements in potato tuber (P, Mg, Na and Mn) and four elements in potato leaves (N, Mg, Fe and B) were higher in organically fertilized field. Lampkin (1990) stated that organic farming system have higher protein contents in the cereals. Hattab *et al.* (1998) found that application of 25 per cent organic N as a basal and 75 per cent inorganic as top dressing proved its efficiency in influencing the various quality parameters *viz.* total amylose, insoluble amylose, crude protein, lipid content, optimal cooking time and gruel loss. Among the organic sources compared, the easily degradable green manures *Sesbania rostrata* and *S. aculeata* were observed to be the most suitable for improvement in quality parameters. Beaulah (2001) reported that secondary and micronutrients (Ca, S and Fe) and macronutrients (NPK) content of leaves and pods were superior under poultry manure + neem cake + *Panchagavya* treatments. Higher nutrient uptake and more nutrient use efficiency in both main and ratoon crops of annual moringa were also observed. Similarly, the quality parameters *viz.*, crude fibre, protein, ascorbic acid, carotene content and shelf life in annual moringa were also higher under organic manure applied with *Panchagavya* as spray (Beaulah *et al.*, 2002). Therefore, this study was conducted to assess the effect of organic manures application and *Panchagavya* spray on nutrient composition of raw rice.

Materials and Methods

A field experiment was conducted during summer season (January–May) of 2004 at Tamil Nadu Agricultural University, Coimbatore, India. The soil of the experimental field was deep, moderately drained, clay loam, low in available N (267 kg ha^{-1}), medium in available P (15.9 kg ha^{-1}) and high in available K (530 kg ha^{-1}). The soil pH was 7.9 and electrical conductivity (EC) of soil was 0.41dsm^{-1} The rice variety ADT 43, with field duration of 118 days, was used in the trial. The seedlings were transplanted on 13 February and harvested on 17 May 2004. The transplanting in the main field was done manually with a spacing of 15 × 10 cm @ two seedlings hill. The experiment was laid out in split plot design with three replications. The experiment consisted of two main plot treatments *i.e.* with *Panchagavya* spray and without *Panchagavya* spray and seven subplot treatments *i.e.* different combinations of farm yard manure, composted poultry manure, composted coir pith and green leaf manure (*Sesbania aculeata*). A control treatment wherein recommended rates of N, P and K (120 : 38 : 38 kg ha^{-1}) was applied as urea (46 per cent N), single super phosphate (16 per cent P_2O_5) and muriate of potash (60 per cent K_2O), respectively was also tried for comparison purpose. The full dose of P_2O_5 and K_2O were applied basally and nitrogen was applied in four split dose of 15 days intervals.

Composted coir pith was prepared by mixing with urea @ 5 kg and 5 bottles *Pleurotus sajorcaju* tpn^{-1} coir waste, making alternate layer of coir waste – urea – coir waste – *Pleurotus* and finally made up a heap like structure. For composting poultry manure, bits of chopped rice straw was mixed with poultry manure @ 1 : 10 on dry weight basis and made into a heap like structure. These heaps were composted for 75 days under aerobic conditions with adequate moisture. Nitrogen content of composted coir pith, composted poultry manure, farm yard manure and green leaf manures on the dry weight basis were analysed and found as 1.34 per cent, 2.74 per cent, 0.64 per cent and 2.80 per cent, respectively. The recommended dose of N was substituted through combination of two organics with 50 per cent recommended N from each. Based on N content the quantum of organic manures applied in different treatments were 9836 kg ha^{-1} of green leaf manure, 11320 kg ha^{-1} of farm yard manure, 6122 kg ha^{-1} of composted coir pith and 3030 kg ha^{-1} of composted poultry manure.

Panchagavya stock solution (20 litres) was prepared using ingredients *viz.* Cow dung (5 kg), Cow's urine (3 litres), Cow's milk (2 litres), Cow's curd (2 litres), and Cow's clarified butter/ghee (1 litre). In addition, sugarcane juice (3 litres), tender coconut water (3 litres) and riped banana (1 kg) were also added to accelerate the fermentation process (Natarajan, 1999). All the materials were added to a wide mouthed mud pot and kept open under shade and stirred twice a day for about 20 minutes, both in the morning and evening to facilitate aerobic microbial activity. The *Panchagavya* stock solution was ready after 10 days. Three per cent solution was sprayed on 30th and 50th days after transplanting (DAT).

Two kg of paddy grains from each experimental plot were milled to get raw rice and analysed. Statistical analysis of the data was done as per the methods suggested by Panse and Sukhatme (1967). Wherever the result was significant, critical difference was worked out at five per cent probability level.

Results and Discussion

Panchagavya spray and organic manures application exerted marked influence on nutrient composition of raw rice. *Panchagavya* spray recorded significantly higher phosphorous content (104.38 mg 100 g^{-1}), potash (87.92 mg 100 g^{-1}), calcium (25.58 mg 100 g^{-1}), magnesium (17.68 mg 100 g^{-1}), sodium (1.08 mg 100 g^{-1}), iron (6.17 mg 100 g^{-1}), copper content (0.351 mg 100 g^{-1}), zinc (1.126 mg 100 g^{-1}) and manganese (1.427 mg 100 g^{-1}) content in raw rice as compared to without *Panchagavya* spray (99.98, 84.70, 24.00, 16.34, 0.98, 5.57, 0.284, 1.038, and 1.284 mg 100 g^{-1}, respectively) (Tables 24.1–24.3). Significantly higher carbohydrate content (79.73 per cent), protein (7.677 per cent), fat (0.564 per cent) and ash (0.861 per cent) content in raw rice was found with *Panchagavya* spray as compared to without *Panchagavya* spray (79.61 per cent, 7.581 per cent, 0.555 per cent and 0.845 per cent respectively) (Table 24.4). The micro-organisms present in the *Panchagavya* (foliar spray) would have helped in the fixation of atmospheric nitrogen in leaves and its adsorption and utilization for protein synthesis. Increased nitrogen content would increase the protein content. These results are in line with earlier findings of Roy and Seth (1971) in radish, Venter (1979) in carrot, Bome *et al.* (1987) in cabbage, Kohil *et al.* (1992) in pea, Beaulah (2001) in moringa, Somasundaram (2003) in maize, sunflower and green gram and Boomiraj (2003) in Bhendi. There are several reasons for increased nutritive value in rice due to spraying of *Panchagavya*. The quantities of IAA and GA present in *Panchagavya* (Somasundaram, 2003), when sprayed two times, could have created the stimuli in the plant system and in turn increased the production of growth regulator in the cell system. Kalarani (1991) reported that, the action of the growth regulators in plant system stimulated the necessary growth and development in plants, leading to better yield. Beaulah (2001) reported that secondary and micro-nutrients (Ca, S and Fe) and macro-nutrients (NPK) content of leaves and pods were superior under poultry manure + neem cake + *Panchagavya* treatments.

Application of recommended NPK through fertilizers (S_7) recorded highest phosphorous (108.36 mg 100 g^{-1}), potash (92.51 mg 100 g^{-1}) and manganese (1.537 mg 100 g^{-1}) content in raw rice, which was on par with 50 per cent N through composted poultry manure + 50 per cent N through green leaf manure (S_5) (108.07, 92.22 and 1.522 mg 100 g^{-1}, respectively). However, S_5 recorded highest amount of calcium (28.59 mg 100 g^{-1}), magnesium (20.60 mg 100 g^{-1}), sodium (1.28 mg 100 g^{-1}), iron (6.52 mg 100 g^{-1}), copper (0.357 mg 100 g^{-1}) and zinc (1.332 mg 100 g^{-1}) content in raw rice, which was significantly superior to S_7 (Tables 24.1–24.3). The maximum amount of raw rice carbohydrate (80.31 per cent) and fat (0.577 per cent) content was found in S_7, in which carbohydrate content is significantly higher than other treatments and fat content was on par with S_5. However, highest amount of protein (7.733 per

Table 24.1: Effect of Organic Manures and *Panchagavya* Spray on Phosphorus, Potash and Calcium Content of Raw Rice (mg 100 g^{-1})

Treatments	Phosphorous		Mean	Potash		Mean	Calcium		Mean
	M_1	M_2		M_1	M_2		M_1	M_2	
S_1	108.91	103.25	106.08	92.45	88.02	90.23	25.60	23.40	24.50
S_2	98.06	94.79	96.42	81.59	79.55	80.57	23.25	22.03	22.64
S_3	109.84	104.85	107.34	93.37	89.62	91.49	28.30	26.82	27.56
S_4	97.06	93.98	95.52	80.59	78.75	79.67	23.03	22.22	22.62
S_5	110.75	105.39	108.07	94.28	90.16	92.22	29.67	27.51	28.59
S_6	94.85	92.08	93.47	78.38	76.52	77.45	24.43	22.31	23.37
S_7	111.22	105.49	108.36	94.75	90.26	92.51	24.79	23.73	24.26
Mean	104.38	99.98	102.18	87.92	84.70	86.31	25.58	24.00	24.79
Source	*SEd*	*CD (p=0.05)*		*SEd*	*CD (p=0.05)*		*SEd*	*CD (p=0.05)*	
M	0.09	0.40		0.34	1.46		0.35	1.50	
S	0.27	0.55		0.23	0.48		0.24	0.50	
M at S	0.36	0.80		0.46	1.50		0.47	1.55	
S at M	0.38	0.78		0.33	0.68		0.35	0.71	

S_1: Farm yard manure + Composted poultry manure; S_2: Farm yard manure + Composted coir pith; S_3: Farm yard manure + Green Leaf manure; S_4: Composted poultry manure + Composted coir pith; S_5: Composted poultry manure + Green leaf manure; S_6: Composted coir pith + Green Leaf manure; S_7: Recommended NPK; M_1: With *Panchagavya* spray; M_2: Without *Panchagavya* spray.

Table 24.2: Effect of Organic Manures and *Panchagavya* Spray on Magnesium, Sodium and Iron Content of Raw Rice (mg 100 g^{-1})

Treatments	Magnesium		Mean	Sodium		Mean	Iron		Mean
	M_1	M_2		M_1	M_2		M_1	M_2	
S_1	17.40	16.65	17.02	1.26	1.04	1.15	6.15	5.59	5.87
S_2	14.69	13.22	13.95	0.91	0.84	0.87	5.52	5.01	5.27
S_3	20.61	19.08	19.85	0.87	0.78	0.83	6.52	6.01	6.26
S_4	15.36	14.67	15.01	1.06	0.98	1.02	5.89	5.01	5.48
S_5	21.71	19.50	20.60	1.35	1.21	1.28	6.77	6.26	6.52
S_6	16.44	15.33	15.88	0.99	0.91	0.95	5.78	4.96	5.37
S_7	17.54	15.95	16.75	1.14	1.08	1.11	6.53	6.06	6.30
Mean	17.68	16.34	17.01	1.08	0.98	1.03	6.17	5.57	5.87
Source	*SEd*	*CD (p=0.05)*		*SEd*	*CD (p=0.05)*		*SEd*	*CD (p=0.05)*	
M	0.23	0.98		0.01	0.05		0.07	0.28	
S	0.26	0.53		0.03	0.06		0.05	0.11	
M at S	0.41	NS		0.04	NS		0.10	0.30	
S at M	0.37	NS		0.04	NS		0.08	0.16	

S_1: Farm yard manure + Composted poultry manure; S_2: Farm yard manure + Composted coir pith; S_3: Farm yard manure + Green Leaf manure; S_4: Composted poultry manure + Composted coir pith; S_5: Composted poultry manure + Green leaf manure; S_6: Composted coir pith + Green Leaf manure; S_7: Recommended NPK; M_1: With *Panchagavya* spray; M_2: Without *Panchagavya* spray.

Table 24.3: Effect of Organic Manures and *Panchagavya* Spray on Copper, Zinc and Manganese Content of Raw Rice (mg 100 g⁻¹)

Treatments	Copper		Mean	Zinc		Mean	Manganese		Mean
	M_1	M_2		M_1	M_2		M_1	M_2	
S_1	0.353	0.303	0.328	1.303	1.100	1.202	1.363	1.220	1.292
S_2	0.330	0.223	0.277	0.953	0.900	0.927	1.203	1.170	1.187
S_3	0.340	0.270	0.305	0.913	0.843	0.878	1.423	1.283	1.353
S_4	0.360	0.303	0.332	1.100	1.043	1.072	1.343	1.243	1.293
S_5	0.393	0.320	0.357	1.393	1.270	1.332	1.680	1.363	1.522
S_6	0.340	0.280	0.310	1.033	0.970	1.002	1.400	1.210	1.305
S_7	0.343	0.290	0.317	1.183	1.140	1.162	1.573	1.500	1.537
Mean	0.351	0.284	0.318	1.126	1.038	1.082	1.427	1.284	1.356

Source	SEd	CD (p=0.05)	SEd	CD (p=0.05)	SEd	CD (p=0.05)
M	0.001	0.006	0.011	0.047	0.014	0.059
S	0.004	0.009	0.030	0.063	0.021	0.042
M at S	0.006	0.013	0.041	NS	0.030	0.076
S at M	0.006	0.012	0.043	NS	0.029	0.060

S_1: Farm yard manure + Composted poultry manure; S_2: Farm yard manure + Composted coir pith; S_3: Farm yard manure + Green Leaf manure; S_4: Composted poultry manure + Composted coir pith; S_5: Composted poultry manure + Green leaf manure; S_6: Composted coir pith + Green Leaf manure; S_7: Recommended NPK; M_1: With *Panchagavya* spray; M_2: Without *Panchagavya* spray.

Table 24.4: Effect of Organic Manures and *Panchagavya* Spray on Carbohydrate, Protein, Fat and Ash Content of Raw Rice (%)

Treatments	Carbohydrate		Mean	Protein		Mean	Fat		Mean	Ash		Mean
	M_1	M_2		M_1	M_2		M_1	M_2		M_1	M_2	
S_1	79.64	79.55	79.60	7.683	7.593	7.683	0.567	0.557	0.562	0.863	0.843	0.853
S_2	79.47	79.40	79.44	7.640	7.530	7.585	0.557	0.547	0.552	0.853	0.827	0.840
S_3	79.73	79.59	79.66	7.733	7.657	7.695	0.573	0.567	0.570	0.873	0.863	0.868
S_4	79.57	79.45	79.51	7.603	7.477	7.540	0.550	0.540	0.545	0.847	0.830	0.838
S_5	79.84	79.72	79.78	7.773	7.693	7.733	0.580	0.570	0.575	0.883	0.870	0.877
S_6	79.46	79.37	79.42	7.550	7.447	7.498	0.540	0.530	0.535	0.843	0.823	0.833
S_7	80.41	80.21	80.31	7.750	7.673	7.713	0.580	0.573	0.577	0.867	0.857	0.862
Mean	79.73	79.61	79.67	7.677	7.581	7.629	0.564	0.555	0.559	0.861	0.845	0.853

Source	SEd	CD (p = 0.05)	SEd	CD (p = 0.05)	SEd	CD (p = 0.05)	SEd	CD (p = 0.05)
M	0.001	0.004	0.002	0.007	0.000	0.002	0.001	0.002
S	0.009	0.018	0.008	0.016	0.001	0.004	0.002	0.004
M at S	0.012	0.024	0.010	0.022	0.002	0.005	0.003	0.006
S at M	0.013	0.026	0.011	0.023	0.003	0.006	0.003	0.006

S_1: Farm yard manure + Composted poultry manure; S_2: Farm yard manure + Composted coir pith; S_3: Farm yard manure + Green Leaf manure; S_4: Composted poultry manure + Composted coir pith; S_5: Composted poultry manure + Green leaf manure; S_6: Composted coir pith + Green Leaf manure; S_7: Recommended NPK; M_1: With *Panchagavya* spray; M_2: Without *Panchagavya* spray.

cent) and ash (0.877 per cent) content in raw rice were recorded by S_5, which was significantly superior to other treatments (Table 24.4). Lampkin (1990) also stated that organic farming system have higher protein contents in the cereals. Similarly, Warman and Howard (1998) also reported higher micronutrient content in potato tuber produced from organically fertilized field than conventionally grown field.

Generally, interaction effect was significant for rice chemical composition. Application of 50 per cent N through composted poultry manure + 50 per cent N through green leaf manure along with *Panchagavya* spray recorded significantly higher Ca, Fe, Cu, Zn, Mn content whereas, protein content was on par with recommended NPK through fertilizers along with *Panchagavya* spray.

The results of the study revealed that *Panchagavya* spray significantly increased the P, K, Na, Ca, Mg, Fe, Cu, Zn, Mn, carbohydrate, protein, fat and ash content in raw rice as compared to without *Panchagavya* spray. Application of recommended NPK through fertilizers recorded significantly higher P, K, Mn, moisture content, carbohydrate and fat content, followed by application of 50 per cent N through composted poultry manure + 50 per cent N through green leaf manure, which did not differ significantly with each other. However, higher Na, Ca, Mg, Fe, Cu, Zn, protein and ash content was recorded in 50 per cent N through composted poultry manure + 50 per cent N through green leaf manure, which was significantly superior to other treatments. From this study it is concluded that application of 50 per cent N through poultry manure and 50 per cent N through green leaf manure along with foliar spray of *Panchagavya* (3 per cent) on 30th and 50th day is the viable organic approach to increase the nutritive value and quality of rice produced.

References

Beaulah, A. (2001). Growth and development of Moringa (*Moringa oleifera* Lam.) under organic and inorganic systems of culture. *Ph.D. Thesis*, Tamil Nadu Agric. Univ., Coimbatore.

Beaulah, A., Vadivel, E. and K.R. Rajadurai (2002). Studies on the effect of organic manures and inorganic fertilizers on the quality parameters of Moringa (*Moringa oleifera* Lam.) cv. PKM 1. In: *Abstracts of the UGC sponsored, National Seminar on Emerging Trends in Horticulture*, Department of Horticulture, Annamalai University, Annamalai Nagar, Tamil Nadu, p. 127–128.

Borne, V., Eid, K. and A. Krarus (1987). Nitrogen fertilization of white cabbage for saucer kraut production. *Soil and Fertil.*, 50: 9492.

Boomiraj, K. (2003). Evaluation of organic sources of nutrients, *Panchagavya* and botanicals spray on Bhendi (*Abelmoschus esculentus* Monech). *M.Sc. Thesis*, Tamil Nadu Agric. Univ., Coimbatore.

FAO (2001). *Medium Term Projection of the World Rice Economy: Major Issues at Stake*. Food and Agriculture Organization. Int. Rice Comm. Newsletter, 50: 1–6.

Hattab, K.O., Natarajan, K. and A. Gopalaswamy (1998). Effect of organics in combination with inorganic nitrogen on the quality of rice. *Oryza*, 35: 343–346.

Kalarani, M.K. (1991). Senescence regulation in Soybean [*Glycine max* (L.) Merill]. *M.Sc. Thesis*, Tamil Nadu Agric. Univ., Coimbatore.

Kohil, V.K., Thakur, I.K. and Y.R. Shukla (1992). A note on response of pea (*Pisum sativum* L) to P and K application. *Hort. J.*, 5(1): 59–61.

Lampkin, N. (1990). *Organic Farming*. Ipswich, U.K., Farming Press Books, p. 701–710.

Roy, R.N.J. and Seth (1971). Nutrient uptake and quality of radish (*Raphanus sativus* L.) as influenced by levels of N, P and K and methods of their application. *Indian J. Hort.*, 28(2): 144–149.

Somasundaram, E. (2003). Evaluation of organic sources of nutrients and *Panchagavya* spray on the growth and productivity of maize-sunflower-green gram system. *Ph.D. Thesis*, Tamil Nadu Agric. Univ., Coimbatore.

Venter, F. (1979). Nitrate contents in carrots (*Daucus carota*) as influenced by fertilization. *Acta. Hort.*, 93: 163–172.

Warman, P.R. and K.A. Howard (1998). Yield, vitamin and mineral contents of organically and conventionally grown potatoes and sweet corn. *Agric. Ecosyst. Environ.*, 68: 207–216.

Chapter 25
Microbial Indicators in Mice Exposed to Pesticides

P. Dhasarathan, A.J.A. Ranjitsingh** and N. Sukumaran****

**Microbial Biotech Unit, Department of Biotechnology,*
Sri Kaliswari College, Sivakasi – 626 130
***Department of Biology, Sri Paramakalyani College, Alwarkurichi – 627 412*
****Centre for Environmental Science, M.S. University, Alwarkurichi – 627 412*

ABSTRACT

Bacteria associated with animals as indicators of pollution stress are less explored. Bacteriotoxic effect of pesticides had eliminated microbes that are leading a symbiotic association in the gut of mice and indirectly affected the feeding of the mice. In the rectum region of control mice, the different groups of bacteria were found to occur in good proportions. *Viz.*, amylolytic (50.00 per cent), gelatinolytic (53.30 per cent), caseinolytic (66.67 per cent) and lipolytic (46.67 per cent). Similarly oesophagus, stomach and intestinal region of mice physiological bacterial group were present in high amount. In the mice exposed to the pesticides quinolphos and acepate, the different physiological groups of bacteria associated with rectums, oesophagus, stomach and intestine were found reduced.

***Keywords**: Mice, Gut microflora, Pesticide toxicity and Physiological grouping.*

Introduction

Bacteria may be considered as the first, and important, line of defence the environment has to combat the effects of xenobiotic chemicals (Painter, 1993). Pesticide contaminants in water affect the bacterial population present in fishes living in that water (Dhasarathan and Ranjitsingh, 2000). Numerous reports are available to understand the biochemical, histological and other mechanisms underlying the chronic effects of pesticides in animals (Anees, 1978, Murthy, 1986, Deminti, 1994, Dutta *et al.*, 1995, Ranjitsingh, *et al.*, 1996 and Sambasiva Rao, 1999). However the utility of bacteria associated with animals as indicators of pollutional stress is less explored (Painter, 1993). Studies on the effect of pesticides on the symbiotic and parasitic microbes that have close association with fish are also scanty. Pesticides were reported to affect the bacterial flora in the gut of fish (Tanasomwang and Muruga, 1988, Dhasarathan *et al.*, 2000). In the present investigation attempts were made to find out whether the changes in the gut microflora in mice due to pesticide stress could be used as bacteriological indicators of pesticide toxicity?

Materials and Methods

The percentage occurrence of different physiological groups of bacteria *viz.*, amylolytic, gelatinolytic, lipolytic and caseinolytic in rectums, oesophagus, stomach and intestine of the control and pesticide treated mice was studied. From the laboratory acclimatized stock of the mice Swiss albino, individuals in the weight range 15–17 g were recruited for the experiments. Using a statistic bioassay test, toxicity of an organophosphate (quinolphos and acephate) was evaluated. From these toxicity values sub-lethal doses were computed. The experimental animals were reared in these sub-lethal concentrations (quinolphos, 3 ppm and acephate 6 ppm) for 30 days. During the experiment the mice were fed *ad libitum* with pellets (Lipton, Bombay). 'The drinking water was changed on alternate days without giving much disturbance to the mice. Simultaneously a control set of mice was maintained in a separate cage for comparative study.

After the experimental regimes, the control and pesticide treated mice were brought to the laboratory in living condition for counting the total heterotrophic bacterial population. The mice were sacrificed and the rectums, oesophagus, stomach and intestine were aseptically dissected out for studying bacterial population. All instruments used to dissect out the tissues, different sets of sterilized instruments were used. The aseptically excised tissues (1 g) were placed in separate Petriplates. Microbial analysis was made individually for rectums, oesophagus, stomach and intestine. Pour plate method was employed for the microbial population analysis. One gram of dissected out tissues were homogenized separately using a known volume of sterile per cent peptone water. Then the homogenate were made to 100 ml using 1 per cent sterile peptone water. Further serial dilutions were done using 9 ml of same 1 per cent sterile peptone water. One ml aliquots of serially diluted homogenates were taken out into sterile Petriplates. About 20 ml sterile molten agar of different types of *viz.*, starch agar, casein agar, gelatine agar and tween agar were poured aseptically into the Petriplates. The plates were rotated in clockwise and in anti-clockwise directions and the nutrients agar medium was allowed to solidify. For each agar plates duplicates were maintained.

Results and Discussion

The percentage distribution of different physiological groups of bacteria in the rectum, oesophagus, stomach and intestine of control and pesticide treated mice and presented in Table 25.1. In the rectum region of control mice, the different groups of bacteria were found to occur in good proportions *viz.*, amylolytic (50.00 per cent), gelatinolytic (53.30 per cent), caseinolytic (66.67 per cent) and lipolytic (46.67 per cent). In oesophageal region of control mice, the presence of different bacterial group was *viz.*, amylolytic (60.00 per cent), gelatinolytic (63.30 per cent), caseinolytic (50.00 per cent) and lipolytic (53.33 per cent). The percentage occurrence of different physiological group in the stomach region of control mice indicated the high presence of gelatinolytic group (66.67 per cent) followed by amylolytic (60.00 per cent), caseinolytic (56.67 per cent) and lipolytic group (46.67 per cent). In the intestinal region of control mice the amylolytic bacterial group was present high amount (73.33 per cent) followed by caseinolytic (70.00), gelatinolytic (60.00 per cent) and lipolytic (56.67 per cent) bacteria.

Table 25.1: Changes in the Distribution of Different Physiological Groups of Bacteria in the Rectum, Oesophagus, Stomach and Intestine of Swiss Albino Mice after Exposing to Different Pesticides

Physiological Type	*Organs*	*Control*		*Quinolphos*		*Acephate*	
		No. of +ve Isolates	*Mean ± SD*	*No. of +ve Isolates*	*Mean ± SD*	*No. of +ve Isolates*	*Mean ± SD*
Amylolytic	Rectum	15	18.3 ± 2.7	12 (20.0)	9.5 ± 1.5 (48.0)	8 (46.7)	8 ± 2.5 (56.1)
bacteria	Oesophagus	18		9 (50.0)		6 (60.7)	
	Stomach	18		8 (55.6)		12 (33.3)	
	Intestine	22		9 (59.1)		6 (72.7)	
Gelatinolytic	Rectum	16	18.3 ± 2.7	15 (6.3)	12.7 ± 1.5 (30.1)	11 (31.3)	9.5 ± 1.7 (48.0)
bacteria	Oesophagus	19		13 (31.6)		9 (52.7)	
	Stomach	20		12 (40.0)		11 (45.0)	
	Intestine	18		11 (38.9)		7 (61.1)	
Caseinolytic	Rectum	20	16.5 ± 2.9	9 (55.0)	8.0 ± 1.2 (51.5)	8 (60.0)	6.3 ± 1.1 (62.1)
bacteria	Oesophagus	15		9 (40.0)		5 (66.6)	
	Stomach	17		8 (64.7)		6 (64.7)	
	Intestine	14		8 (61.9)		6 (71.4)	
Lipolytic	Rectum	14	15.3 ± 1.3	8 (42.7)	9.3 ± 2.6 (39.3)	5 (64.3)	5.5 ± 1.5 (67.2)
bacteria	Oesophagus	16		6 (62.5)		4 (75.0)	
	Stomach	14		10 (28.6)		8 (42.8)	
	Intestine	17		13 (23.5)		5 (70.6)	

Values in parenthesis are percentage change over control; mean and standard deviation are given. Total numbers of cultures tested were thirty.

In the mice exposed to the pesticides quinolphos and acephate, the different physiological groups of bacteria associated with rectums, oesophagus, stomach and intestine were found reduced (Table 25.1). The reduction of amylolytic bacteria in various organs of quinolphos treated mice was 47.95 per cent whereas it was 56.16 per cent in acephate treated mice. The percentage changes in the presence of gelatinolytic bacteria in different organs of quinolphos and acephate treated mice were 30.14 per cent and 47.95 per cent respectively. When compared to control mice, the decrease in caseinolytic bacterial genera in the rectums, oesophagus, stomach and intestine of quinolphos and acephate treated mice were 51.52 per cent and 62.12 per cent. Lipolytic bacterial genera were greatly reduced in different organs studied in quinolphos treated mice (67.21 per cent). In acephate treated mice the reduction of lipolytic bacterial genera in all the organs tested was 39.34 per cent. From the results it was inferred that both the pesticide quinolphos and acephate were toxic to the different physiological group of bacteria living in the organs like rectums, oesophagus, stomach and intestine. Microbiological assay of experimental mice indicated the toxicity of the pesticides. Of the two pesticides tested acephate was found more toxic to all the four types of bacteria studied than quinolphos. The reduction in the occurrence of different bacterial genera in the gut was reported to affect the digestive ability in fishes (Dhasarahan and Ranjitsingh, 2000). Thus it was clear that the bactriotoxic effect of pesticides had eliminated microbes that are leading a symbiotic association in the gut of the mice and indirectly affected the feeding and energetic value of the mice. Several pesticides were reported to affect the feeding and energy budget in animals (Pandian and Bhaskaran, 1983., Vasanthi *et al.*, 1990 and

Ramakrishnan *et al.*, 1997. The reported reduction of gut micro flora in mice under pesticide stress was one of the reasons for poor feeding, digestion and a fall in energetic value.

Acknowledgement

The authors are thankful to the Dean and Professors, Department of Immunology, Stanley Medical College, Chennai for providing facilities to carry out the work.

References

Anees, M.A. (1978). Hepatic pathology in the fresh water teleost *Channa punctatius* (Block), exposed to sub-lethal and chronic levels of three organophosphorous insecticides. *Bull. Environ. Contam. Toxicol.*, 19: 524–527.

Deminti, B. (1994). Ocular effects of organophosphates: A historical perspective of saku disease. *J. Appl. Toxicol.*, 14: 119–129.

Dhasarathan, P and A.J.A. Ranjitsingh (2000). Xenobiotic stress on the alimentary tract microflora of the fish, *Cyprinus carpio. J. Ad. Bios.*, 11: 47–54

Dhasarathan, P., Palaniappan, R and A.J.A. Ranjitsingh (2000). Effect of acephate and butachlor on the digestive enzyme and proximate composition of the fish *Cyprinus carpio. Indian J. Environ. and Ecoplan.*, 3: 611–614.

Duttai, H.M. (1995). A comparative approach for evaluation of the effects of pesticides in fish. In: *Fish Morphology and Horizons of New Research*, (Eds.) Datta, J.S and D. Datta Hiran. Oxford IBH, New Delhi, p. 249–277.

Painter, A. (1993). A review of tests for inhibition of bacteria. In: *Ecotoxicology Monitoring*, (Ed.) Mervyn Richardson. VCH, New York, p. 17–35.

Pandian, T.J. and R. Bhaskaran (1983). Food utilization in the fish *Channa striatus* exposed to sub-lethal concentrations of DDT and methyl parathion. *Proc. Indian Acad. Sci.*, 92: 475–481.

Ramakrishnan, M., Arunachalam, S. and S. Palanichamy (1997). Sub-lethal pesticides on feeding energetic in *Cyprinus carpio. Ecotoxicol. Environ. Monit.* 1: 59–64.

Ranjitsingh, A.J.A., Haniffa, M.A and C. Padmalatha (1996). Organophosphate pesticide toxicity to the bioenergetic of a fresh water snail *Indoplanorbis exustus. Poll. Res.*, 5: 89–93.

Sambasica Rao, K.R.S. (1999). *Pesticide Impact on Fish Metabolism*. Discovery Publishing House, New Delhi, pp. 233.

Tanasomwang, V. and K. Muruga (1988). Intestinal microflora of larval and juvenile stages in lapinase Flounder (*Paralicthis olivaceus*). *Fish Pathol.*, 23: 77–83.

Vasanthi, R., Bhaskaran, P. and S. Palanichamy (1990). Influence of carbafuron on growth and protein conversion efficiency in some fresh water fishes. *J. Ecobiol.*, 2: 85–88.

Chapter 26
Effect of Leachate Contamination on Soil Properties

C. Bala Murali Krishna, R.K. Yaji** and S. Shrihari***

**Department of Civil Engineering, NMAMIT, Nitte, Udipi, Karnataka, India*
***Department of Civil Engineering, NITK, Surathkal, D.K., Karnataka, India*

ABSTRACT

Unscientific methods of disposal of solid waste as landfill on low lying areas causes serious environmental and geotechnical problems. The leachate generated from the decomposition of solid waste causes the pollution of soil layers. In this study an attempt has been made to investigate the effect of leachate on soil properties by laboratory tests. The tests included the Liquid Limit, Plastic Limit, Shrinkage Limit, Plasticity Index and UCC strength of the soil. From the test results it was observed that the leachate considerably influences the properties of soil. The results of this study is presented in this article.

Introduction

Due to population escalation, improved living standards and rapid industrial growth, the disposal of solid waste on land is becoming increasingly more common. Leachate from landfills pollute the soil and ground water. All types of pollution have direct or indirect effects on air and water as well as on soil properties. In recent years, majority of the research in the area of environmental geotechnology is related to the land disposal of wastes. Soils exposed to chemicals over time display changes in their properties. Depending on the intensity, duration and the type of contaminant, the type of interaction and reactions that occur with in the soil water system, many properties of the soil are subject to change. At present, design and construction of most geotechnical project is based on test results following ASTM and Indian Standards. The standards are based on controlled conditions at room temperature with distilled water as pore fluid. Since field situations and standard control conditions are significantly different, premature or progressive failures frequently occur. Expensive damage to the floors, pavements and foundations of a light industrial building in a fertilizer plant in Kerala state was reported by Sridharan *et al.* (1981). Rao *et al.* (1995) reported that phosphate contamination has a marked effect on the heave and collapse characteristics of unsaturated laboratory soil. Efforts were made to study the influence of contaminants such as caustic soda, urea, phosphoric acid, waste I.C. engine oil, sugar mill liquid waste and coconut oil on soil properties by Yaji *et al.* (1996). They reported that soil properties have been modified by the interaction of soil with these contaminants. Shivapullaiah *et al.* (1996) reported that the effluent from fertilizer plant significantly alters the index and permeability characteristics of black cotton soil and has little effect on the properties of red earth. Some of the other important works on these aspects have been reported by Kumapley *et al.* (1985), Srivastava *et al.* (1994), Natraj *et al.* (1995), Sitaram *et al.* (1995), Shoba *et al.* (1996). In conventional soil mechanics, it is assumed that the Atterberg limits remain constant for the given soil. However, the above mentioned recent studies have indicated that these properties change when pore fluid changes. The detrimental effects of leachate contamination on soil properties has not been adequately addressed. It is hence the intention of this study to examine the effect of solid waste leachate on consistency limits of soil and UCC strength of the soil.

Materials and Methodology

In order to study the effect of leachate on soil properties a soil column study has been conducted in the Civil Engineering laboratory, NMAMIT, Nitte. Eight soil column setups were used in this study. Figure 26.1 shows a pair of soil column setup. The experimental column setup shown in the Figure 26.1 consists of a PVC pipe of 1.2 meter height and 15.24 cm (or 6 inch) in diameter. One side of the soil column setup was closed by pipe end made from PVC and the another side was kept open for the application of pore fluid (synthetic leachate or distilled water). In each column soil was filled in three layers up to a depth of one meter with the soil collected from solid waste disposal site at Vamanjoor village (where the Mangalore city municipal solid waste is being dumped from the past three decades). The three soil layers (making 1 meter depth) filled in the column are in the same sequence of soil layers as in the field and corresponding to the field densities. The top, middle and bottom soil layer field densities were 1.76 gm/cm^3, 1.783 gm/cm^3, 1.891 gm/cm^3 respectively. The soil column consists of top and middle layers of 30 cm depth each and the bottom layer of depth 40 cm. The top, middle and the bottom soil layers are denoted as soil sample A, soil sample B and soil sample C respectively. Drainage facilities were provided for all the soil column setups at the bottom of the each soil layer. To study the effect of leachate on soil properties synthetic leachate was prepared in the laboratory according to prevalent literature (Chauhan *et al.* (1998), Khan *et al.*(1994)). Table 26.1 shows the quality of synthetic leachate prepared. Synthetic leachate (400 ml on soil each column) was applied to four soil columns from the top everyday and the four leachate applied soil column (L–S Column) are denoted as L1, L2, L3 and L4. Similarly distilled water (400 ml on each soil column) has been applied to four soil columns from the top everyday and the four distilled water applied soil column (DW–S Column) are denoted as D1, D2, D3 and D4. Arrangements were made to distribute both synthetic leachate and distilled water drop by drop on soil surface in the column with the help of separating funnel. The soil columns L1 and, D1, L2 and D2, L3 and D3 and L4 and D4 were dismantled after two, four, six and twelve months respectively. Each layer of soil samples were collected separately and analyzed for various soil consistency limits and UCC strength of soil. Importance has been accorded to qualitative magnitude of effect of leachate contamination rather than the mechanism due to which this happens in this article.

Table 26.1: The Quality of Synthetic Leachate Prepared

Sl. No.	Parameter	Unit	Expected Value	Observed Value
1.	pH	–	7	6.5–7.05
2.	Conductivity	ms/ppm	1	1.4–1.5
3.	Hardness	mg/l	500	480–540
4.	Chlorides	mg/l	100	101–125
5.	Alkalinity	mg/l	300	300–380
6.	Sulphates	mg/l	120	95–125
7.	Nitrates	mg/l	10	9.18–12.41
8.	Ammonia	mg/l	15	9.27- 18.75
9.	Phosphates	mg/l	5	0.11–0.15
10.	Sodium	mg/l	50	44.5–57.5
11.	Potassium	mg/l	30	28.08–39.4
12.	COD	mg/l	125	125

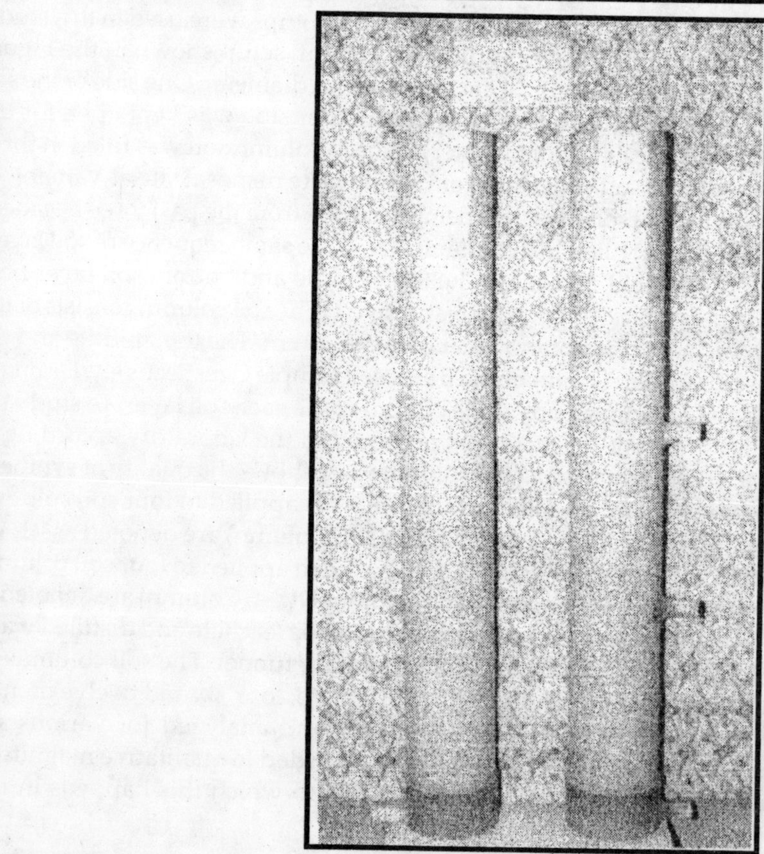

Figure 26.1: The Experimental Column Setup

The consistency limits of the soil were determined for the soil fraction passing 425 micron sieve. The liquid limit of the soil was determined by the single point penetration method using cone penetrometer. The plastic limit, shrinkage limit and plasticity index were determined as per the Indian standards. The UCC strength of the soil was determined by conducting the UCC test.

Results and Discussion

The observed values of soil consistency limits and UCC strength of soil at different time periods of application of leachate in L–S Column and application of distilled water in DW–S Column are presented in Tables 26.2 and 26.3 respectively. Comparison of soil properties before the application of leachate and after the application of leachate as presented in Tables 26.2 and 26.3 show that there has been a significant increase in liquid limit, plastic limit, shrinkage limit and plasticity index of soil and decrease in UCC strength of the soil. The variation of soil properties with time, at different depths of soil column due to leachate application and by the application of distilled water (as control test) are presented in the form of Figures 26.2–26.31.

Table 26.2: Effect of Leachate Contamination on Properties of Soil at Different Time

Test	*Initial*	*After 2 Months*	*After 4 Months*	*After 6 Months*	*After 12 Months*
Soil Sample–A					
Liquid limit (per cent)	60.28	62.14	67.32	70.71	72.93
Plastic limit (per cent)	48.04	52.36	54.48	57.92	58.57
Plasticity index (per cent)	12.24	9.78	12.84	12.79	14.36
Shrinkage limit (per cent)	31.14	32.16	35.52	38.34	42.76
UCC strength (KN/m^2)	55.0	52.5	49.2	47.4	48.6
Soil Sample–B					
Liquid limit (per cent)	58.99	62.25	67.47	73.64	71.92
Plastic limit (per cent)	38.08	38.34	40.16	42.39	43.45
Plasticity index (per cent)	20.91	23.91	27.31	31.25	28.47
Shrinkage limit (per cent)	31.21	35.78	42.1	41.62	42.36
UCC strength (KN/m^2)	63.9	61.7	55.8	52.2	54.3
Soil Sample–C					
Liquid limit (per cent)	67.66	63.92	70.32	76.42	78.54
Plastic limit (per cent)	46.64	46.36	46.52	49.16	51.84
Plasticity index (per cent)	21.02	17.56	23.8	27.26	26.7
Shrinkage limit (per cent)	29.23	32.11	35.14	36.72	36.24
UCC strength (KN/m^2)	56.5	54.6	51	47.5	49.4

The variation in liquid limit of soil samples A, B and C *i.e.* the top, middle and the bottom soil layers respectively from L–S Column with time are shown in Figures 26.2 to 26.4. The Figures 26.5 to 26.7 show the variation in liquid limit of soil samples A, B and C from DW–S Column with time. The observed liquid limit in the soil samples A, B and C taken from both the soil columns is found to be increasing with time. Considerable variation in liquid limit was observed in all the soil samples from L–S Column as compared to DW–S Column. At the end of the study period the observed increase in

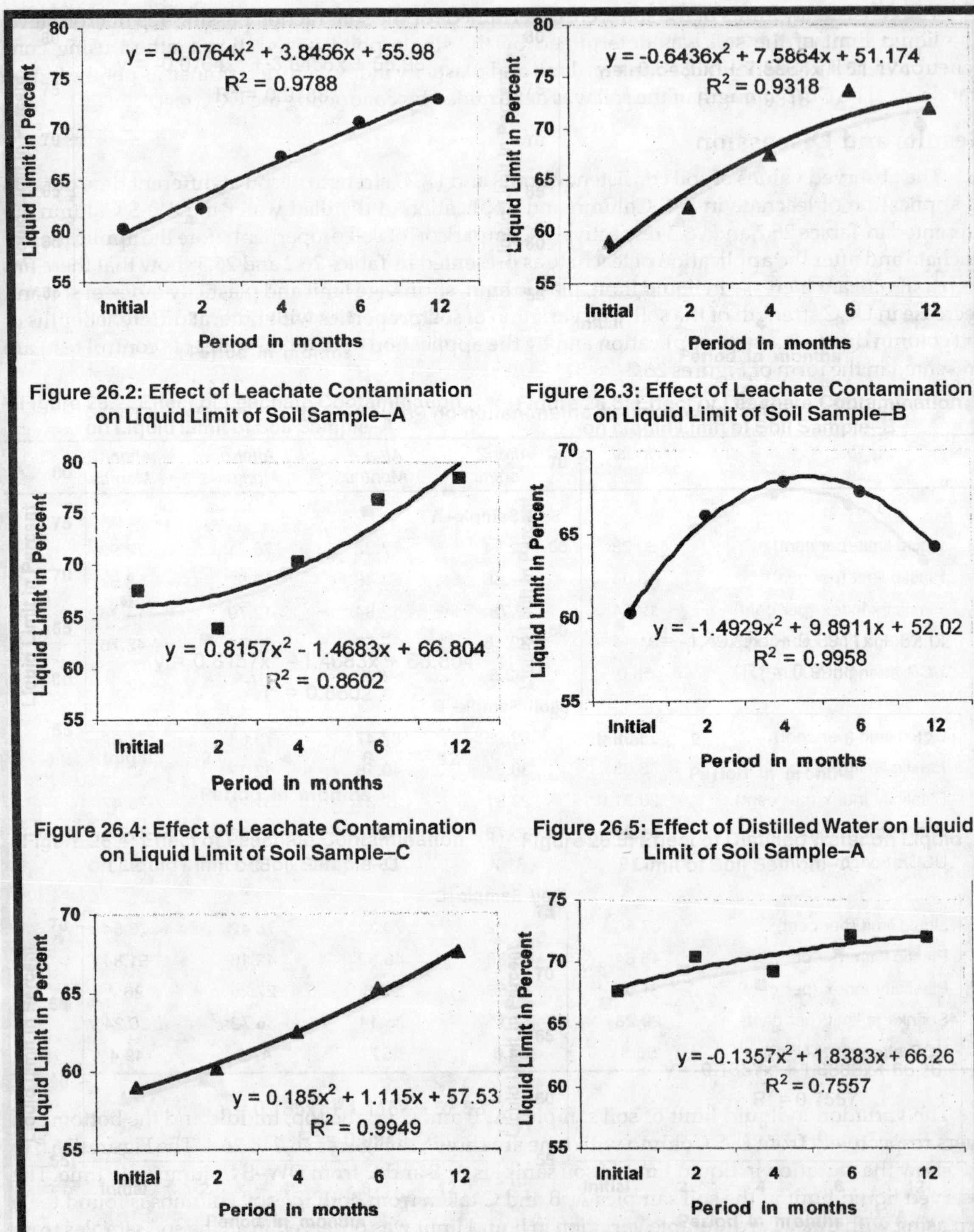

Figure 26.2: Effect of Leachate Contamination on Liquid Limit of Soil Sample–A

Figure 26.3: Effect of Leachate Contamination on Liquid Limit of Soil Sample–B

Figure 26.4: Effect of Leachate Contamination on Liquid Limit of Soil Sample–C

Figure 26.5: Effect of Distilled Water on Liquid Limit of Soil Sample–A

Figure 26.6: Effect of Distilled Water on Liquid Limit of Soil Sample–B

Figure 26.7: Effect of Distilled Water on Liquid Limit of Soil Sample–C

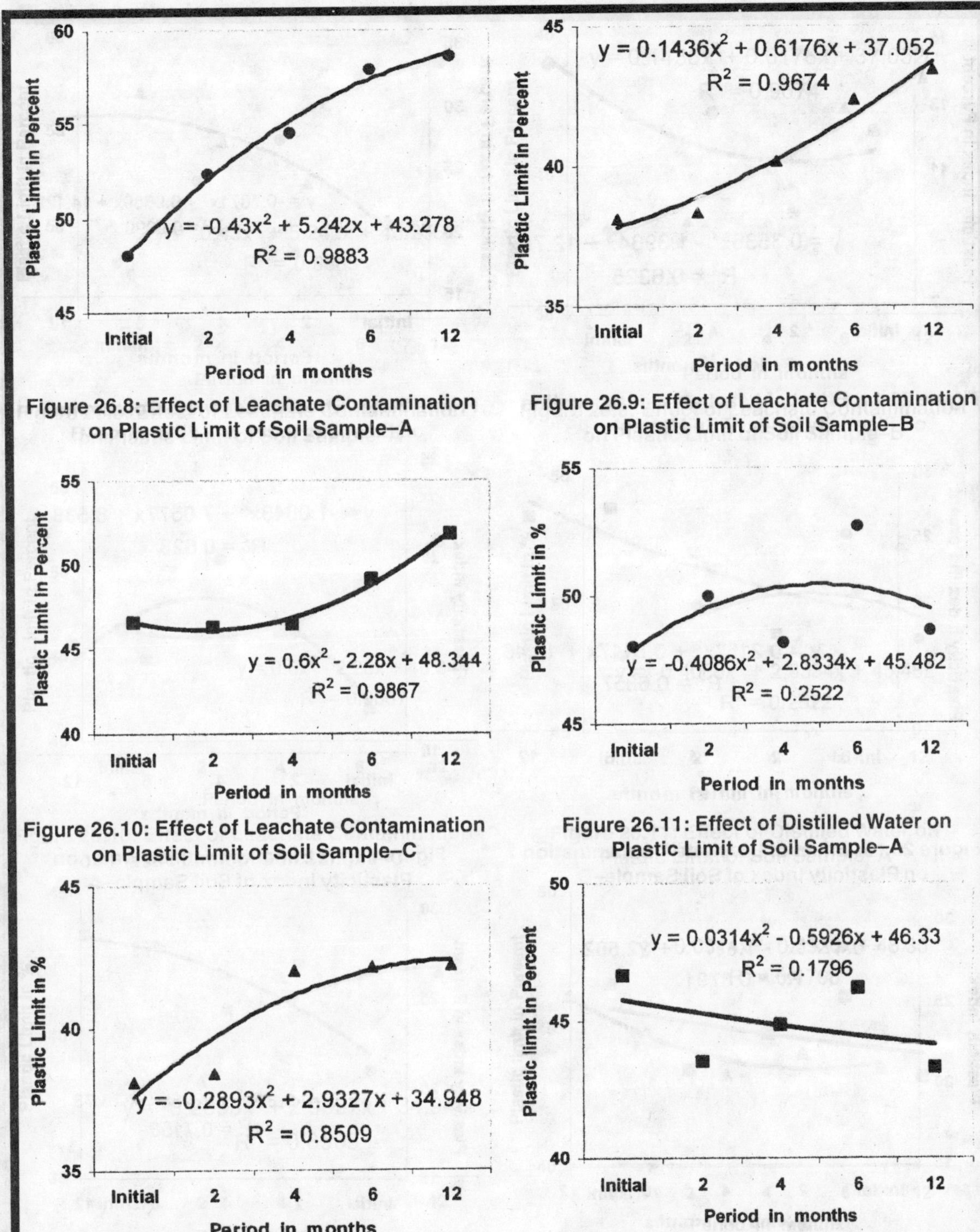

Figure 26.8: Effect of Leachate Contamination on Plastic Limit of Soil Sample–A

Figure 26.9: Effect of Leachate Contamination on Plastic Limit of Soil Sample–B

Figure 26.10: Effect of Leachate Contamination on Plastic Limit of Soil Sample–C

Figure 26.11: Effect of Distilled Water on Plastic Limit of Soil Sample–A

Figure 26.12: Effect of Distilled Water on Plastic Limit of Soil Sample–B

Figure 26.13: Effect of Distilled Water on Plastic Limit of Soil Sample–C

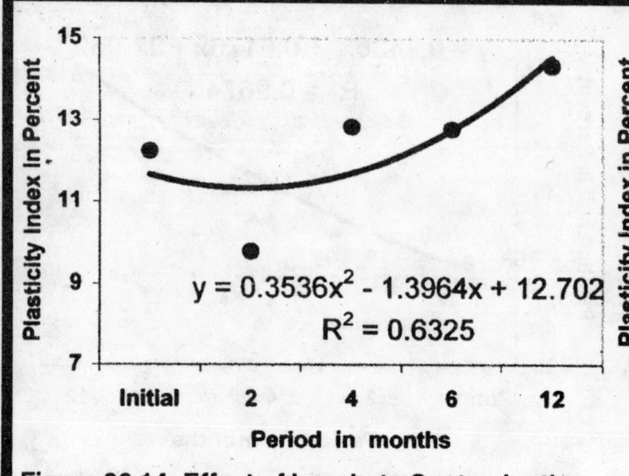

Figure 26.14: Effect of Leachate Contamination on Plasticity Index of Soil Sample–A

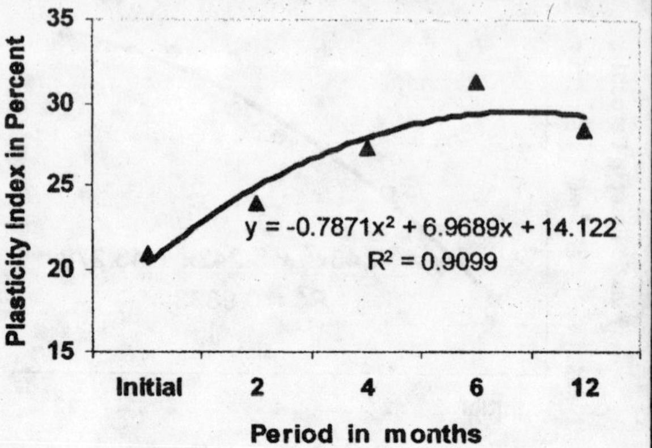

Figure 26.15: Effect of Leachate Contamination on Plasticity Index of Soil Sample–B

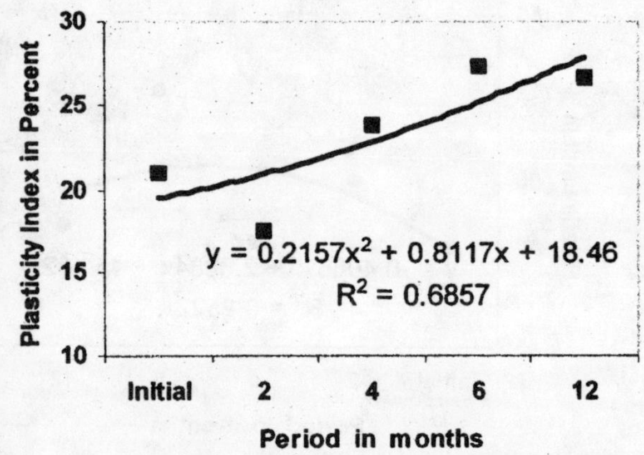

Figure 26.16: Effect of Leachate Contamination on Plasticity Index of Soil Sample–C

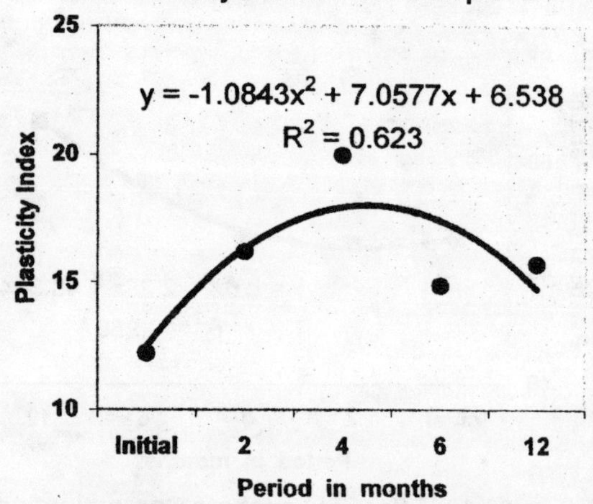

Figure 26.17: Effect of Distilled Water on Plasticity Index of Soil Sample–A

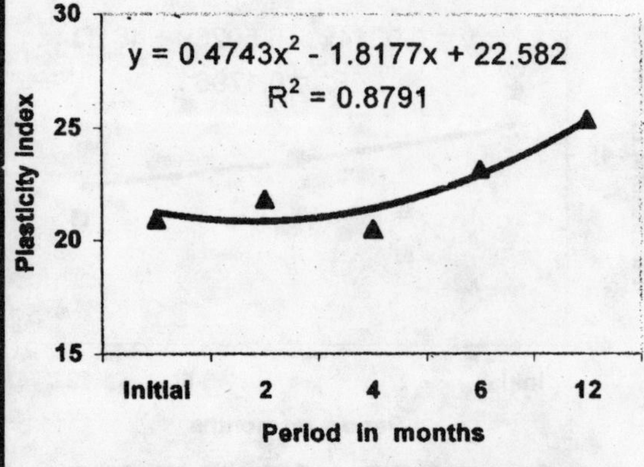

Figure 26.18: Effect of Distilled Water on Plasticity Index of Soil Sample–B

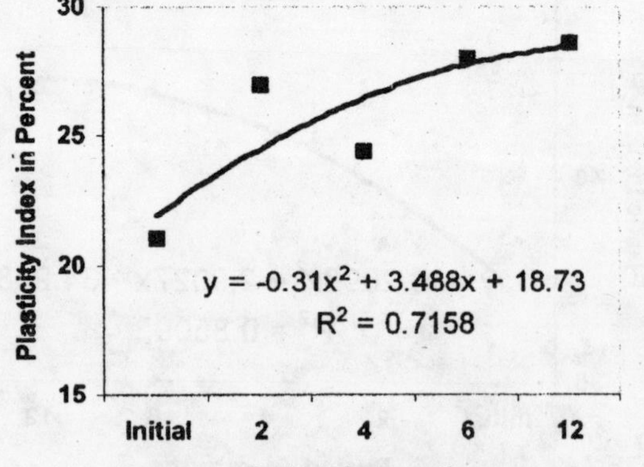

Figure 26.19: Effect of Distilled Water on Plasticity Index of Soil Sample–C

liquid limit (comparing with initial liquid limit value) is 20.98 per cent, 21.92 per cent and 16.08 per cent in the soil samples A, Band C respectively from L–S column. In DW–S Column the increase in liquid limit observed is 6.60 per cent, 14.54 per cent and 6.29 per cent in the soil sample A, B and C respectively. The maximum increase in liquid limit is found in the soil sample B from both the soil columns.

Table 26.3: Effect of Distilled Water on Properties of Soil at Different Time

Test	*Initial*	*After 2 Months*	*After 4 Months*	*After 6 Months*	*After 12 Months*
		Soil Sample–A			
Liquid limit (per cent)	60.28	66.14	68.16	67.52	64.26
Plastic limit (per cent)	48.04	49.96	48.17	52.68	48.59
Plasticity index (per cent)	12.24	16.18	19.99	14.84	15.67
Shrinkage limit (per cent)	31.14	30.39	31.42	33.5	36.72
UCC strength (KN/m²)	55.0	52.4	51.3	50.7	52.4
		Soil Sample–B			
Liquid limit (per cent)	58.99	60.18	62.54	65.27	67.57
Plastic limit (per cent)	38.08	38.38	42.04	42.13	42.19
Plasticity index (per cent)	20.91	21.8	20.5	23.14	25.38
Shrinkage limit (per cent)	31.21	34.71	35.26	38.76	32.62
UCC strength (KN/m²)	63.9	60.5	61.6	56.6	58.2
		Soil Sample–C			
Liquid limit (per cent)	67.66	70.44	69.23	72.16	71.92
Plastic limit (per cent)	46.64	43.52	44.86	46.18	43.29
Plasticity index (per cent)	21.02	26.92	24.37	27.98	28.63
Shrinkage limit (per cent)	29.23	34.68	38.32	38.29	36.42
UCC strength (KN/m²)	56.5	54.7	53.3	52.8	53.4

Except in the case of soil sample C from DW–S Column, considerable variation in plastic limit is observed in all the soil samples from L–S Column as compare to DW–S Column. At the end of the study period the observed increase in plastic limit is (comparing with initial plastic limit value) 21.92 per cent, 14.1 per cent and 11.15 per cent in the soil sample A, B and C respectively from L–S column. No proper trend is observed in variation in plastic limit of soil sample C from DW–S Column. The observed increase in plastic limit of soil samples A and B are 1.14 per cent and 10.79 per cent respectively. The variation in plastic limit of soil sample A, B and C with time from L–S Column are shown in Figures 26.8 to 26.10 and from DW–S Column are shown in Figures 26.11 to 26.13.

The plasticity index of different soil samples were calculated from the observed liquid limit and plastic limit values. The variation in plasticity index of soil sample A, B and C from L–S Column with time are shown in Figures 26.14 to 26.16 and from DW–S Column with time are shown in Figures 26.17 to 26.19. The plasticity index were increasing in all the soil samples from both the soil columns with increasing in contamination time. At the end of the study period the observed increase in plasticity index (comparing with initial plasticity index value) is 17.32 per cent, 36.15 per cent and 27.02 per

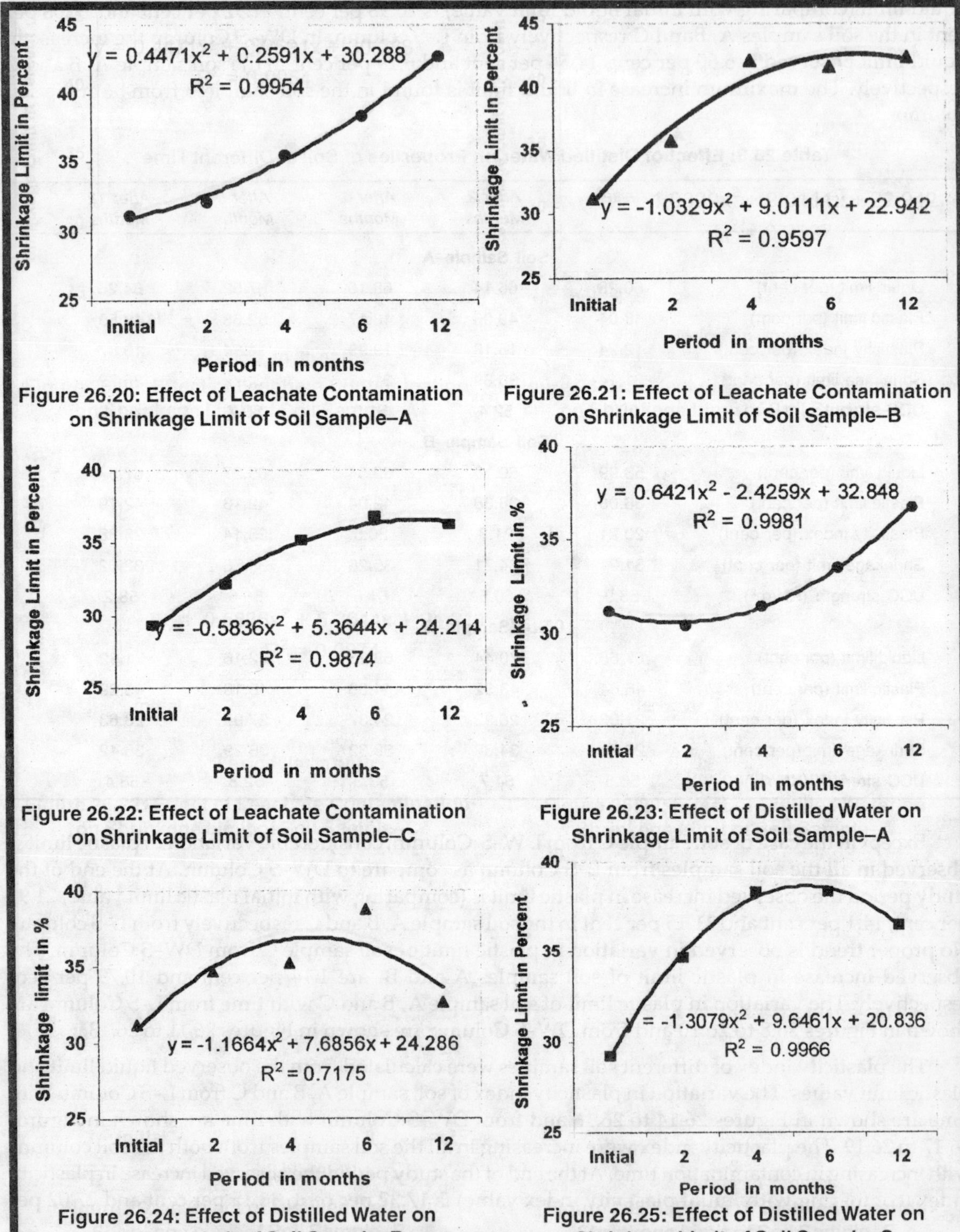

Figure 26.20: Effect of Leachate Contamination on Shrinkage Limit of Soil Sample–A

Figure 26.21: Effect of Leachate Contamination on Shrinkage Limit of Soil Sample–B

Figure 26.22: Effect of Leachate Contamination on Shrinkage Limit of Soil Sample–C

Figure 26.23: Effect of Distilled Water on Shrinkage Limit of Soil Sample–A

Figure 26.24: Effect of Distilled Water on Shrinkage Limit of Soil Sample–B

Figure 26.25: Effect of Distilled Water on Shrinkage Limit of Soil Sample–C

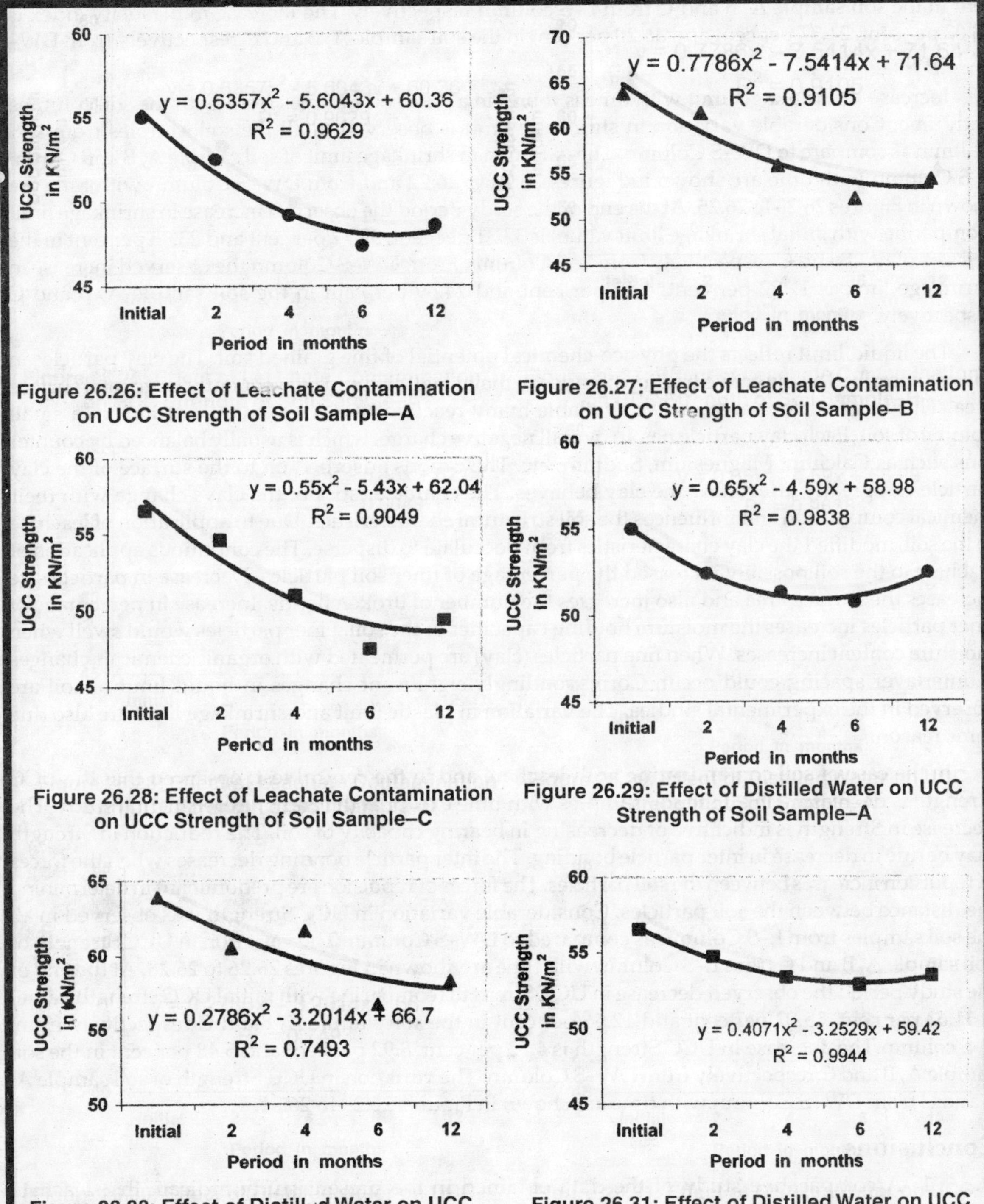

Figure 26.26: Effect of Leachate Contamination on UCC Strength of Soil Sample–A

Figure 26.27: Effect of Leachate Contamination on UCC Strength of Soil Sample–B

Figure 26.28: Effect of Leachate Contamination on UCC Strength of Soil Sample–C

Figure 26.29: Effect of Distilled Water on UCC Strength of Soil Sample–A

Figure 26.30: Effect of Distilled Water on UCC Strength of Soil Sample–B

Figure 26.31: Effect of Distilled Water on UCC Strength of Soil Sample–C

cent in the soil sample A, B and C from L–S column respectively. The increase in plasticity index is 28.02 per cent, 21.37 per cent and 36.20 per cent in the soil sample A, B and C respectively from DW–S Column.

Increase in shrinkage limit with time is found in all the soil samples from both the soil columns with time. Considerable variation in shrinkage limit is observed in all the soil samples from L–S Column as compare to DW–S Column. The variation in shrinkage limit of soil sample A, B and C from L–S Column with time are shown in Figures 26.20 to 26.22 and from DW–S Column with time are shown in Figures 26.23 to 26.25. At the end of the study period the observed increase in shrinkage limit (comparing with initial shrinkage limit value) is 37.31 per cent, 35.72 per cent and 23.98 per cent in the soil sample A, B and C respectively from L–S Column. From DW–S Column the observed increase in shrinkage limit is 17.92 per cent, 4.52 per cent and 24.59 per cent in the soil sample A, B and C respectively.

The liquid limit reflects the physico-chemical potential of fine grained soil. The clay particles in soil are described as the active fraction because of their small size and consequently large total surface area. This large surface area makes available many reactive sites for exchange of ions in a small volume of soil. Each clay particle has an overall negative charge, which is usually balanced by counter ions such as Calcium, Magnesium, Sodium, etc. The cations adsorbed on to the surface of the clay particle can greatly affect how the clay behaves. The characteristics of the clays change with their chemical composition and influences the soil structural characteristics. Due to application of leachate to the soil, modified the clay characteristics from flocculate to disperse. The continuous application of leachate to the soil possibly increased the percentage of finer soil particles. Decrease in particle size increases the surface area and also increases the number of broken bonds. Increase in percentage of finer particles increases the moisture holding capacities of the soil. Finer particles would swell when moisture content increases. When fine particles (clay) are permeated with organic chemicals changes in interlayer spacing could occur. Correspondingly significant changes in liquid limit of soil are observed in the experimental studies. The variation in plastic limit and shrinkage limit are also due same reason.

In the case of soil contaminating with leachate and in the control test, observed that the UCC Strength is decreasing in all the soil samples with time except at the end of experimental study. The decrease in Strength is indicative of decreasing in bearing capacity of soil. The reduction in Strength may be due to decrease in inter particle bonding. The inter particle bonding decreases when the forces of repulsion increases between the soil particles. The forces of repulsion are predominant in determining the distance between the soil particles. Considerable variation in UCC Strength was observed in all the soil samples from L–S Column as compared to DW–S Column. The variation in UCC Strength of soil sample A, B and C from L–S Column with time are shown in Figures 26.26 to 26.28. At the end of the study period the observed decrease in UCC Strength (comparing with initial UCC strength value) is 11.63 per cent, 15.02 per cent and 12.56 per cent in the soil sample A, Band C respectively from. L–S column.The decrease in UCC Strength is 4.72 per cent. 8.92 per cent and 5.48 per cent in the soil sample A, B and C respectively from DW–S Column. The variation in UCC Strength of soil sample A, B and C from DW–S Column with time are shown in Figures 26.29 to 26.31.

Conclusions

1. A comparative study of the data obtained in the present study indicate that leachate contamination modified the soil properties.
2. Consistency limits of soil were modified by the application of leachate.
3. The UCC Strength of soil decreases with increase in contamination time with leachate.

References

Chauhan, B.S., Khan, B., Karri, A. and R.P. Singh (1998). Characteristics and groundwater pollution of solid waste leachate. *Asian J. Chem.*, 10(4): 824–827.

Gabr, M.A. and S.N. Valero (1995). Geotechnical properties of municipal solid waste. *J. Geotechnical Testing*, 18(2): 241–251.

Khan, S.A., Rao, C.U. and M. Bandyopadhyay (1994). Characteristics of leachates from solid wastes. *Indian J. Environ. Hlth.*, 36(4): 248–257.

Kumaply, N.K. and Ishola (1994). The effect of chemical contamination on soil strength. *Proc. of 11th ICSMFE*, Sanfrancisco, p. 1199–1201.

Shivapullaiah, P.V., Raju, K.V.B. and A. Sridharan (1996). Effect of leachate on the properties of compacted clays. *Proc. IGC*, Madras, p. 520–522.

Shivapullaiah, P.V., Srinivasa Raghavan, R. and V.K. Stalin (1998). Effect of lime content on the volume change behaviour of soil. *Proc. IGC*, New Delhi, p. 51–54.

Shoba Cyrus and M. Thomas Roy (1996). Effect of tannery waste on the behaviour of soil. *Proc, IGC*, Madras, p. 544–546.

Sitaram, N., Ravindranath, C. and A. Vijay (1995). Prediction of settlement characteristics of shedi soil of D.K. District. *Proc. IGC*, Bangalore, 1: 13–15.

Sridhran, A, Nagaraj, T.S. and P.V. Shivapullaiah (1981). Heaving of soil due to acid contamination. *Proc. of 10th, ICSMFE*, Stockholm, 2: 383–386.

Srivastava, R.K., Singh, M. and R.P. Tiwari (1994). Laboratory study of soil industrial wastewater interaction behaviour. *Proc. of 13th ICSMFE*, New Delhi, p. 1553–1556.

Sudhakar, M. Rao and P. Mohan Rami Reddy (1995). Collapse behaviour of a laboratory contaminated soil. *Proc, IGC*, Bangalore, 1: 245–248.

Yaji, R.K. and Ramakrishna Gowda (1995). Effect of contamination by some chemicals on the engineering behviour of shedi soil. *Proc. IGC*, Bangalore, 1: 241–244.

Yaji, R.K., Ramakrishna Gowda, C. and Sandeep Jha (1996). Influence of contamination on the behviour of shedi soil. *Proc. IGC*, Madras, p. 540– 543.

Chapter 27

Bioactive Potentiality of *Catharanthus roseus*

Padma Chatterjee and Sucharita Das Gupta*

Plant Physiology and Biochemistry Laboratory, Department of Botany, University of Kalyani, Kalyani – 741 235, West Bengal

**E-mail: schatterjeecal@yahoo.co.in*

ABSTRACT

The bioactive potentiality of *Catharanthus roseus* was examined n terms of its antifungal and antibacterial action. The crude 50 per cent aqueous ethanolic extract of *Catharanthus roseus* was tested for antibacterial effect by agar cup method and for antifungal effects by disc-diffusion method (Bauer *et al.*, 1966). It was revealed in this study that it could inhibit the growth of *Staphylococcus aureus, E. coli, Fusarium oxysporum, Curvularia* sp. and *Aspergillus niger.* The inhibitory effect was highest in the dose 70 mg/ml.

Introduction

An ecofriendly approach of linking bioactive properties of phyto-chemicals with synthetic drug replacement tendency has led to identification of active principles of various plants which can be used as non-toxic compounds of medicinal and economic values.

Catharanthus roseus G. Don is a shrub, member of family Apocyanaceae and has been reported to have immense bioactive properties. Antimicrobial activity of indol alkaloid obtained from *Catharanthus roseus* against some bacterial and fungal strains was shown by Rozas (1979). Anna and Bridget showed anticancer properties of Vinca alkaloids. Hill (2001) *C. roseus* was also shown to have anticancer property by Muriel (2004).

The present investigation was carried out with the aim to study the antibacterial and antifungal activity of *C. roseus*.

Materials and Methods

Mature plants of *Catharanthus roseus* was collected from Kalyani of West Bengal during the month of June–July. Bacterial cultures of *S. aureus* and *E. coli* and fungal cultures of *Fusarium oxysporum*, *Curvularia* sp. and *Aspergillus niger* were taken from the department of Botany, Kalyani University.

500 gm sundried, powdered, leaves of *C. roseus* were soaked in 3 liter of 50 per cent aqueous ethanol at room temperature for 4 days. It was charcoalised and filtered. The filtrate was concentrated under reduced pressure and a brown residual solid was obtained.

The antibacterial assay was done following agar cup medium plates.1 ml bacterial suspension each of *E. coli* and *Staphylococcus aureus* was uniformly spread. On the seeded agar plates, wells of 0.8 cm diameter were bored with sterile cork borer. Three concentrations (30, 50, 70 mg/ml) of test compound were loaded in the wells. The petridishes were incubated at 37°C and after 24 hours the diameters of the inhibition zones surrounding the point of application of different dilutions were measured. The tests were done in triplets and mean values were recorded.

The antifungal assay was done following disc-diffusion method. Sample extracts of respective doses (30, 50, 70 mg/ml) were soaked to at 9 mm-diameter filter paper (Whatman). 20 ml autoclaved P.D.A. media (pH maintained at 6.8–7) was poured into each of the sterile petriplates (9 cm each). Inoculation was done from pathogenic pure culture. 5 mm diameter of inoculum was taken by a cork borer, for individual inoculation. After inoculation all the soaked filter papers were placed on inoculated media and the extract was allowed to diffuse. A control was maintained with distilled water. All the plates were incubated at 30°C ± 1 temperature and the diameter of the zone inhibition were measured after 7 days.

Results and Discussion

Table 27.1 represents the effect of the extract of *Catharanthus roseus* on the colony growth of *A. niger*, *F. oxysporum* and *Curvularia* sp. The results indicates that the 50 per cent aqueous ethanolic extract of *C. roseus* has an inhibitory effect on fungal colony growth. The most effective dose was 70 mg/ml.

Table 27.1: Studies on Antifungal Efficacy of Crude Extracts of *C. roseus* Against *A. niger* (A.N), *Fusarium oxysporum* (F.O) and *Curvularia* sp. (C. sp.) by Disc-diffusion Method

Dose (mg/ml)	Diameters of Inhibitory Zone (cm)			Inhibition (%)		
	A.N.	F.O.	C.sp	A.N	F.O	C.sp
Control	0	0	0	0	0	0
30	1.14 ± 0.6	1.14 ± 0.57	1.14 ± 0.6	12.72	12.73	12.73
50	2 ± 0.99	2.08 ± 1.0	1.88 ± 0.94	18.18	18.9	16.36
70	2.84 ± 1.4	2.4 ± 1.2	2.48 ± 1.2	25.82	21.82	22.55

The effects of the 50 per cent aqueous ethanolic extract of *C. roseus* on the growth of *S. aureus* and *E. coli* are presented in Table 27.2. The results indicate that the 50 per cent aqueous ethanolic extract of *C. roseus* has an inhibitory effect on the growth of the bacterial colony. The inhibition was most effective in 70 mg/ml.

Table 27.2: Studies on Antibacterial Efficacy of Crude Extracts of *C. roseus* on *Staphylococcus aureus* (S.A.) and *E.coli* (E.C.) by Agar Cup Method

Dose (mg/ml)	*Diameters of Inhibitory Zone (cm)*		*Inhibition (%)*	
	S.A	*E.C*	*S.A*	*E.C*
Control	0	0	0	0
30	1.56 ± 0.53	1.61 ± 0.81	14.18	14.64
50	2.1 ± 1.03	2.85 ± 1.43	19.09	25.91
70	2.86 ± 1.43	3 ± 1.5	26	27.27

References

Anna, Krucznskim and T. Hill Bridget (2001). Anticancer properties of Vinca alkaloids. *Critical Reviews in Oncology/Hematology*, 40(2): 159–173.

Bauer, A.W., Sherris, T.M. and W.H.M. Kirby (1996). Antibiotic susceptibility testing by a standardized single disk method. *A.M.J. Clin. Path. Pathol.*, 45: 493.

Muriel, J. Montbriand (2004). *Herbs or Natural Products that Decrease Cancer Growth*, Part One of a Four-Part Series, 31(4).

Rojas, Harnandez N.M. (1979). Antimicrobial activity of indol alkaloids from *Catharanthus roseus* against some bacterial and fungal strains. *Rev. Cubana Med. Trop.*, 31(3): 199–204.

Chapter 28
Trace Elemental Concentrations in Vegetative Parts of Weed Plant *Achyranthes aspera* L. by SEM-EDS Method

N. Ramamurthy, J. Subashini* and M. Parthasarathy***

**Department of Physics, **Physics Wing, DDE, Annamalai University, Annamalainagar – 608 002, Tamil Nadu, India*

ABSTRACT

The vegetative parts, leaves, stem and roots of weed plant *Achyranthes aspera* Linn are analysed using SEM-EDS. The microphotograph obtained from Scanning Electron Microscope and weight percentage of specific elemental concentration of Na, Mg, P, S, CI, K, Ca, Mn, Fe, Cu, Zn and Pb obtained from Energy Dispersive X-Ray Spectrometer attached to SEM are analysed sample wise. The results are reported and discussed.

Keywords: SEM-EDS, Trace elements, Vegetative parts, Weed, Achyranthes aspera L.

Introduction

Each micronutrient play a specific biochemical role and if present in sufficient quantity, the growth and hence the yield attributes may be seriously reduced (Kanwar and Randhawa, 1967).

Achyranthes aspera Linn (Family: Amaranthaceae) is a small genius of stiff herbs found in tropical and sub-tropical regions. *Achyranthes aspera* occurs commonly as a weed in the drier wastelands (Anonymous, 1985).

The purpose of our present study is to find the microphotograph of the structure and the short review in the trace elements accumulation in the vegetative parts like leaf, stem and root of weed plant *Achyranthes aspera* Linn. The microphotograph obtained from SEM, supported by energy dispersive X-ray (EDS) microanalysis device, to identify the elements like Sodium (Na), Magnesium (Mg), Phosphorus (P), Sulphur (S), Chloride (Cl), Potassium (K), Calcium (Ca), Manganese (Mn), Iron (Fe), Copper (Cu), Zinc (Zn), and Lead (Pb) in the vegetative parts of weed *Achyranthes aspera* L., collected from Nakkaravandangudi Village, Chidambaram, Tamil Nadu.

Materials and Methods

Roots, stem and leaves were carefully collected from some *Achyranthes aspera* L. weed plants. Fresh plants were thoroughly washed with ample water and then distilled water, to remove clay sands, dusts and associated algae. The vegetative parts leaves, stem and roots were carefully separated from the plant. Cleaned vegetative parts were shade dried and then dried in oven at 60°C for four hours to remove moisture content. Dried samples were ground to fine powder. These samples are used for the SEM-EDS analysis.

The microphotographs of these samples were recorded using SEM JEOL model, JSE–5610 LV available at Centralised Instrumentation and Services Laboratory (CISL), Annamalai University, with an accelerating voltage of 20 keV, at high vacuum (HV) mode and Secondary Electron Image (SEI). The maximum magnification possible in the equipment is 3,00,000 times with a resolution of 3 nm., typically setting at a magnification of X1500 (leaf and stem), X500 for root samples of study.

The semi-quantification elemental analyses to identify the weight percentage of major and minor elements present in the samples were done using the OXFORD INCA Energy Dispersive X-ray Fluorescence Spectrometer (EDS).

Results and Discussion

Figure 28.1 shows the microphotograph (morphological structure) and corresponding EDS spectrum of the vegetative parts like leaf, stem and root sample of weed plant *Achyranthes aspera* L. Figure 28.2 gives the relative distribution of trace elements of vegetative parts, in the form of a graph in which the originate and abscissa are given in the terms of percentage and types of elements respectively. The result of weight percentage of elements distribution in the samples is furnished in Table 28.1.

Table 28.1: The Percentage of Trace Elements Present in the Leaf, Stem and Root Sample of Weed Plant *Achyranthes aspera* L.

Characteristic Elements	*Elemental Concentration in %*		
	Leaf	*Stem*	*Root*
Na	5.82	14.48	4.16
Mg	11.84	6.16	11.92
P	4.27	0.81	7.74
S	2.92	7.43	3.09
Cl	2.24	2.69	5.19
K	20.31	23.90	17.30
Ca	19.12	15.88	8.26
Mn	4.43	–	0.06
Fe	11.98	0.32	0.35
Cu	5.21	16.99	21.49
Zn	5.28	0.45	0.05
Pb	6.58	10.89	20.39

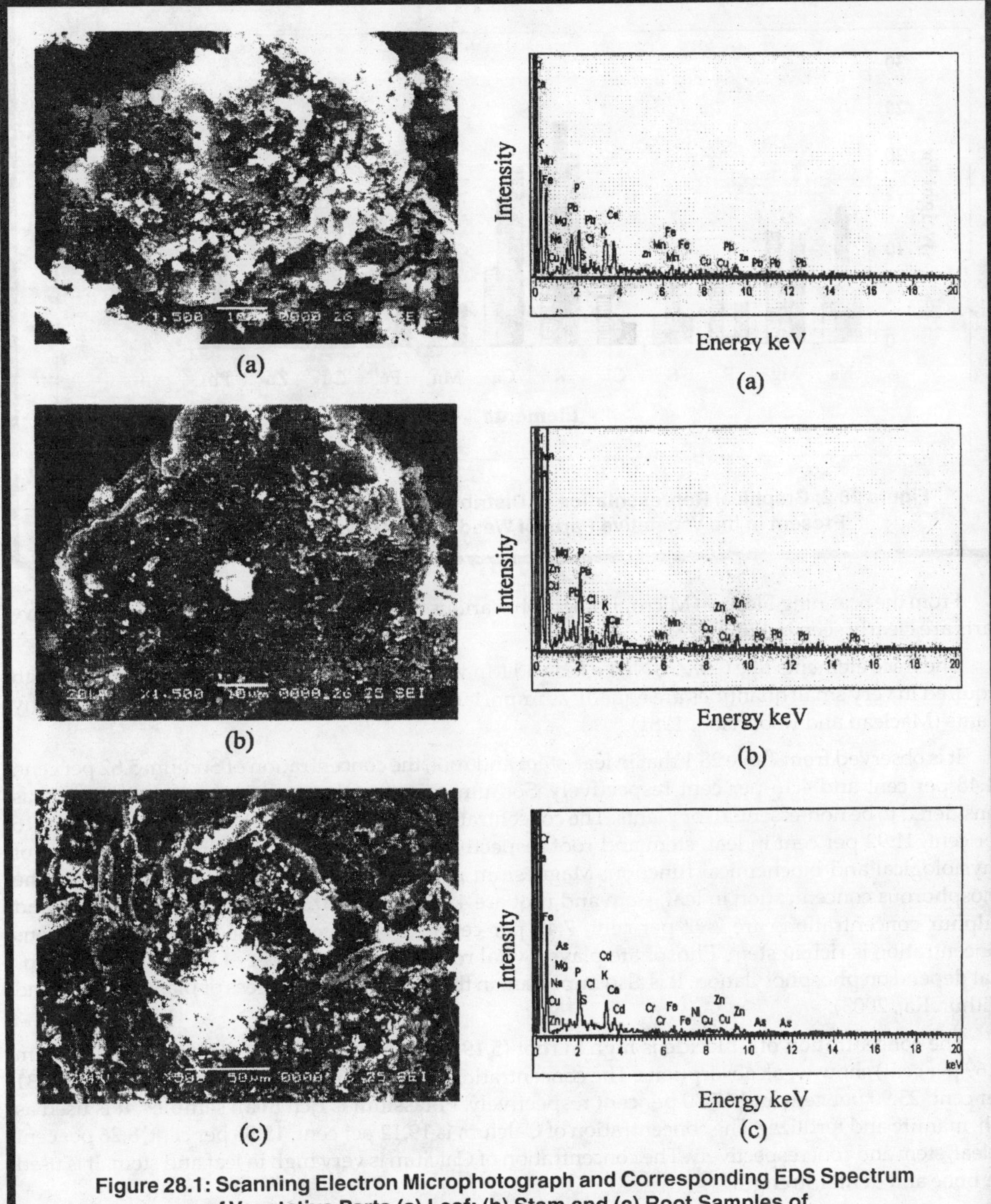

Figure 28.1: Scanning Electron Microphotograph and Corresponding EDS Spectrum of Vegetative Parts (a) Leaf; (b) Stem and (c) Root Samples of Weed Plant *Achryanthes aspera* L.

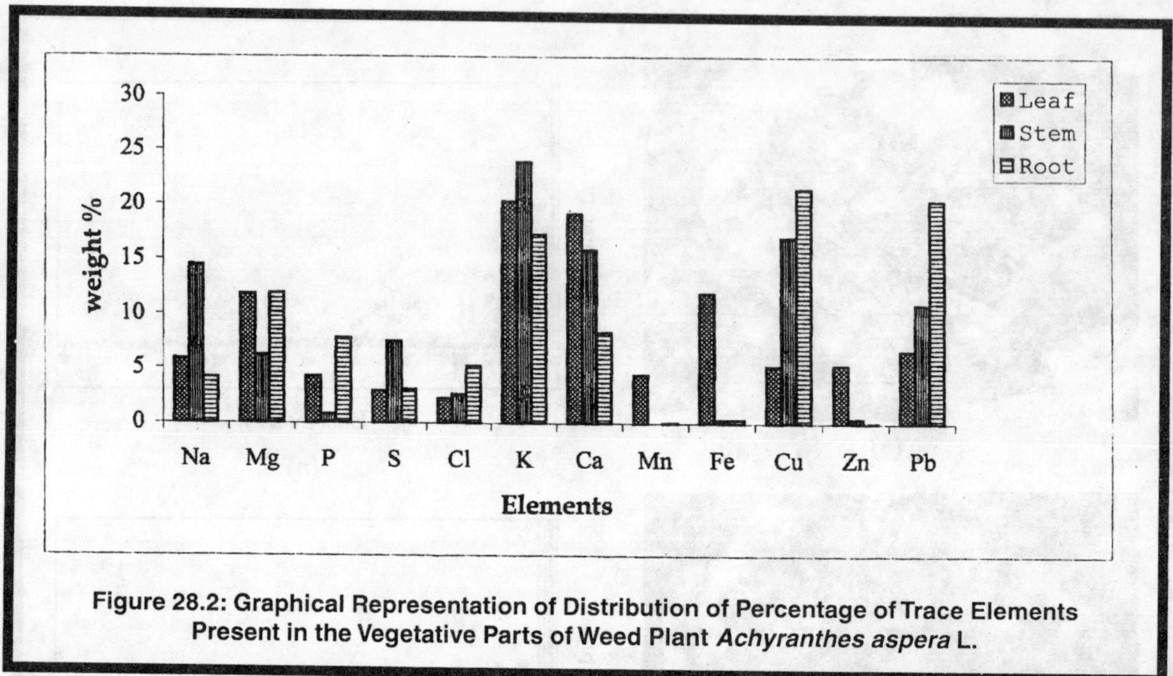

Figure 28.2: Graphical Representation of Distribution of Percentage of Trace Elements Present in the Vegetative Parts of Weed Plant *Achyranthes aspera* L.

From the Scanning Electron Microphotograph, various morphological cell structures of vegetative parts are clearly seen in Figure 28.1.

The trace elements like Cu, Zn, Ni, Fe, Cr and Mn which are essential for plant growth although required in very small quantities are equally as important as the major elements in producing healthy plants (Maclean and Robertson, 1981).

It is observed from Table 28.1 that in leaf, stem and root, the concentration of Sodium 5.82 per cent, 14.48 per cent and 4.16 per cent respectively. Sodium concentration is rich in stem. Sodium was considered to be non-essential for plants. The concentration of Magnesium value is 11.84 per cent, 6.16 per cent, 11.92 per cent in leaf, stem and root respectively. Magnesium is involved in a number of physiological and biochemical function. Magnesium is rich in leaf and root plant than stem. The Phosphorous concentration in leaf, stem and root are 4.27 per cent, 0.81 per cent, 7.74 per cent and Sulphur concentrations are 2.92 per cent, 7.43 per cent, 3.09 per cent respectively. Phosphorous concentration is rich in stem. Phosphate plays a vital role in a large number of enzymatic reactions that depend on phosphorylation. It is also important in the reproductive process of plants (Murali and Mithun Raj, 2003).

The concentration of Chloride is high in root (5.19 per cent). Both leaf (2.24 per cent) and stem (2.69 per cent) show weak absorptions. The concentration of Potassium in leaf, stem and root are 20.31 per cent, 23.90 per cent and 17.30 per cent respectively. Potassium is rich in all samples. It is used as ash, manure and fertilizer. The concentration of Calcium is 19.12 per cent, 15.88 per cent, 8.26 per cent in leaf, stem and root respectively. The concentration of Calcium is very high in leaf and stem. It is used for bone and dental strength.

Concentration of Copper is maximum as root is 21.49 per cent and in stem it is 16.99 per cent and concentration of copper in leaf is 5.21 per cent. Copper provides metabolic control over auxin synthesis and is involved in protein metabolism. Zinc is an essential trace element for the normal functioning of cells, because it takes part in the synthesis of DNA and thus cell growth. Zinc is one of the essential nutrients for plant growth. Hence, Zinc is essential for the normal, functioning of the cells including protein synthesis, carbohydrates metabolism, cell growth and cell division (Kudesia, 1980).

In this present investigation of Zinc records 5.28 per cent, 0.45 per cent, 0.05 per cent in leaf, stem and root respectively. The percentages of Lead are 6.58 per cent, 10.89 per cent, 20.39 per cent in leaf, stem and root respectively. Concentration of lead is maximum in root (20.39 per cent). The percentages of Iron are 11.98 per cent, 0.32 per cent, 0.35 per cent in leaf, stem and root respectively. Concentration of Iron is maximum in leaf plant (11.98 per cent) than stem and root. The trace elements like Fe and Mn, which are essential for plant growth although required in very small quantities are equally important in producing healthy plants.

The leaf, stem and root samples of *Achyranthes aspera* Linn contain all essential elements Na, Mg, P, S, CI, K, Ca, Mn, Fe, Cu and Zn in abundance. The elements K and Ca are very much high compared to other elements. The elements Mn and Zn, which are essential for metabolic activity of the plant are present in abundance in the leaves than in other parts. P and S, which are highly reactive elements, are present in ample quantity in all parts. The toxic element Pb is present in abundance in all parts. From SEM-EDS analysis, it is concluded that *Achyranthes aspera* L. absorbs all essential nutrients from the soil and hence, act as weed among the main plants.

References

Anonymous (1985). *The Wealth of India*, PID, CSIR, New Delhi, IA: 24.

Kanwar, J.G. and N.S. Randhawa (1967). *Micronutrient Research in Soil and Plants in India: A Review*, New Delhi.

Kudesia, V.P. (1980). *Pollution (Every Where)*. Pragati Prakashan, Begam Bridge, Meerut.

Murali, M. and Mithun Raj (2003). Some studies on effect of integrate low cost sanitation on soil at Marripalem, Visakhapatnam. *Indian J. Env. Prot.*, 23(9): 1068–1073.

Chapter 29

Effect of Intensity of Puddling on the Transport and Transformation of Urea Nitrogen in Rice Soil

A.K. Dash, R. Nayak** , B.K. Mishra** and B.B. Behera***

**Training Associate, K.V.K., G. Udayagiri, Kandhamal – 762 100*
***Department of Soil Science and Agricultural Chemistry,*
Orissa University of Agriculture and Technology, Bhubaneswar – 751 003, Orissa, India

ABSTRACT

The rice field was puddled with different puddling intensities *i.e.*, T_1–Twice puddling by wooden country plough; T_2–Once puddling by zig zag puddler; T_3–Twice puddling by zig zag puddler; T_4–Once puddling by zig zag puddler after dhinacha incorporation and T_5–Once puddling by zig zag puddler after addition of FYM @ 5 t ha^{-1} so as to study the transportation and transformation of nitrogen in rice soil. Leachate samples at a soil depth of 15 cm, 30 cm and flood water sample were collected from the microplot (225 cm × 225 cm). No NH_2–N and NO_3–N was detected either from leachate or from flood water sample. It has been observed that NH_4^+–N appeared in flood water soon after application of fertilizer. The amount of NH_4^+–N leached through the flood water was maximum in T_1, intermediate in T_2 and T_3 and minimum in T_4 and T_5. The leachate sample collected from 15 cm soil depth indicated that maximum concentration of NH_4^+–N was observed with T_5 (16.50 mg L^{-1}) followed by T_4 (12.0 mg L^{-1}), T_1 (9.6 mg L^{-1}), T_2 (6.75 mg L^{-1}) and T_3 (6.51 mg L^{-1}). In presence of organic matter/green manure with less intensity of puddling provided more amount of NH_4^+–N to the crop plant for higher production.

Keywords: *Country plough, Zig zag puddler, Flood water.*

Introduction

Puddling is the mechanical reduction of apparent specific volume (Budman and Rubin, 1968) and it is an important tillage practice for transplanted rice. When urea is broadcasted in rice field, it get dissolved in water resulting in high concentration of amide form (NH_4^+–N) and transported by diffusion (Savant and De Datta, 1979). Mahajan *et al.* (1991) reported that leaching loss of nitrogen from urea was more under saturated condition than due to intermittent wetting and drying. Mishra *et al.* (1999) reported that downward movement of NH_4^+–N was more than lateral movement. Motousch (1982) reported that increasing bulk density from 1.72 to 1.75 mg m^{-3} decreases the percolation rate which also greatly reduces the leaching loss of nitrogen. Many other workers have been studied the leaching loss of nitrogen and percolation loss of water with same puddling intensity. But present investigation was carried out in order to study the effect of different intensities of puddling on the transport and transformation of urea nitrogen in soil.

Materials and Methods

A field experiment was carried out at Central Farm of Orissa University of Agriculture and Technology, Bhubaneswar so as to evaluate the efficiency of different intensities of puddling on the transport and transformation of nitrogen present in prilled urea (PU) applied to rice soil. The soil of the experimental site was sandy clay loam in texture (62 per cent sand, 16 per cent silt and 22 per cent clay) with pH 8.5, organic carbon 5.3 g kg^{-1} and bulk density (B.D.) 1.77 mg m^{-3}. The experiment consisted of 5 tillage treatments in 5 strips each replicated thrice in a randomised block design so as to produce different intensities of puddling in the rice field. The treatments are as follows:

T_1–Twice puddling by wooden country plough (Conventional practice)

T_2–Once puddling by zig zag puddler

T_3–Twice puddling by zig zag puddler

T_4–Once puddling by zig zag puddler after incorporation of dhaincha as *in situ* green manure.

T_5–Once puddling by zig zag puddler after incorporation of FYM @ 5 t ha^{-1}.

After imposing the treatments, rice seedling (Cv. Lalat) of 21 days old were transplanted in the experiment field with recommended dose of fertilizer. A common dose of 40 kg each of P_2O_5 and K_2O per hectare as single super phosphate and muriate of potash respectively was applied as basal. The prilled urea (PU) was applied in 3 splits *i.e.* at transplanting, tillering and at panicle initiation stage. One microplot of 0.225 × 0.225 m^2 was prepared in the middle of each treatment plot of second replication by inserting a tin plate upto a depth of 0.20 m. Inside the microplot soil solution sampler were installed at a soil depth of 0.15 m and 0.30 m so as to collect leachate sample with specific time interval. PU was applied @ 20 kg N ha^{-1} in the main field as well as inside the microplot at the time of transplanting. The leachate samples were collected at an interval of 3 h for 1 day, 4 h for 1 day and 24 h for 15 days alongwith flood water sample from each microplot. These samples were analysed for NH_2–N, NH_4^+–N and NO_3–N. The NH_2–N of leachate was determined by modified diacetyl monoxime method (Mulvaney and Bremner, 1979), NH_4^+–N was determined colorimetically by Indophenol Blue Method (Keeney and Nelson, 1982) and NO_4^-–N in the leachate was determined by 2,4 Disulphonic Acid Method (Jackson, 1973).

Results and Discussion

The concentration of NH_2–N, NH_4^+–N and NO_3^-–N in the leachate as well as in the flood water was determined. No NH_2–N and NO_3^-–N was observed in leachate sample as well as flood water sample. So the BTC for these two ions are not presented here. The graphical presentation of concentration (mg L^{-1}) of NH_4^+–N correspondence to time is known as Break Through Curve (BTC) of particular ions.

Concentration of NH_4^+–N in the Flood Water

The BTCs of NH_4^+–N stemming from flood water due to application of 20 kg N ha^{-1} in form of PU with respect to different intensities of puddling throughout the growth period of crop was presented in Figure 29.1. It has been revealed that NH_4^+–N appeared in flood water soon after application of PU in the plot. The time required to reach the maximum concentration was 22.5 h, 35 h, 30.5 h, 35 h and 11 h for T_1, T_2, T_3, T_4 and T_5 respectively. The maximum content of NH_4^+–N was observed to be 19, 10, 14, 9.5 and 13.5 mg L^{-1} in the flood water with respect to T_1, T_2, T_3, T_4, and T_5 respectively. The NH_4^+–N concentration in flood water disappeared gradually over a period of 150 h (*i.e.* 6.25 days), The area under BTC indicated that the amount of NH_4^+–N in the flood water was maximum for T_1, intermediate for T_2 as well as T_3 and minimum in T_4 alongwith T_5. The differential concentration of NH_4^+–N with different intensities of puddling possibly due to, low dispersability of clay associated with low NH_4^+–N adsorption in clay micelle for which more NH_4^+–N observed in flood water in T_1. When intensities of puddling was increased in T_2 and T_3 the clay despertion was more associated with more adsorption of NH_4^+–N in these soil resulting less amount of NH_4^+–N in the flood water in other treatments as compared to T_1. But presence of organic matter or green manure also induced more clay dispersion as well as improved soil structure which induced movement of more NH_4^+–N to lower layer resulting reduction of NH_4^+–N content in the flood water in T_4 and T_5 as compared to T_1, T_2 and T_3.

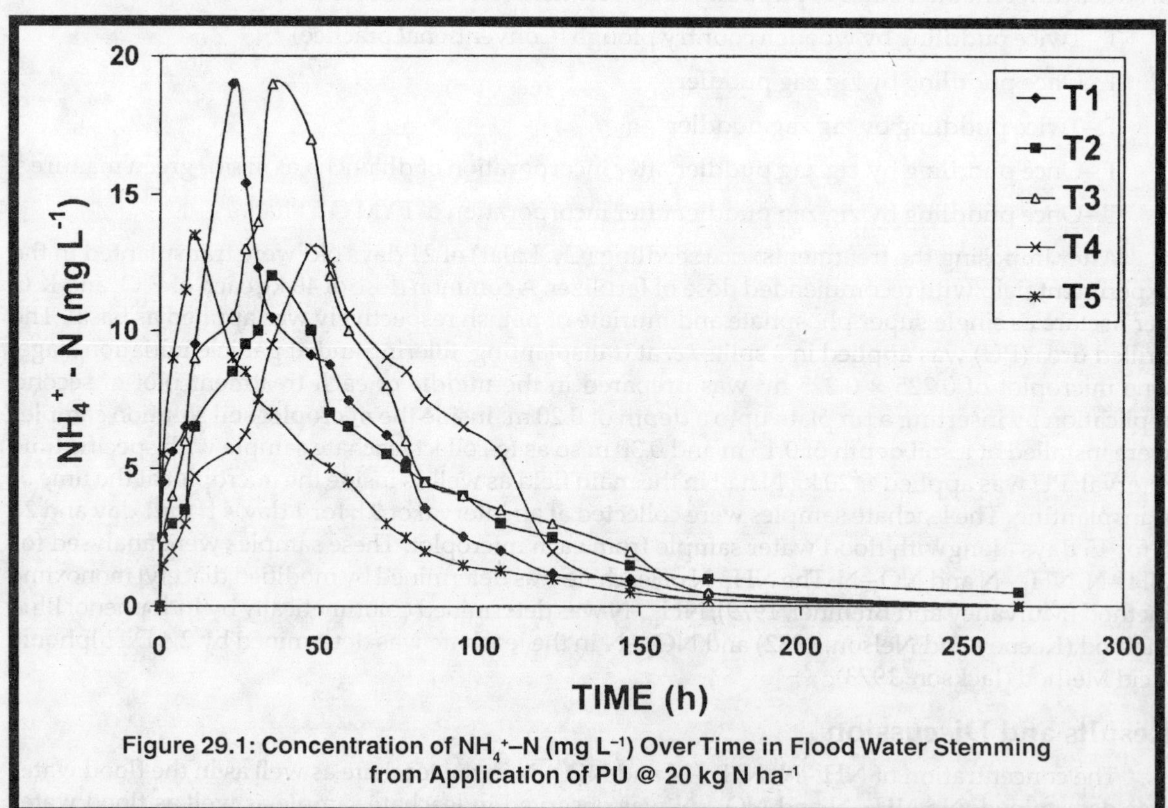

Figure 29.1: Concentration of NH_4^+–N (mg L^{-1}) Over Time in Flood Water Stemming from Application of PU @ 20 kg N ha^{-1}

Concentration of NH_4^+–N at 15 cm Soil Depth

The concentration of NH_4^+–N at 15 cm soil depth stemming from the application PU @ 20 kg N ha^{-1} was presented in Figure 29.2. These BTCs indicated appearance of NH_4^+–N was delayed in the leachate as compared to flood water and the maximum concentration in the leachate was observed at about 26 to 36 h of its application. The maximum concentration of NH_4^+–N was 9.6, 6.75, 6.51, 12.0 and 16.50 mg L^{-1} at 15 cm soil depth with respect to T_1, T_2, T_3, T_4 and T_5 respectively. It has been revealed that amount of NH_4^+–N was maximum in T_5 and T_4, intermediate in T_1 and minimum in T_2 and T_3. This might be due to improved pore size distribution in presence of organic matter which induced high leaching of NH_4^+–N from surface to 15 cm soil depth. On the other hand, the leaching of NH_4^+–N was minimum in T_2 and T_3 because of puddling by zig zig puddler resulting in high degree of clay dispersion as well as more retention of NH_4^+–N in puddled layer. As more of NH_4^+–N was observed in the flood water in T_1, T_2 and T_3 than T_4 and T_5 obviously the amount of NH_4^+–N detected in 15 cm layer was lower in T_1, T_2 and T_3. Further leaching loss of NH_4^+–N in T_3 was less than T_1 at 15 cm soil depth revealed that increasing intensity of puddling (Twice by zig zag puddler) restricted the movement of NH_4^+–N to a soil depth of 15 cm. However in presence of organic matter (T_4 and T_5) the NH_4^+–N was observed for a longer period in leachate sample at a soil depth of 15 cm. It can be concluded that in the presence of organic matter less amount of NH_4^+–N found in the flood water and that amount was more in 15 cm soil depth which can be utilised by the crop plant efficiently for a longer period.

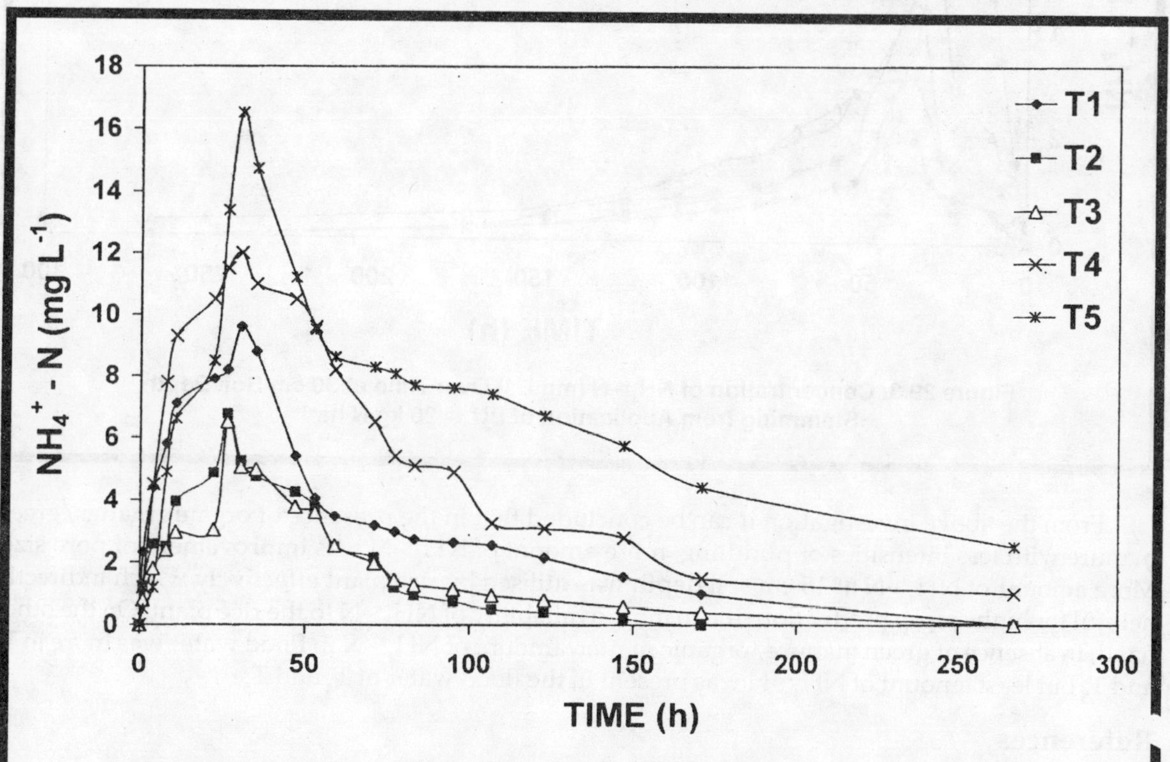

Figure 29.2: Concentration of NH_4^+–N (mg L^{-1}) Over Time at 15 cm Soil Depth Stemming from Application of PU @ 20 kg N ha^{-1}

Concentration of NH_4^+–N at 30 cm Soil Depth

The concentration of NH_4^+–N in the leachate collected at a soil depth of 30 cm was presented in Figure 29.3. From the result it was revealed that maximum concentration of NH_4^+–N was exceeded from 10 mg L^{-1} at 30 cm soil depth. It has been observed that amount of NH_4^+–N found at 30 cm soil depth was low as compared to NH_4^+–N at 15 cm soil depth irrespective of treatment. The result indicated that a lot of NH_4^+–N was absorbed in soil surface resulting drastic reduction in concentration at a soil depth of 30 cm irrespective of intensities of puddling.

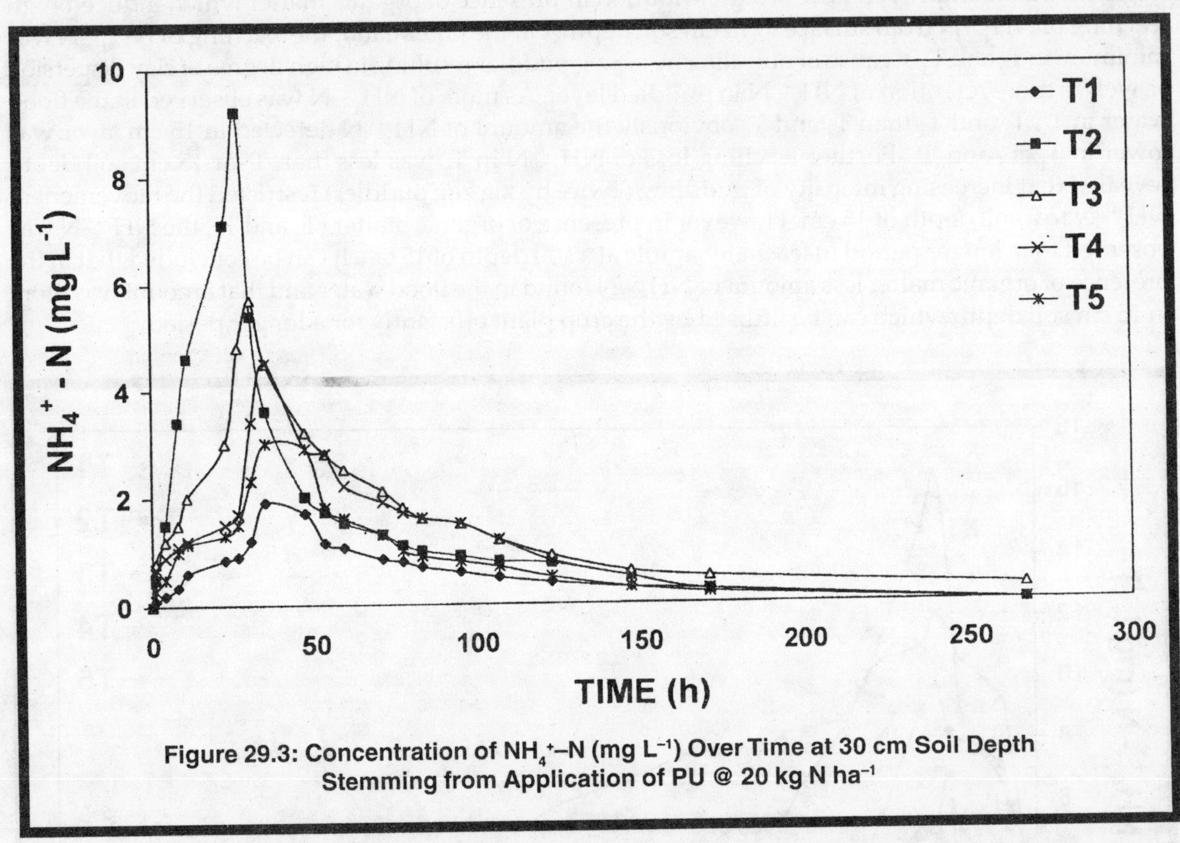

Figure 29.3: Concentration of NH_4^+–N (mg L^{-1}) Over Time at 30 cm Soil Depth Stemming from Application of PU @ 20 kg N ha^{-1}

From the above investigation it can be concluded that in the presence of organic matter/green manure with less intensities of puddling, more amount of NH_4^+–N was improvement of pore size. More amount of NH_4^+–N at 15 cm soil depth was utilised by rice plant effectively which indirectly helpful for higher crop production due to more availability of NH_4^+–N to the rice plant. On the other hand, in absence of green manure/organic matter, amount of NH_4^+–N in flood water was more in T_2 and T_3 but least amount of NH_4^+–N was present in the flood water of T_4 and T_5.

References

Budman, G.B. and Rubin (1968). Soil puddling. *Soil Soc. American Proc.*, 13: 27–36.

Jackson, M.L. (1973). *Soil Chemical Analysis*. Prentice Hall of India Pvt. Ltd., New Delhi.

Keeney, D.K. and D.W. Nelson (1982). *Methods of Soil Analysis*, 2nd Edition. Number 9 (Part 2) in the series of Agronomy, American Society of Agronomy, Soil Science Society of America, Medison, Wisconsin, USA, p. 643–698.

Mahajan, K.K. and B.R. Tripathy (1991). Leaching loss of nitrogen from different forms of Urea in columns of Alfisols. *Ind. J. Soil Sci.*, 34(4): 618–624.

Mishra, B.K., Dash, A.K., Das, A.K., Jena, D. and S.K. Swain (1999). Evaluation of placement methods for urea super granules in wetland rice (*Oryza sativa*) Soil. *Indian J. of Agron.*, 44(4): 710–716.

Motousch, C. (1982). Nitrogen form in fertilizers and their movement in soil. *Rostlina Vyroba*, 28(10): 1011–1018.

Mulvaney, R.L. and J.M. Bremner (1979). A modified diacetyl monoxime method for colorimetric determination of urea. *Commun. Soil Sci. Plant Anal.*, 10: 1163–1170.

Savant, N.K. and S.K. De Datta (1979). Nitrogen release patterns from deep placement of urea in a wetland rice soil. *Soil Sci. Soc. Am. J.*, 43: 131–134.

Chapter 30

Ecological Impact and its Consequence to Interrelationship Between Harmful and Beneficiary Insect/Pest of Paddy

S.R. Singh, P. Rukamani Devi*, N.B. Devi*, W.K. Devi* and N.S. Devi***

**Post Graduate Studies Centre, HRDRI, Canchipur –795 003, Imphal*
***Botany Department, Imphal College, Imphal, Manipur*

ABSTRACT

Agro-ecosystem plays an important role as environment changing would be consequent to changes in vast ecosystem. Eventually the ecological factors and insect pest diversity on paddy field has been formulated on selective and two varieties *viz.*, Leimaphou (CV_1) and Phoudongba (CV_2) during 2001–2002 to facilitate an understanding of the insect pest and farmer's friend at three different phases of crop growth and the corresponding influence of meteorological factors and their co-relatives. The finding high lighted the agro-ecological aspects of paddy the basics of ecologists, environmentalist, agriculturist in the wet land of Imphal valley. The farmer's friend insects of Dragonfly, Nepa, Water stider, Water treader, Spider, Belostoma, Trichogramma, Telenomus etc. and in harmful pest of green leaf hopper, Rice case worm, White stem borer, Grasshopper, Whorl maggot, Carausius mantis, Army worm, Gundhi bug, Brown plant hopper, Leaf hopper, Mole cricket, Green horn caterpillar, Cut worm moth, Hispa etc. were detected. The ratio of defenders to beneficial insects remains dominated (2 : 1) in all the investigated phases. All pest of rice are below the given Economic Injury Level (EIL) and Economic Threshold Level (ETL). The HYV have higher susceptible to pests and have higher number of other organisms.

Keywords: *Agro-ecosystem, Farmer's friend, Economic injury level, Economic threshold.*

Introduction

Many changes have occurred and agricultural productivity is much greater than formerly, but even today cereal yields affected by many factors outside the direct human control, especially in the long term cereal agro-ecosystem, a very complex and dynamic process with many new and important biological renovation and other still be covered by further broad based research.

The changes due to the time with the development impact influence much especially by plant breeding and mechanization to a dramatically increased yields. In India the yield of paddy exponently increased from 668 kg/ha in 1950–51 to 1013 to 1740 kg/ha in the year 1990's however the rate of changes has recently decreased quite sharply and there is now no significant upward trend from the conventional paddy varieties (Anonymous, 2000; Singh, 2002). Similar trend also existed in Britain (Cooke, 1967; Attwood, 1970). Pesticide and agro-chemical inputs generally have continued to increases exponentially (Potta and Vickerman, 1974), so that yields appear to have reached an asymptote. It seems therefore that many changes in crop husbandry must have been counter productions. Some of this underlying reason appears to be biological problems of investigation and recently new attempts have been made to alleviate this simplification which this entails by mixing varieties by the use of break crops and by imposed hygiene, but none of this systems has been carried out for long enough to permit any long-term production. In Manipur such work have not yet counted/calculated and included even in agriculture department. For this purpose, the present work have been taken under the following objectives:

1. To determine the ecological impact to insect and pest population.
2. To analyse the ratio of beneficiary and harmful insects in the agro-ecosystem of Imphal valley.
3. To determine the feasibility of IPM in ecosystem.

Materials and Methods

Investigation on the agro-ecological studies on paddy fields was conducted in farmers' field during two kharif crop seasons *viz.*, 2001 and 2002 at "Kitna Panung Loukon" situated in the Imphal East District of Manipur (24°C′ to 25°16′N latitude and 33°37′ to 94°15′ E longitude). The corresponding meteorological parameters during the course of study was scheduled from Tulihal Airport, Imphal and ICAR, Lamphel.

The investigation covered the 3 phases *viz.*, vegetative, reproductive and ripening phases of test crop varieties Leimaphou (CV_1) and Phoudongba (CV_2) comprising the determination of harmful and beneficiary insects. An area of 1 m² in the field was undertaken as sample unit and such units of 20 numbers/ha. were selected and repeatedly visited as required to complete the observation for harmful and beneficiary insect/pests. Determination for harmful and farmers' friend or beneficiary insects during three consecutive phases *viz.*, vegetative, reproductive and ripening phases of paddy includes detection, identification and population estimation at an interval of at least one per four days in vegetative phase, at least two per seven days in reproductive phase and three per ten days in ripening phase by using insect catching net and adopting other insect trapping techniques.

Results and Discussion

Table 31.1(a) shows counting of harmful and beneficiary organisms during 2001. On 4th June in CV_1 certain harmful and beneficiary insects was collected among them 40 numbers were harmful pest (*i.e.* 30 grasshoppers, 10 green leafhoppers) and 20 numbers were beneficiary, it includes 10 spiders and 10 water striders. On the next two observations same numbers of harmful and beneficiary organisms were collected. Beneficiary consists 20 spiders and 10 water strider and harmful consists 30 grasshoppers, 20 green leafhoppers. In the next observation, harmful organisms were increased but same number of beneficiary insects was observed upto 30 numbers and in the next period beneficiary are haphazardly occurred but harmful insects are amazingly increased. During this period of observation, maximum numbers of helpful insects was collected with 70 numbers on 24th day of which there are 30 spiders, 20 dragonflies and 20 striders. And among the harmful insects, 30 white stem borers, 10 army worms, 10 stick insects, 20 rice case worms and 60 grasshoppers.

Table 30.1(a): Harmful and Helpful Insects/Pests Present During Vegetative Phase (in per sq.m.)

		Y_1 (2001)									
Type of Paddy	Date of Obser-vation	Total No. of Org. $(O_1)m_2$	Helpful/ Beneficia-ries Nos. $(H_1)m_2$	Harmful Nos. $(H_1)mg$	Max. Temp. $(Tx_1)°C$	Min. Temp. $(T_1)°C$	Mean Temp. $(Tm_1)°C$	RH Max. $(RHX_1)\%$	RH Min. $(RH_1)\%$	RH Mean $(RHn_1)\%$	Rain-fall $(Rf_1)mm$
Laimaphou CV_1	4th June	60	20	40	26.9	20.6	23.75	91	90	90.5	8.4
	8th June	80	30	50	27.5	21.1	24.3	94	75	84.5	18.4
	12th June	80	30	50	30.3	20.1	25.2	84	75	79.5	10.4
	16th June	130	30	100	27.4	21.3	24.35	82	82	82	24.3
	20th June	100	20	80	28.7	21	24.85	80	62	71.5	3.4
	24th June	200	70	130	30.1	22.4	26.25	72	70	71	1.2
	28th June	180	40	140	31.6	21.8	26.7	73	65	69	1.2
		N = 830	N = 240	N = 590							
Phoudongba CV_2	4th June	80	20	60	26.9	20.6	23.75	91	90	90.5	8.4
	8th June	100	40	60	27.5	21.1	24.3	94	75	84.5	18.4
	12th June	120	30	90	30.3	20.1	25.2	84	75	79.5	10.4
	16th June	40	10	30	27.4	21.3	24.35	82	82	82	24.3
	20th June	150	50	100	28.7	21	24.85	80	62	71.5	3.4
	24th June	220	40	180	30.1	22.4	26.25	72	70	71	1.2
	28th June	250	90	160	31.6	21.8	26.7	73	65	69	1.2
		N = 960	N = 280	N = 680							

In CV_2 maximum number of beneficiary insects was observed on 28th days and least numbers of insects was observed on 16th day with only 10 numbers and highest number of harmful was observed on 24th days with 180 numbers. Helpful and beneficiary insects was observed as an haphazardly nature *i.e.* increase from 20 to 40 numbers then decreases from 30 to 10 and increases from the 10 to 50 and also again decreases from 50 to 40 and increases from 40 to 50 numbers. In these helpful organisms, spiders and dragonflies are dominant during 8th to 20th days. Among the harmful insects grasshoppers was observed as a dominant followed by stem borer, green horned caterpillar, whorl maggot etc. During the present investigation the maximum number of harmful insects was recorded as 590 numbers in CV_1 and 680 numbers in CV_2.

Table 30.1(b) revealed the analysis result of harmful and beneficiary insects/pests during 2002. In CV_1, same type of beneficiary insect (*viz.*, spiders, striders, butterflies) was collected on 4th, 12th and

16th day with 20 numbers, it includes 12 spiders and 8 water spiders but different species was occurred on 20th and 24th day with same numbers of organism (*i.e.* 50 each), it includes 10 dragonflies, 20 spiders and 20 striders. Maximum collection of harmful insects was on 24th, 12th and 20th day with 150, 110 and 100 numbers respectively. It includes green leafhopper, rice case worm, army worm, stem borer and grasshopper on 28th days. Both harmful and beneficiary insects was recorded with 110 numbers on 28th day in CV_1 during 2002. In beneficiary insects it includes 50 butterflies, 30 spiders and 30 water spiders and in harmful, 50 stem borers (30 whites and 20 yellows) 30 grasshoppers, 20 green hoppers, 10 army worms. In case of CV_2 maximum beneficiary was recorded on 28th and 8th day respectively. On 8th day it observes 80 dragonflies, 10 spiders, and 10 striders. The maximum numbers of harmful types were recorded on 28th day with 180 numbers it includes 50 green leafhoppers, 20 maggots, 30 stem borers, 10 stick insects and 70 grasshoppers. From the above analysis it shows that maximum harmful insect in CV_1 as 610 numbers and 560 numbers in CV_2.

Table 30.1(b): Harmful and Helpful Insects/Pests Present During Vegetative Phase (in per sq.m.)

					Y_2 (2002)						
Type of Paddy	*Date of Observation*	*Total No. of Org. (O_2)*	*Helpful/ Beneficiaries Nos. (H_2)*	*Harmful Nos. (Ha_2)*	*Max. Temp. (Tx_2)°C*	*Min. Temp. (T_2)°C*	*Mean Temp. (Tm_2)°C*	*RH Max. (RHx_2)%*	*RH Min. (RH_2)%*	*RH Mean (RHn_2)%*	*Rainfall (Rf_2)mm*
Laimaphou CV_1	4th June	50	20	30	30.6	19.9	25.25	70	66	68	nil
	8th June	90	40	50	30.6	19.5	25.05	68	66	67	nil
	12th June	130	20	110	28.3	20.7	24.5	95	82	88.5	41.5
	16th June	80	20	60	30.7	21.6	26.15	90	86	88	4.8
	20th June	150	50	100	31.5	21.5	26.5	80	65	72.5	5.5
	24th June	180	50	150	28.6	21.5	25.05	83	81	82	10.8
	28th June	220	110	110	30.1	22.4	26.25	79	63	71	0.1
		N = 920	N = 310	N = 610							
Phoudongba CV_2	4th June	70	30	40	30.6	19.9	25.25	70	66	68	nil
	8th June	60	10	50	30.6	19.5	25.05	68	66	67	nil
	12th June	80	20	60	28.3	20.7	24.5	95	82	88.5	41.5
	16th June	120	40	80	30.7	21.6	26.15	90	86	88	4.8
	20th June	130	40	90	31.5	21.5	26.5	80	65	72.5	5.5
	24th June	80	20	60	28.6	21.5	25.05	83	81	82	10.8
	28th June	280	100	180	30.1	22.4	26.25	79	63	71	0.1
		N = 820	N = 260	N = 560							

Table 30.2(a) shows the harmful and beneficiary insects in reproductive phase during 2001. In CV_1 maximum harmful insects were recorded on 47th day with 230 numbers in 2001 and gradually decreases on the next two observations *i.e.* 54th and 61st day with 220 and 210 numbers. The important species were 170 grasshoppers, 20 army worms, 20 gundhi bugs, 10 cut worm moths, 10 green horn caterpillars, 50 rice case worm moths, 10 rice gundhi bugs, 30 army worms, 10 cut worm moths, 30 green leafhoppers. Same numbers of species and their composition were recorded on 61st and 68th day, both with 210 harmful pests. Maximum beneficiary insects was recorded on 68th and 89th days old

plant with 130 numbers and their component includes 30 spiders, 20 giant water bugs, 50 dragonflies and 30 water striders. Minimum beneficiary insects was recorded on 40th and 54th days with 40 numbers of 30 water striders and 10 spiders.

Table 30.2(a): Determination of Harmful and Helpful Organism Present During Reproductive Phase (in per s.q.m.)

							Y_1 (2001)				
Type of Paddy	*Date of Obser-vation*	*Total No. of Org.* (O_1)cm	*Helpful/ Beneficia-ries Nos. Org.*(H_1)	*Harmful Nos. Org.* (Ha_1)	*Max. Temp.* (Tx_1)°C	*Min. Temp.* (T_1)°C	*Mean Temp.* (Tm_1)°C	*RH Max.* (RHx_1)%	*RH Min.* (RH_1)%	*RH Mean* (RHn_1)%	*Rain-fall* (Rf_1)mm
Laimaphou CV_1	5th July	250	60	190	32.5	22.1	27.3	71	64	67.5	nil
	12th July	200	40	160	30.8	23.3	27.05	91	94	92.5	2.6
	19th July	300	70	230	30.1	22.9	26.5	87	85	86	9
	26th July	260	40	220	29.8	22.6	26.2	84	82	83	1.2
	2nd Aug.	280	70	210	29.1	22.1	25.6	91	81	86	6.4
	9th Aug.	340	130	210	32.1	21.1	26.6	78	78	78	nil
	16th Aug.	180	50	130	31.2	22	26.6	91	88	89.5	10.2
	23rd Aug.	240	120	120	30.5	21.9	26.2	85	68	76.5	nil
	30th Aug.	280	130	150	30.3	24.3	27.3	82	80	81	nil
	6th Sep.	200	60	140	30.8	22.4	26.6	81	78	79.5	2.5
		N = 2530	N = 770	N = 1760							
Phoudongba CV_2	5th July	260	70	190	32.5	22.1	27.3	71	64	67.5	nil
	12th July	130	20	110	30.8	23.3	27.05	91	94	92.5	2.6
	19th July	180	50	130	30.1	22.9	26.5	87	85	86	9
	26th July	200	80	120	29.8	22.6	26.2	84	82	83	1.2
	2nd Aug.	170	80	90	29.1	22.1	25.6	91	81	86	6.4
	9th Aug.	240	80	160	32.1	21.1	26.6	78	78	78	nil
	16th Aug.	140	30	110	31.2	22	26.6	91	88	89.5	10.2
	23rd Aug.	180	40	140	30.5	21.9	26.2	85	68	76.5	nil
	30th Aug.	230	60	170	30.3	24.3	27.3	82	80	81	nil
	6th Sep.	200	40	160	30.8	22.4	26.6	81	78	79.5	2.5
		N = 1930	N = 550	N = 1380							

In case of CV_2 same numbers of beneficiary insects was recorded on consecutive three periods of observation *i.e.* 54th, 61st and 68th days with 80 numbers of types it includes 40 dragonflies, 20 striders and 20 spiders. Least number of beneficiary was recorded on 40th day old plant with 20 nos./sq.m. Maximum harmful insects were recorded on 33 days old plant with 190 numbers it includes 70 green leafhoppers, 30 stick insects, 80 grasshoppers and 10 army worms. Minimum number was recorded on 61st day with 90 nos./sq.m. It includes 30 maggots, 20 green leaf hoppers, 20 green horned caterpillars and 20 grasshopper. Following this analysis it shows that maximum harmful insect in CV_1 is 1760 numbers in comprising with 1380 nos. in CV_2 and beneficiary or farmer's friend as 70 numbers as against 550 numbers in CV_2.

Table 30.2(b) depicts harmful and helpful/beneficiary/farmer's friend presents during reproductive phase, 2002. In CV_1 during 2002 maximum numbers of harmful insects was recorded on 89th days old plant with 200 numbers/sq.m. It includes 30 green horned caterpillar moths, 20 green horned caterpillars, 30 rice gundhi bugs, 80 grasshoppers, 20 stock insects and 20 rice case worms. Minimum harmful insects was recorded on 33rd days old plant with 100 nos./sq.m. Maximum beneficiary was recorded on 82nd days old plant with 100 nos./sq.m. It consists of 40 dragonflies, 40 spiders, and 20 striders. Least numbers of beneficiary insect was recorded on 40th and 61st days with 30 nos./sq.m. It includes 20 striders and 10 spiders. In CV_2, during 2002, least number of beneficiary insects was recorded on 40th and 68th days with 20 nos./sq.m. and maximum was on 82nd day with 90 nos./sq.m. The types includes 30 jumping spiders, 20 water treaders, 40 predatory damsels. Maximum numbers of harmful insects was recorded on 47th days with 200 nos./sq.m. in CV_2 (2002). It includes 40 stem borers, 30 rice case worms, 30 green leafhoppers, 20 brown plant hoppers, 10 mole crickets and 70 grasshoppers. On 82nd days old plant same numbers of harmful and beneficiary insects was recorded with 90 numbers per sq.m. in CV_2 (2002). From the present investigation it shows that maximum attainment of harmful insects was observed with 1630 numbers in CV_1 and 1480 numbers in CV_2 and the number of helpful/beneficiary/farmer's friend in CV_1 as 680 and 470 in CV_2 respectively during 2002.

Table 30.3(a) revealed that determination of harmful and beneficiary insects/pest in ripening phase during 2001. In CV_1, maximum numbers of harmful insects was recorded on 125 day old plant with 60 nos./sq.m. and the second highest was on 145th clay with 40 nos./sq.m. and the rest during ripening phases they are lesser than 40 nos./sq.m. and most of them are dragonflies (2001). On 115th and 125th days, same types and same numbers of helpful/beneficiary/farmer's friend insects was recorded with 120 nos./sq.m. and highest was on 135th days with 150 nos./sq.m. in CV_1 during 2001. In this phase grasshoppers dominated among the harmful insects. In CV_2 (2001) maximum harmful insects was recorded on 135th day with 50 nos./sq.m. and minimum on 155th day with 20 nos./sq.m.; it consists of grasshoppers and green leafhoppers. Maximum beneficiary insects was recorded on 145th days with 70 nos./sq.m. which of them are dragonflies, spiders, striders, water treaders. From the present analysis it shows that maximum harmful insects in CV_1 as 170 nos. and 180 nos. in CV_2, farmer's friend in CV_1 as 510 numbers and 270 numbers in CV_2. In CV_1, during 2002 maximum numbers of harmful insects was recorded as 50 nos./sq.m. on 125th day it includes grasshoppers and green leafhoppers and beneficiary was recorded on 115th days with 110 nos./sq.m. It consists of dragonfly and jumping spider. In CV_2, maximum numbers of harmful pest was recorded on 125th day with 120 nos./sq.m. Present investigation shows that number of beneficiary/farmer's friend/helpful insect in CV_1 as 360 numbers and 430 numbers in CV_2 and harmful insects in CV_1 as 170 numbers and 230 numbers in CV_2 during 2002. During the investigation period in the consecutive crop seasons of 2001 and 2002 a number of harmful and beneficiary insects were detected and recorded. The harmful and beneficiary insects in the test and surrounding fields during the experimental period *i.e.* Kharif crop seasons of 2001 and 2002 includes army worm, gundhi bug, cut worm, green horn, caterpillar, grasshopper, rice case worm moth, whorl maggot, stock insect, white stem borer, mole cricket, green leafhopper etc. in harmful pest and spider, water strider, dragonfly, nepa, water treader, belostoma, egg parasites of borers and defoliators (*Trichogramma, Telenomus*) etc. in beneficiary are the important insect/pests species. Some of these species were always scarce but were nevertheless widespread [Table 30.3(b)].

Table 30.2(b): Determination of Harmful and Helpful Organism Present During Reproductive Phase (in per sq.m.)

		Y_2 (2002)									
Type of Paddy	Date of Obser-vation	Total No. of Org. (O_2)cm	Helpful/ Beneficia-ries Nos. Org.(H_2)	Harmful Nos. Org. (Ha_2)	Max. Temp. (Tx_2)°C	Min. Temp. (T_2)°C	Mean Temp. (Tm_2)°C	RH Max. (RHx_2)%	RH Min. (RH_2)%	RH Mean (RHn_2)%	Rain-fall (Rf_2)mm
Laimaphou CV_1	5th July	180	80	100	27.7	22.8	25.25	86	83	84.5	23.3
	12th July	210	30	180	27.3	22.9	25.1	84	80	82	3.7
	19th July	280	90	190	29.6	20.4	25	85	73	79	nil
	26th July	270	90	180	28.5	21	24.75	82	71	76.5	nil
	2nd Aug.	180	30	150	29.8	22.7	26.25	86	73	79.5	4
	9th Aug.	130	40	190	31.5	23.7	27.6	76	71	73.5	nil
	16th Aug.	190	80	110	28.8	22.5	25.65	93	84	88.5	20
	23rd Aug.	240	100	140	29.7	21.1	25.4	83	82	82.5	nil
	30th Aug.	280	80	200	28.5	20.6	24.55	72	70	71	nil
	6th Sep.	250	60	190	29.4	22.5	25.95	77	73	75	nil
		N = 2210	N = 680	N = 1630							
Phoudongba CV_2	5th July	130	30	100	27.7	22.8	25.25	86	83	84.5	23.3
	12th July	180	20	160	27.3	22.9	25.1	84	80	82	3.7
	19th July	250	50	200	29.6	20.4	25	85	73	79	nil
	26th July	250	80	170	28.5	21	24.75	82	71	76.5	nil
	2nd Aug.	170	30	150	29.8	22.7	26.25	86	73	79.5	4
	9th Aug.	210	20	190	31.5	23.7	27.6	76	71	73.5	nil
	16th Aug.	140	30	110	28.8	22.5	25.65	93	84	88.5	20
	23rd Aug.	180	90	90	29.7	21.1	25.4	83	82	82.5	nil
	30th Aug.	200	60	140	28.5	20.6	24.55	72	70	71	nil
	6th Sep.	230	60	170	29.4	22.5	25.95	77	73	75	nil
		N = 1940	N = 470	N = 1480							

During vegetative phase, maximum number of organisms in terms of helpful/beneficiary/farmer's friend was recorded as 70 numbers in CV_1 in 2001. The period considers with the meteorological parameters of temperature 22.4°C in minimum and 30.1°C in maximum, relative humidity 72–70 per cent and rainfall 1.2 mm. And in terms of harmful pest it was recorded as 140 numbers. The period consider with the meteorological parameters of temperature 21.8°C in minimum and 31.6°C in maximum relative humidity 65–73 per cent and rainfall 1.2 mm.

In CV_1 (2002) maximum number of helpful/beneficiary/farmer's friend was recoded as 110 numbers with the corresponding meteorological parameters of 22.4°C in minimum and 30.1°C in maximum temperature, relative humidity 63–79 per cent and rainfall 0.1 mm. Maximum harmful pest was recorded as 150 numbers with corresponding meteorological parameters of 21.5°C in minimum and 28.6°C in maximum temperature, relative humidity 81–83 per cent and rainfall 10.8 mm. In CV_2, maximum number of organisms in terms of helpful/beneficiary/farmer's friend was recorded as 90 numbers. The period considers with the meteorological parameters of temperature 21.8°C in minimum and 31.6°C in maximum, relative humidity 65–73 per cent and rainfall 1.2 mm in 2001 (Table 31.1). But in case of 2002, the maximum numbers of the farmer's friend was 100 numbers and harmful insects as 180 numbers and corresponding meteorological parameters of temperature 22.4°C as minimum and 30.1°C in maximum, relative humidity 63–79 per cent and rainfall 0.1 mm (Table 31.1). The present analysis shows that in vegetative phase, the highest harmful pest was occurred in CV_2 with 180 numbers. The finding was in agreement with that of Oseto (2000), Pathak *et al.* (1998), Mangan *et al.* (2000), Velasco (1999) and De Beon (1999).

Table 30.3(a): Determination of Helpful/Beneficiary/Farmer's Friend and Harmful Organism During Ripening Phase of Rice (CV)

Y_1 (2001)												
Type of Paddy	*Date of Obser-vation*	*Total No. of Org. $(O_1)m_2$*	*Helpful/ Beneficia-ries Nos. Org.(H_1)*	*Harmful Nos. Org. (Ha_1)*	*Max. Temp. (Tx_1)°C*	*Min. Temp. (T_1)°C*	*Mean Temp. (Tm_1)°C*	*RH Max. (RHx_1)%*	*RH Min. (RH_1)%*	*RH Mean (RHn_1)%*	*Rain-fall (Rf_1)mm*
Laimaphou CV_1	16th Sept.	100	80	20	26	20.6	23.3	85	75	80	nil
	26th Sept.	150	120	30	29.5	21	25.25	85	75	80	nil
	6th Oct.	180	120	60	28.5	20.5	24.5	78	72	75	nil
	16th Oct.	170	150	20	30.2	19.6	24.9	88	85	86.5	nil
	26th Oct.	80	40	40	30.1	15	22.55	78	71	74.5	nil
		N = 680	N = 510	N = 170							
Phoudongba CV_2	16th Sept.	80	50	30	26	20.6	23.3	85	75	80	nil
	26th Sept.	70	40	30	29.5	21	25.25	85	75	80	nil
	6th Oct.	90	40	50	28.5	20.5	24.5	78	72	75	nil
	16th Oct.	110	60	50	30.2	19.6	24.9	88	85	86.5	nil
	26th Oct.	100	80	20	30.1	15	22.55	78	71	74.5	nil
		N = 450	N = 270	N = 180							

Maximum number of harmful pest in CV_1 during reproductive phase (2001) was recorded as 230 numbers with meteorological parameters of 22.9°C in minimum and 30.1°C in maximum temperature relative humidity 85–87 per cent and rainfall 9 mm and the maximum number of helpful/beneficiary/farmer's friend as 130 numbers with the meteorological parameters of 21.1°C in minimum and 29.7°C in maximum temperatures, relative humidity 78 per cent and rainfall nil. In CV_1 (2002) highest number of organisms in terms of farmer's friend was recorded as 100 numbers with corresponding meteorological parameters of 21.1°C in minimum and 29.7°C in maximum temperatures relative humidity 82–83 per cent and rainfall nil. Harmful pest (CV_1 in 2002) as 200 numbers with the meteorological parameters of 20.6°C in minimum and 28.5°C in maximum temperature, relative humidity 70–72 per cent (Table 30.1). In case of CV_2 (2001), maximum number of harmful pest was recorded as 190 numbers with meteorological parameters of temperature 22.1°C in minimum and 32.5°C in maximum, relative humidity 64–71 per cent and rainfall nil. Helpful/beneficiary/farmer's friend as 80 numbers, the corresponding meteorological parameters of temperature 22.6°C in minimum and

29.8°C in maximum, relative humidity 82–84 per cent and rainfall 1.2 mm. In 2002, maximum number of farmer's friend was recorded as 90 numbers with corresponding meteorological parameters of temperature 21.1°C in minimum and 29.7°C in maximum, relative humidity 82–83 per cent. Harmful pest (CV_2 in 2002) as 200 numbers with corresponding meteorological parameters of temperature 20.4°C in minimum and 29.6°C in maximum and relative humidity 85–73 per cent (Table 30.1). From the above analysis it shows that environmental factors affect the outbreak of insect pest in the paddy field. The finding was in agreement with that of John *et al.* (2000–2001), Mello (1997), De Kraker *et al.* (2001), Hiroi (1999), Chaiyawat and Khan (1990), Hix (2000), Mochida *et al.* (1990), Heong (1997), Oseto (2000), Guo *et al.* (2000) and Rapusas and Heong (1995).

Table 30.3(b): Determination of Helpful/Beneficiary/Farmer's Friend and Harmful Organism During Ripening Phase of Rice (CV)

		Y_2 (2002)										
Type of Paddy	*Date of Observation*	*Total No. of Org. (O_2)*	*Helpful/ Beneficiaries Nos. Org. (H_2)*	*Harmful Nos. Org. (Ha_2)*	*Max. Temp. (Tx_2)°C*	*Min. Temp. (T_2)°C*	*Mean Temp. (Tm_2)°C*	*RH Max. (RHx_2)%*	*RH Min. (RH_2)%*	*RH Mean (RHn_2)%*	*Rainfall (Rf_2)mm*
Laimaphou CV_1	16th Sept.	70	50	20	29	22.8	25.9	81	75	78	1.1
	26th Sept.	130	80	50	29.6	20.4	25	85	80	82.5	nil
	6th Oct.	150	110	40	30.8	20.6	25.7	80	76	78	nil
	16th Oct.	100	80	40	22.1	18.3	20.2	96	91	93.5	5.2
	26th Oct.	160	40	20	27.7	3.4	20.55	80	68	74	nil
		N = 510	N = 360	N = 170							
Phoudongba CV_2	16th Sept.	120	60	60	29	22.8	25.9	81	75	78	1.1
	26th Sept.	80	30	50	29.6	20.4	25	85	80	82.5	nil
	6th Oct.	150	120	30	30.8	20.6	25.7	80	76	78	nil
	16th Oct.	130	110	20	22.1	18.3	20.2	96	91	93.5	5.2
	26th Oct.	100	50	50	27.7	3.4	20.55	80	68	74	nil
		N = 660	N = 430	N = 230							

The highest number of harmful pest in CV_1 during ripening phase (2001) was recorded as 60 numbers on 6th October with meteorological parameters of temperature 20.5°C in minimum and 28.5°C in maximum, relative humidity 78–72 per cent and rainfall nil. The maximum helpful insect was recorded as 150 numbers in CV_1 (2001). The period corresponds with the meteorological parameters of temperature 19.6°C in minimum and 30.2°C in maximum, relative humidity 85–88 per cent. In CV_2, maximum number of harmful insect was recorded as 50 numbers with corresponding meteorological parameters of temperatures 20.5°C in minimum and 28.5°C in maximum, relative humidity 72–78 per cent. The highest helpful/beneficiary/farmer's friend insects were recorded as 80 numbers with corresponding meteorological parameters of temperature 15°C in minimum and 30.1°C in maximum and relative humidity 71–78 per cent (CV_1 in 2001) (Table 30.3). The present studies suggest that the abundance of cereal pests may often be affected by predators of defenders. Predators in general are polyphagous species. However detailed studies are required and are in need. At present such works are in progress with keen observation. Even though the study was a preliminary the present work have laid a firm stepping stone with scattered evidences on pest outbreaks when the predator pressure is reduced. The finding was in agreement with that of Paith (1999), Smith (1986), Pimental (1971). On the other hand prey species survive better when predators are experimentally removed (Pimental, 1971; Raob *et al.*, 1984). Now a days, it is clear that predator are satisfactorily used in biological control (Smith, 1986; Raob *et al.*, 1984).

Table 30.4: Ratio of Harmful Pest and Helpful/Beneficiary/Farmer's Friend in the Experimental Paddy Field at Different Phases of Crop Growth in Crop Seasons 2001 and 2002

Type of Paddy	*Y_1 (2001)*				*Y_2 (2002)*			
	Name of Phases	*Harmful Pest*	*Helpful/ Beneficiaries/ Farmer's Friends*	*Ratio of Harmful and Farmer's Friend*	*Name of Phases*	*Harmful Pest*	*Helpful/ Beneficiaries/ Farmer's Friends*	*Ratio of Harmful and Farmer's friend*
CV_1	Vegetative	590	240	2 : 1	Vegetative	610	310	1.9 : 1 or 2 : 1
CV_2	phases	680	280	2.4 : 1	phases	560	260	2 : 1
CV_1	Reproductive	1760	770	2.3 : 1	Reproductive	1630	680	2.3 : 1
CV_2		1380	550	2.5 : 1		1480	470	3 : 1
CV_1	Ripening	510	170	3 : 1	Ripening	360	170	2 : 1
CV_2	phases	350	200	1.6 : 1	phases	430	230	1.8 : 1

The present investigation on agro-ecology of paddy fields clearly indicates the harmful and helpful ratio as 2 : 1 in vegetative phase in CV_1 for both 2001 and 2002 crop seasons; 2.4 : 1 and 2 : 1 in CV_2 for 2001 and 2002 crop seasons and respectively in reproductive phase the ratio occurred as 2.3 : 1 in CV_1 and 2.5 : 1 in CV_2 in 2001 crop seasons and in 2002, for CV_1 as 2.3 : 1 and 3 : 1 for CV_2. In case of ripening phase 3 : 1 as CV_1 in 2001, 2 : 1 in 2002 and 1.6 : 1, 1.8 : 1 in CV_2 during 2001 and 2002 respectively (Table 30.4). The finding was in corroborative with that of other workers (Oseto, 2000; Pathak *et al.*, 1998; Velasco, 1999 and De Beon, 1999). They also stated the efficiency of interpreting in predator-prey ratio, rather than the absolute number of predators, so as to determine the extent of predation. Similar finding was reported from various works (King, 1971; Pearson, 1966). The present finding suggest an immediate need for deep analysis of agro-ecosystem of widely cultivated paddy crop in Imphal valley as well as different agro-climatic area wise basis for a region so that large scale practice of IPM may be practice successfully and without disturbing environmental hazards. The present investigation clearly shows that the rice crop is attacked by a number of insect pests, diseases, nematodes and weeds right from the seedling stage to harvest resulting in significant crop losses. Proper studies of agro-ecosystem and absorption of sound practices can enable to raise healthy rice crop without use of unjudicious pesticides, weedicides and other chemicals. In this regards, Smith (1986) rightfully stated that the concept of IPM is not a disjunct development in crop protection, but is an evolutionary stage on the story of pest control. It is significantly different from strategies.

References

Annonymous (2000). *India 2000: A Reference Annual*. Publication Division, Ministry of Information and Broadcasting, Government of India.

Attwood, P.J. (1970). *Proc. 10th Br. Weed Control Conf.* The Cereal Farmers Needs, p. 873–880.

Chaiyawat and Khan (1990). The effectiveness of silica with relation to the tolerance to the yellow stem borer.

Cooke, G.W. (1967). *The Control of Soil Fertility*. Crosby Lockwood, London.

De Leon, T.P. (1999). Magnificent result of the farmer's cognitive blocks on integrated pest management. In: *Opportunity and Initiative in Philippines Rice RD&E. Proc. of the Philippines Rice RD&E Scholar's Forum*, July 26–27, 1999, Phil Roce, Mdigaya Mang, Nueva Ecija, Philippines: DA-Phil, Rice, 199, SB210, P5 P436.

De Kraker, J. Van Huis, A., Van Lenderen, J.C., Heong, K.L. and R. Rabbinge (2001). The effect of prey and predator density on predation of rice leaf folder eggs by the cricket Metioch Vittaticoltes and confirmed the positive correlation.

Guo, Y.J., Wang, N.Y. and J.W. Jiang (2001). Ecological significance of neutral insects as nutrient bridege for predators in irrigated rice with report community. *Chin. J. Biol. Cont.*, 11(1): 5–9.

Heong, K.L. (1997). Works of rice IPM networks. *Proc. 3–5 November, 1992*, Seminar Room Chandler Hall, IRRI, 1992, SB206, D26M5A.

Hiroi, K. (1999). The integrated pest management measurement of IPM and pesticide risk reduction in various countries. *J. Agric. Sci.*, 54(1): 26.

Hix, R.L. (2001). Development of an IPM monitoring programme for adult rice water weevils and reported the feasibility of applicability the programme. *Tueson, Ariz*, pp. 102.

John, G.C., Pheny Sophia, Khico Bunnarith and Polchanity (2001–2001). The ecological characterization of biotic constraints of rice in Combodia.

King, C.M. (1971). Studies on the Ecology of the Weasel (Mustela nivalis). *Ph.D. Thesis*. University of Oxford.

Mangon, J. and M.S. Mangan (1998). The comparison of two IPM. Training strategies in China along with the importance of concepts of the rice ecosystem for sustainable insect pest management etc. 15(3): 209–221.

Mello, J. (1997). The environment impact on the production of irrigated rice. Itoyai, Sc., Brazil: Empresade pesquisa Agropecuariae Extensao Rural de santa catarina. SB 210, B7 R4, p. 81–85.

Mochida, O. (1990). Occurrence of insect pests and their control in tropical Asia. In: *Environmental Impact and Ecosystem in Agriculture and Forestry*, p. 147–157.

Oseto, C.Y. (2000). *The Physical Control of Insects in Rice Field*. Boca Raton, Fa., Lewis Publisher, SB 933.3.153, p. 25–100.

Paith (1999). Present status of paddy insect pest diversity in Goa. *Bionature*, Bhopal Department of Zoology, Goa, University, Goa, 40 and 206: 19(1), 27–30.

Pathak, M.D., Roa, Y.R.U.J., Kameswar Raw, K.V. and S.K. Mukhopadhyay (1998). *The Integrated Pest Management in Rice*. p. 259–300.

Pearson, O.P.J. (1996). The prey of carnivores during one cycle of mouse abundance. *Anim. Ecol.*, 35: 217–233.

Pimental, D. (1977). Ecological basis of insect pest, pathogen and weed problems. p. 3–31. In: *Origin of Pest, Parasite, Disease and Weed Problems*, (Eds.) J.M., Cherrett and G.R. Sagar. Blackwell Scientific Publication, Oxford, p. 413.

Potta, G.R. and G.P. Vickerman (1974). In: Symp. Brit. Ecol. Soc., *Degraded Environments and Resources Renewal*. Arable ecosystems and the use of agrochemicals.

Raob, R.L., De Foliart, G.K. and G.G. Kennedy (1984). An ecological approach to managing insect populations. In: *Ecological Entomology*, (Eds.) C.B. Huffakar and R.L. Raob. Wiley, New York, p. 697–728.

Rapusas, H.R. and K.L. Heong (1995). Workshop report reducing early season insecticide used for leaf folder control in rice: Impact, economics and risks, Los Banos.

Singh, S.R. (2002). *Handbook of Agriculture in Manipur*. Research Farm and Research Publication Division, HRDRI, Canchipur.

Smith, R.F. (1986). History and complexity of IPM. In: *Pest Control Strategy*, (Eds.) E.H. Smith and David Pimental. Academic Press Inc., New York, pp. 334.

Van Enden, H.F. and G.F. Williams (1974). Insect stability and diversity in agro-ecosystem. *Annual Rev. Entomol.*, 19: 455–475.

Velasco, L.R.I. (1999). The extension approaches to rice with relation to integrated pest management in the Philippines. *Philip Agric. Sci.*, 82(3): 285–295.

Chapter 31
Mineral Composition and Salinity Tolerance of Mangroves in Different Habitats of Kerala

K. Murugan

Department of Botany, University College, Thiruvananthapuram, Kerala – 695 034, India

ABSTRACT

Six species of mangroves from the coastal belt of Kerala were analysed for distribution of major inorganic constituents like sodium, potassium, calcium, phosphorus, magnesium and chloride under varying habitats such as tide–dominated, basin and riverine mangroves. On the basis of ion uptake and retention, mangroves are categorised in to: exclusion–*Rhizophora apiculata, Bruguiera cylindrica;* extrusion–*Avicennia officinalis, Acanthus ilicifolius;* accumulation–*Lumnitzera racemosa, Excoecaria agallocha. Bruguiera* species exclude salt but have poorly developed potassium uptake possibly due to this, it is restricted distributionally. Sodium ions are abundant in mangrove waters, but the plants have efficient mechanisms for potassium uptake. The salt accumulating species are least dominant and *L. Racemosa* is disappearing fastly. *Avicennia* species have an efficient salt extrusion mechanism which allows them to dominate near the shore. The importance of potassium is confirmed by studies on senescent leaves. The capacity of plants to maintain a high cytosolic $Na^+ : K^+$ ratio is likely to be one of the key determinants of plants to salt tolerance. Excess sodium in the substratum affects calcium and magnesium uptake, and this is reflected in additional uptake of calcium and withdrawal of magnesium from senescent leaves in fresh water conditions. It appears that mangroves suffer from insufficiency of calcium and magnesium in metabolic tissues. Salt accumulating species develop senescent leaves which accumulate sodium and chloride ions. They provide potassium and phosphorus to other organs. When the plant is exposed to fresh water conditions for longer period, sodium and chloride uptake is reduced and uptake of calcium is increased. Fresh water condition initially invades mangrove associates later fresh water sedges. Mangroves grow well when salt is present in the soil water.

Keywords: *Tide-dominated, Basin, Riverine mangroves, Exclusion, Extrusion, Accumulation, Salinity, Senescent, Sedges, Atomic spectroscopy, Cations.*

Introduction

Although the history of mangroves can be traced to antiquity, its ecological importance was recognized only recently. Mangroves are characterized by their high fidelity to the ecotone influenced by tides. Mangroves are tropical wet land plants encountered inadvertently in habitats *viz.*, seashore lines, sheltered estuaries, lagoons and river deltas inundation with a gradual transition to terrestrial vegetation. This forest consists of taxonomically diverge flora ranging from ferns to flowering plants which share a suit of convergent adaptation towards salinity and anoxic/hypoxic habitat (Upadhyay, 2002). Mangroves possess an exclusive ability to function in a saline environment. Even though salinity is just one inimical factor of the environment to plant life, mangrove plants have unique geological, histo-morphological and physiological adaptations of their own to survive in the saline environments. These interesting but complex features have been ably reviewed by Walsh (1974). There have been numerous investigations of mineral metabolism in mangroves. Currently mangroves are classified into three strata namely tide-dominated, basin and riverine mangroves based on tidal range, salinity and nutrient flux (Hogarth, 1999). Tide-dominated mangroves are characterized by high tidal range, shallow inter tidal zone, considerable sediment and high salinity; basin mangroves are sheltered from wave action, infrequent touch of tides, little turbulence of water currents and variance in salinity; riverine mangroves have low tidal range, strong fresh water influx and heavy load of sediments.

Mangroves are found where salinity ranges from 0–90 ppt (Ball and Munns, 1992). The physiological process which allow mangroves to live in the constantly changing environment where the land meets the sea, are unique among plants (Ball, 1998). To cope with such fluctuating environments mangroves have developed high levels of plasticity both at the individual level and within species and ecotypes (Sugihara *et al.*, 2000). Significant entry of Na^+ will result in severe growth reduction in glycophytic species, will lead to mild toxicity in salt tolerant species, but may benefit halophytes (Maathuis and Amtmann, 1999). The capacity of plants to counteract salinity stress will strongly depend on the status of their K^+ nutrition (Marschrev, 1995; Tsukamoto and Nakanishi, 1998). This article reports major elemental composition in relation to salt tolerance in mangroves growing under various ecological conditions such as tide-dominated, basin and riverine mangroves of Kerala.

Materials and Methods

Ayiramthengu is one of the undisturbed potential basin mangrove area located 3 km away from Oachira along the Arabian sea at Quilon district (80°53′ N, 76°35′ E) Kallai river (11° 14′ N; 75° 47′E) in Calicut also supports patchy riverine mangrove vegetation. The confluence of Kallai river with the sea is near Panniankara, about 2 km west of Kallai railway station. Pathiramanal (90° N; 75.5° 6′ E) a rich sea shore mangrove vegetation of Alapuzha district. Kumarakom a basin mangrove patch is situated west of Kottayam on the eastern bank of Vembanad estuary. The above mentioned areas were representatives of mangroves from extremely saline to nearly fresh water environment. Species studied were *Rhizophora apiculata* Blume, *Bruguiera cylindrica* Blume, *Aegiceras corniculatum* Linn, *Avicennia alba* Blume, *Excoecaria agallocha* Linn, *Acanthus ilicifolius* Linn and *Lumnitzera racemosa* wild. Plant materials were collected from all the above said regions. The first two were interior or basin mangroves, the second a seashore environment and the last face fresh water influx. Known quantity of sea water, soil and leaf samples were collected seasonally *i.e.*, pre-monsoon, monsoon and post-monsoon. The materials were surface sterilized, dried and powdered for atomic spectroscopy following the procedure of Bharagava and Raghupathi (1993). Major elements were detected and quantified. The experiments were repeated thrice and the results were analysed statistically.

Results and Discussion

Atomic spectroscopy was adopted to quantify the mineral nutrients in the soil, water and leaves (Table 31.1). Analysis was done for six species from each of the habitat. No significant variation was observed in mineral nutrients except in the level of micronutrients. In Table 31.2, values for the major inorganic constituents in the mangroves at Pathiramanal, Kumarakam and Kallai were given. It is clear from the tables that mangroves accumulate Na^+ and Cl^- ions when grown in a salt rich environment. However, in take of these two ions is greatly curtailed under fresh water conditions. The data also shows that the $Na^+ : K^+$ ratio in Pathiramanal is nearly 37 in sea water, 8 in soil and is less in tissues. Comparatively in Kallai 20 in sea water, 0.75 in soil and much less in tissues. Mangroves grow in an environment of salinity whose salinity is between that of fresh water and sea water and absorb sodium and chloride ions. Mangroves deploy a variety of means to cope with this unpromising environment. The principal mechanisms are exclusion of salt by the roots, tolerance of high tissue salt concentrations and elimination of excess salt by secretion. The interplay between these is complex and is not clearly understood (Ball, 1988 a). Based on the concept of Tomlinson (1986), the mangroves are categorized into salt excluding, salt excreting and salt accumulating. Of these salts excluding type, *Rhizophora* form dominant patches along the sea shore whereas *Bruguiera* have a very narrow distribution among the mangrove flora of Kerala. Salt secreting species such as *Avicennia, Aegiceras* form dense flora along the coastal belt and have efficient salt extrusion mechanism by salt glands. The lower surface of *Avicennia* is densely covered with hairs, which raise the secreted droplet of salty water away from the leaf surface, preventing the osmotic withdrawal of water from leafy tissues. Salt accumulating types are less dominant, and of these *Lumnitzera racemosa* is fastly disappearing from the interior swamps of mangroves of Kerala. It appears that for survival in the mangrove swamps either exclusion or extrusion mechanisms are more efficient than accumulation. Salt regulation in mangroves is achieved by an efficient selective absorption mechanism. In salt excluding species like *Rhizophora* have ultra filters in roots. Similarly Na^+ and Cl^- ions are deposited in the bark of old stem and roots approximately 2 to 3 times higher concentrations than those of xylem (Table 31.3). This may help to remove salt from metabolic tissues. Mangroves are evergreen plants and usually do not develop large number of senescent leaves. In the present investigation, analyses were made of both in green and senescent leaves of salt accumulating species like *Excoecaria agallocha* and *Lumnitzera racemosa.* Each of these species accumulates ions. It is clear from the bar diagram that the older leaves were depositories of Na^+ and Cl^- ions and also served as reservoirs from which potassium, magnesium and phosphorus were withdrawn (Figure 31.1). Redistribution of minerals is well known in plants. Williams (1955) reported essential elements are more readily available from old leaves than from rooting medium. Seth and Wareing (1967) demonstrated that phosphorus compounds and other metabolites are mobilized and transported from mature organs to developing and reproductive structures. They further showed that senescence of mature organs is associated with loss of metabolites. The present analysis on mangroves are quiet similar and suggest that potassium and phosphorus are essential elements which are difficult to obtain from the soil and hence are redistributed. Accumulation of calcium in senescent leaves is possibly related to development of the abscission zone in accordance with physiological processes found in plants in general. Mangroves accumulate sodium and chloride when grown in salt rich environment. It is clear that, under riverine conditions, there was an increased calcium uptake than sodium and chloride in relation to tide-dominated conditions. Potassium uptake was not reduced. The present investigations at Ayiramthengu face competition from fresh water marshy plants. It was observed that, due to fresh water influx, there was invasion by *Cyperus, Fimbristylis, Caesalpinia, Premna, Hygrophila* and other fresh water marsh plants and that the mangroves are gradually being replaced by these plants. The area flourished by *Rhizophora, Bruguiera* was gradually replaced by *Derris trifolia, Clerodendrum inerme* and later by other fresh water sedges. The present results suggest that mangroves prefer a non-competitive saline environment than non-saline ones. Walsh (1974) summarized the view on growth of mangrove under saline and non-saline conditions and indicated that mangroves grow well when salt is present in the soil water.

Table 31.1: Mineral Components of Left Tissues in Mangroves at Ayiramthengu (Interior/Basin). Values in sea water are expressed as g/liter; in soil as g/100g air-dried soil; and in leaves as g/100 g of dry tissue.

	Ayiramthengu					
	Na	*K*	*Ca*	*Cl*	*Mg*	*Na : K*
Water	1.3	0.05	0.03	1.99	0.67	26
Soil	0.23	0.07	025	0.39	0.06	3.28
Avicennia	0.56	0.69	0.36	1.3	0.88	0.81
Acanthus	2	1.1	0.57	2.2	0.94	1.82
Excoecaria	3.3	1.93	1.83	5.3	0.76	1.71
Rhizophora	2.7	0.44	0.85	0.92	0.89	6.13
Bruguiera	2.1	0.17	0.51	4.9	0.71	12.4
Lumnitzera	2.4	0.92	0.34	2.01	0.77	2.6

Table 31.2: Major Mineral Components in Mangroves at Pathiramanal (Sea Shore), Kumarakom (Interior/Basin) and Kallai (Riverine). Values in sea water are expressed as g/liter; in soil as g/100 g air-dried soil; and in leaves as g/100 g of dry tissue.

	Pathiramanal					*Kumarakom*					*Kallai*				
	Na	*K*	*Ca*	*Cl*	*Na : K*	*Na*	*K*	*Ca*	*Cl*	*Na : K*	*Na*	*K*	*Ca*	*Cl*	*Na : K*
Water	7.1	0.19	0.29	14	37.4	1.1	0.04	0.04	1.98	27.5	0.02	0.001	0.001	0.032	20.0
Soil	0.5	0.06	0.04	0.61	8,3	0.2	0.07	0.03	0.38	2.85	0.03	0.028	0.06	0.029	0.002
Avicennia	2.4	0.88	0.39	3.71	2.72	0.6	0.71	0.37	1.25	0.83	0.38	0.41	0.24	0.68	0.93
Acanthus	2.0	1.12	0.56	2.2	1.78	2.1	1.2	0.58	2.19	2.07	0.62	1.89	1.17	2.94	0.33
Excoecaria	2.3	1.2	2.3	4.22	1.92	3.2	1.95	1.86	5.2	1.64	0.11	0.48	1.7	0.51	0.23
Rhizophora	3.6	0.52	0.98	4.92	6.92	2.8	0.46	0.84	3.89	6.09	1.5	0.45	0.79	2.7	3.3
Bruguiera	3.1	0.21	0.52	5.2	14.8	2.0	0.19	0.49	4.8	10.6	1.4	0.19	0.48	2.8	7.4

Table 31.3: Major Inorganic Constituents in Various Parts of *Rhizophora apiculata*. Values are expressed as g/100 g of fresh tissue.

Plant Part	*Cl*	*Na*	*K*	*ca*
Stem bark	1.17	1.14	0.34	0.14
Stem xylem	0.42	0.37	0.3	0.12
Root bark	1.1	0.94	0.41	0.22
Root xylem	0.55	0.47	0.37	0.2

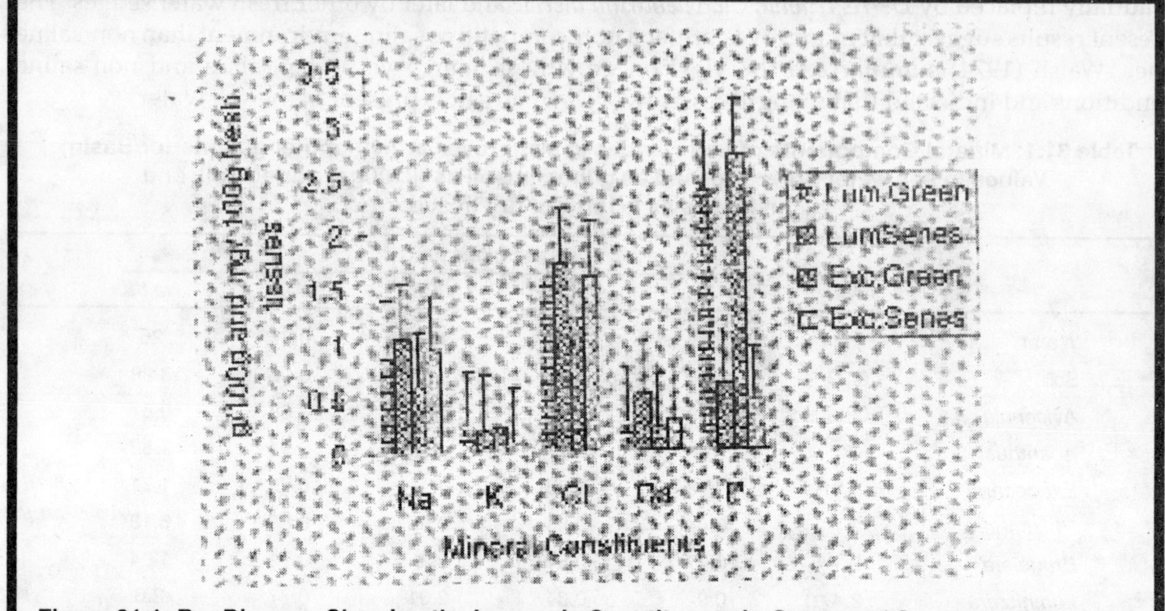

Figure 31.1: Bar Diagram Showing the Inorganic Constituents in Green and Senescent Leaves of *Lumnitzera racemosa* and *Excoecaria agallocha*. Values are expressed in g/100g of fresh tissue except for phosphorus, where values are expressed in mg/100 g of fresh tissue.

Ion uptake by mangroves is affected by excess sodium in soil water. The numerically dominant sodium enters tissues and function as the major cation. Due to great uptake of sodium, the ion most affected is potassium, which must be absorbed against competition with sodium as well as against a concentration gradient. The ratio of sodium : potassium is 37 in sea water, 8 in mangrove soil water, and approximately less in leaf tissues in tide-dominated salt mangrove belts. This can happen only in the presence of an efficient potassium uptake mechanism. Tables 31.1 and 31.2 portrait the preferential potassium uptake in the mangrove habitat in Kerala. Among the mangrove species this mechanism is most active in *Avicennia, Aegiceras, and Excoecaria* and is less efficient in *Bruguiera.* This explains the narrow distribution of *Bruguiera* species along the tide-dominated zones of Kerala. Potassium plays a crucial role in developing salt tolerance as summarized by many authors (Joshi and Mishra, 1970; Hedge and Joshi, 1974). The present study suggests the importance of development of efficient K^+ uptake for effective salt tolerance mechanism in mangroves. Similar results are obtained in *Clerodenron inerme* has greater K^+ uptake in the natural sea shore habitat than under garden conditions (Hedge and Joshi, 1974). It is evidence that increased uptake of sodium and potassium ions affects uptake of calcium and magnesium because the sum total of all cations in a plant, remains constant. The present results indicate calcium insufficiency in mangroves which may be due to excess sodium in the leaf tissues as indicated by Osmond (1966). It appears that reduced amount of Ca^{++} in metabolic tissues of mangroves can affect metabolic phenomena like C_4 photosynthesis. Similarly magnesium concentration in mangroves is not high as those reported from other glycophytes. This may be compensated by withdrawing magnesium from senescent leaves or the older parts of stem/root bark.

References

Ball, M.C. (1988a). Ecophysiology of Mangroves. *Trees*, 2: 129–142.

Ball, M.C. (1998). Mangrove species richness in relation to salinity and water logging: A case study along the Adelaide river flood plain, northern Australia. *Global Ecology and Biogeography Letters*, 7: 73–82.

Ball, M.C. and R. Munns (1992). Plant responses to salinity under elevated atmospheric concentrations of CO_2. *Aust. J. Bot.*, 40: 515–526.

Bhargava and Raghupathy (1993). Analysis of plant material for macro and micro nutrients. In: *Methods of Analysis of Soils, Plants, Water and Fertilizer*, (Ed.) H.L.S. Tandon, FDCO, New Delhi.

Hedge, B.A. and G.V. Joshi (1974). Mineral salt absorption in saline rice variety, Kala Rata. *Plant and Soil*, 41: 421–424.

Hogarth, P.J. (1999). *The Biology of Mangroves*. Oxford University Press, New York.

Joshi, G.V. and S.D. Mishra (1970). Photosynthesis and mineral metabolism in senescent leaves of *Clerodendron inerme* Gaertn. *Indian. Exp. Biol.*, 8: 41–43.

Maathuis, F.J.M. and A. Amtmann (1999). K^+ nutrition and Na^+ toxicity: The basis of cellular K^+/Na^+ ratios. *Annals Bot.*, 84: 123–133.

Marschner, H. (1994). *Mineral Nutrition of Higher Plants*, 2nd edition. Academic Press, London.

Osmond, C.B. (1966). Divalent cation absorption and interaction in *Atriplex. Austral. J. Biol. Sci.*, 19: 37–48.

Seth, A.K. and P.F. Wareing (1967). Hormone directed transport of metabolites and its possible role in plant senescence. *J. Exp. Bot.*, 18: 65–77.

Sugihara, K. Hanagata, N. Dubinsky, Baba, S. and I. Karube (2000). Molecular characterization of cDNA encoding oxygen evolving enhancer protein 1 increased by salt treatment in the mangrove *Bruguiera gymmorhiza. Plant. Cell. Physiol.*, 41: 1279–1285.

Tomlinson, P.B. (1986). *The Botany of Mangroves*. Cambridge University Press, Cambridge.

Tsukamoto, T. and Y. Nakanishi (1998). The salinity of soil water and the chemical elements of leaves of *Rhizophora stylosa* and *Bruguiera gymmorhiza* of the mangrove forest in the coastal area of Irimote island, Okinawa. *Environ. Control. Biol.*, 36: 27–40.

Upadhyay, V.P., Ranjan, R. and J.S. Singh (2002). Human-mangrove conflicts: The way out. *Curr. Sci.*, 83: 1328–1336.

Walsh, G.E. (1974). Mangroves: A Review. In: *Ecology of Halophytes*, (Eds.) R. Reimhold and W. Queen. Academic Press, New York, pp. 51–174.

Williams, R.F. (1955). Redistribution of mineral elements during development. *Ann. Rev. Plant. Physiol.*, 6: 25–42.

Chapter 32

Efficacy of Some Insecticide Sequences Against Pod Borer Complex in Pigeon Pea

J.R. Kadam, G.N. Patil, A.P. Chavan, D.B. Kadam and B.M. Mhaske

Department of Entomology, Mahatma Phule Krishi Vidyapeeth, Rahuri, District Ahmednagar – 413 722, Maharashtra, India

ABSTRACT

The results of the study on the bio-efficacy of sole (full crop coverage) and staggered (covering alternate strip alternately sprays of some insecticide sequences for management of pigeon pea pod borer complex revealed that for protection against the pod borers, the sequence with sole or staggered sprays of fenvalerate-endosulfan-monocrotophos was the most promising followed by endosulfan-monocrotophos-fenvalerate; endosulfan-fenvalerate-monocrotophos and HaNPV-cypermethrin-triazophos were highly promising. These sequences showed 12.10 to 26.20 larvae of pod borers, 30.01 to 51.00 per cent pod damage, 12.41 to 16.19 q/ha grain yield and 6.33 to 14.67 adults of tur plume moth parasite (*Apanteles* sp.)/100 pods as against 43.98 larvae of pod borers, 73.16 per cent pod damage, 7.71 q/ha grain yield, 1 : 4.22 to 1 : 10.66 ICBR and 16.83 parasite adults/100 pods in untreated control. However, staggered sprays conserved significantly and substantially higher population of pod borer parasitoids 11.00 to 14.67/100 pods against 6.33 to 9.33 in the sole sprays of some sequences. These promising sequences avoided 37.80 to 52.50 per cent loss in yield as against the average loss of 41.00 per cent.

***Keyword:** Pod borer complex, Efficacy, Insecticide sequences.*

Introduction

Pigeon pea, *Cajanus cajan* (Linnaeus) Mill sp. is commonly known as red gram or *tur.* About 95 per cent production of pigeon pea is from south Asia. 90 per cent of which belongs to India. Pigeon pea pod borers, *Helicoverpa armigera* (Hubner) and pod fly *Melanagromyza obtusa* (Malloch) are major pest which caused up to 40 per cent (Rai and Singh, 1976), and 21 per cent (Anonymous, 1985) damage respectively. The recommendations as per the Deparment of Agriculture, Maharashtra State are spraying of endosulfan 35 EC 0.07 per cent or phenthoate 50 EC 0.07 per cent or phosalone 35 EC 0.07 per cent or quinalphos 25 EC 0.04 per cent or monocrotophos 36 WSC 0.04 per cent or Neem seed extract 5 per cent or HaNPV 250 LE/ha and dusting of carbaryl 10 per cent or quinalphos 1.5 D or methyl parathion 2 D at the rate of 20 kg/ha (Anonymous, 1992). For the last few years it has been observed and also complained by many farmers that pod borers are not effectively controlled by chemical insecticides. This is the part of the hazards of chemical insecticide with intention to reduce the ill effects, it is arbitrarily advised to use insecticide from various groups a alternately. So far, not much attention has been paid to search out suitable sequences of insecticide for the control of pod borers which could minimise the ill effects. Keeping in view this fact, the present investigation were undertaken to study efficacy of different insecticide sequences applied as sole and staggered sprays against the pod borer complex.

Materials and Methods

Experimental Layout

A field experiment was laid out on bio-efficacy of sole and staggered sprays of some insecticide sequences against pod borer complex of pigeon pea at the instructional farm, Department of Entomology, Mahatma Phule Krishi Vidyapeeth, Rahuri, District Ahmednagar, Maharashtra, India. There were thirteen treatment replicated three times in randomized block design. The gross and net plot size were 4.5 × 4 m, 4.3 × 3.4 m respectively adopting spacing of 30 × 10 cm. The variety used was ICPL–87. The treatment details are given in Table 32.1.

Table 32.1: Efficacy of Sole and Staggered Sprays of Some Insecticide Sequences Against Pod Borer Complex of Pigeon-pea

Sl.No.	*Treatments*	*No. of Pod Borer Larvae per 50 Pods*	*Pod Borer Damage*	*Parasitoid Adults per 100 Pods*	*Grain Yield (q/ha) (Avoidable Losses)*	*ICBR*
1.	E-M-F sole	17.60	35.00 (36.27)*	9.33	13.61 (43.27)	1 : 6.67
2.	E-M-F staggered	14.46	34.66 (36.05)	14.67	14.04 (45.22)	1 : 7.71
3.	F-E-M sole	14.07	30.01 (30.33)	8.67	15.94 (51.56)	1 : 10.28
4.	F-E-M staggered	12.10	41.33 (40.01)	13.33	16.19 (52.50)	1 : 10.66
5.	M-F-E sole	20.07	40.01 (39.29)	6.33	12.86 (39.96)	1 : 6.05
6.	M-F-E staggered	16.70	36.99 (35.60)	11.00	13.25 (41.96)	1 : 6.63
7.	E-F-M sole	18.40	32.99 (35.05)	7.67	13.30 (41.95)	1 : 6.65
8.	E-F-M staggered	14.84	32.67 (34.85)	13.33	13.47 (42.91)	1 : 6.93
9.	E-E-E sole	26.20	51.00 (45.57)	8.33	12.41 (37.80)	1 : 4.22
10.	E-E-E staggered	22.01	49.99 (44.98)	12.33	12.52 (38.57)	1 : 4.38
11.	HaNPV-Cy-Tri sole	16.80	32.99 (35.05)	7.33	15.08 (48.80)	1 : 6.55
12.	HaNPV-Cy-Tri staggered	14.30	32.67 (34.85)	12.00	14.96 (48.59)	1 : 6.46
13.	Untreated control	43.98	73.16 (58.79)	16.83	7.71 (0.0)	–
	S.E. ±	1.281	1.31	0.48	0.26	–
	CD at 5% level	3.69	4.00	1.37	0.71	

Note: E: Endosulfan (0.07 per cent); M: Monocrotophos (0.04 per cent); F: Fenvalerate (0.01 per cent); Cy: Cypermethrin (0.005 per cent), Tri-Trizophos (0.06 per cent).

* Figures in parentheses indicate aresin values.

Method of Sprays

The first spray of the insecticide sequences for protection against *tur* pod borer complex was started at the pod formation stage. Second and third sprays were given subsequently at an interval of 15 days. In the case of sole application to *tur* crop in the respective treatment was given. The procedure for staggered application technique suggested by Kadam (1993) was followed. For this initially 6 row strip of the crop (half plot) was treated with respective insecticide. The remaining strip was treated with the same insecticide after 5 days of the first. The staggered sprays initiated with the sole sprays.

Observation

The efficacy of insecticide sequences was evaluated by selecting five plants randomly from each plot for recording the observation on number of pod borers larvae before each application and at 3rd, 7th, and 10th day of the application and the averages were worked out. At the time of harvest, one hundred pods were randomly separated from each bulk samples of each plot and examined for recording the pod damage. The field of grains obtained from the total plants from each of the net plot was recorded separately. The yield per plot was converted in the yield per hectare. Economics of the treatments was also worked out.

Results and Discussion

Number of Pod Borer Larvae per 50 Pods

The average number of pod borers in the treatments ranged 12.10 to 26.20 against the population of 43.98 in untreated control. The sequences with staggered spray of F-E-M recorded lowest (12.10) population of pod borers. However, it was on par to the sequences with sole spray of F-E-M, HaNPV-Cy-Tri; staggered spray of E-M-F and E-F-M.

Pod Borer Damage

The sequences with F-E-M showed significantly least (30.01) damage. Next highly promising sequences in their descending order of superiority were staggered sprays of HaNPV-Cy-Tri and E-F-M (32.67), E-M-F (34.66), M-F-E (36.99) and sole sprays of HaNPV-Cy-Tri (32.99). The pod damage in remaining sequences was 34.66 to 51.00 per cent as against significantly higher damage of 73.16 per cent in untreated control. The damage in staggered spray sequences was mostly lesser than their sole spray. The indications of these results are in conformity to those reported by Parade (1993) regarding impact of the application on damage inflicted by *E. vitella* on okra crop grown for seed purpose.

Parasitoid Adults per 100 Pods

The staggered spray sequences of insecticide resulted in significantly increasing the population (11.00 to 14.67) of the parasite is compared to that (6.33 to 9.33) in their sole sprays. However, most promising sequences in this regard were the staggered sprays of the sequences with E-M-F, F-E-M and M-F-E as such the staggered application technique helped to conserve and augment the parasite as compared to that in conventional sole application method. The staggered sprays were observed to increase 51.03 per cent population of the parasite as compared to those in sole sprays. Kadam (1993) reported that the staggered sprays increased 35.48 per cent population of entomophagus coccinellids over to that in sole sprays applied for the control of *E. vitella* on okra. Thus, the results of the present study are agreement with those reported by Kadam (1993).

Grain Yield q/ha

The sequences (T2, T4, T6, T8 and T10) applied as staggered sprays resulted in giving significantly higher (13.25 to 16.19 q/ha) yield over those applied as sole sprays (12.86 to 15.08 q/ha). On the other hand the difference in yields of sole staggered sprays of T11 and T12 were non-significant. However, the data showed that the yields in staggered sprays in case of all the treatments gave higher yields than sole sprays. The next highly promising sequences for the yield were sole (15.08 q/ha) and staggered (14.96 q/ha) spray of HaNPV-C-Tri and those of E-M-F (13.61 to 14.04 q/ha). Highest yield of (16.19 q/ha) was recorded in staggered sprays of F-E-M followed by at par yield of (15.94 q/ha) in its sole sprays.

Economics of Different Treatment Tested Against Pigeon-pea Borers

The staggered application technique was slightly more profitable (1 : 4.38 to 1 : 10.66) than sole (1 : 4.22 to 1 : 10.28) application method similar trend of the profitability was also observed in sole and staggered spray of different sequences. However, the most important feature of the staggered application techniques was that it was more efficacious in reducing pest infestation and promising for conservation of natural enemies of the test pest. More or less similar type of observation were reported by Kadam (1993) in respect of the suitability of the staggered application technique evaluated against *E. vitella* on okra.

Considering overall performance of the treatments the sequences with F-E-M was the most promising followed by E-M-F, E-F-M and HaNPV-Cy-Tri were highly promising.

References

Anonymous (1985). *ICRISAT Annual Report*, pp. 200.

Anonymous (1992). *Crop Protection Recommendations*. Published by the Department of Agriculture, Maharashtra State, Pune, pp. 33.

Kadam, J.R. (1993). Management of *Earias vitella* (Fabricious) on okra. *Ph.D. Thesis*, MPKV, Rahuri.

Parade, S.D. (1993). Efficacy of different insecticide against shoot and fruit borer *E. vitella* (Fab) on seed crop of okra. *Msc. (Agri.) Thesis*, MPKV, Rahuri, pp. 52.

Ral, L. and H.K. Singh (1976). Efficacy of malathion plus DDT (25/25 EC) against gram pod borer *Heliothis armigera* Hub. *Pesticides*, 10(1): 42–44.

Chapter 33
Association of Grain Yield with Component Characters in Bread Wheat (*Triticum aestivum* L.) Under Heat Stress Environment

Manmohan Sharma, V.S. Sohu and G.S. Mavi

Department of Plant Breeding, Punjab Agricultural University, Ludhiana – 141 004

ABSTRACT

Grain yield is the most important economical attribute but being an artifact, is influenced by a number of yield components. Selection for traits showing positive correlation with grain yield may improve the yield performance of genotype under heat stress. Ten genetically diverse bread wheat varieties *i.e.* PBW 154, PBW 138, PBW 445, PBW 466, PBW 373, PBW 443, PBW 435, HP 1731, Cetia and Weaver were crossed in diallel fashion. Parents and F_1s (including reciprocals) were evaluated under normal (E_1) and heat stress (E_2) environments. Results indicated significant and positive correlation of grain yield with grains per plant, biological yield, spikes per plant, harvest index and 1000-grain weight under heat stress y environment. Depending upon nature of correlation among these traits under heat stress, selection for harvest index, spikes per plant and grains per plant may be used as criteria for improving yield under heat stress.

Keywords: Correlation, Association, Heat tolerance, Triticum aestivum.

Introduction

High temperatures adversely influence several morphological and physiological traits during the life cycle of wheat and culminate into lower final yield. Grain yield is the most important economical attribute but being an artifact, is influenced by a number of yield components. Therefore, the associations of yield with its components measured through correlation and path coefficients is very helpful in deciding the selection criteria. A judicious selection for component characters can be applied for increasing yield under heat stress environments. Certain criteria in this regard have been set. Selection for traits showing positive correlation with grain yield may improve the yield performance of genotypes under heat stress. The present investigation aimed to study association of grain yield with component traits and suggest suitable selection criteria for improving grain yield of bread wheat under heat stress environment.

Materials and Methods

The material for study comprised 10 genetically diverse bread wheat varieties having variable degrees of heat tolerance *i.e.* PBW 154, PBW 138, PBW 445, PBW 466, PBW 373, PBW 443, PBW 435, HP1731, Cetia and Weaver. They were crossed in all possible combinations including reciprocals to produce 90 crosses. The crosses were evaluated along with parents in a randomised complete block design with three replications in each of two environments (E_1 and E_2) created by staggering the dates of sowing. E_1 represented normal (mid November sowing) and E_2–heat stress (mid December sowing) environments for the crop. Each genotype was sown in row of one meter length with row to row spacing of 23 cm and plant to plant spacing 10 cm. Now, experimental rows were grown to avoid border effects. Data were recorded on five randomly taken plants for each genotype for traits such as plant height, leaf area, days to heading, days to maturity, duration of grain filling, biological yield, grains per plant, spikes per plant, grains per spike, spikelets per spike, 1000-grain weight, hectoliter weight and protein content. Phenotypic coefficients of correlation were computed as suggested by Al-Jaibouri *et al.* (1958) and their significance was tested against 'r' values as given by Fischer and Yates (1963).

Results and Discussion

Results of correlation of grain yield with component characters and for each pair of component characters under normal (E_1) and heat stress (E_2) environments are presented in Tables 33.1 and 33.2, respectively. Grain yield was observed to be significantly and positively correlated with grains per plant, biological yield, spikes per plant, grains per spike, spikelets per spike and plant height. These results were as per the earlier reports of Nanda *et al.* (1980) and Tiwari and Rawat (1993). Under heat stress environment (E_2), significant and positive correlation of grain yield with grains per plant, biological yield, spikes per plant, harvest index and 1000-grain weight was observed. Such observations were in accordance with the findings of Reynolds *et al.* (1994), Rahman *et al.* (1997) and Singh *et al.* (1997). The component characters showing significant and positive correlation with grain yield in E_2 were further investigated for correlation with respect to each other. Results indicated that biological yield showed positive association with grains per plant, spikes per plants and 1000-grain weight. Spikes per plant depicted positive association with component traits being investigated. Grains per plant and 1000-grain weight were found to be negatively correlated but association was non-significant. Similarly, harvest index showed significant and negative association with biological yield. Correlation of both harvest index and 1000-grain weight was positive and significant with grain yield, but association of grains per plant with grain yield was found to be stronger than that of 1000-grain weight with grain yield. Thus it may conclude that for improving wheat yield under heat stress, selection for harvest index, spikes per plant and grains per plant may be an appropriate strategy.

Table 33.1: Phenotypic Correlation Coefficients for Different Pairs of Characters in Bread Wheat Under Non-stress Environment (E_1)

	Plant Height	Leaf Area	Days to Heading	Days to Maturity	Duration of Grain Filling	Bio-logical Yield	Grain Yield	Harvest Index	Grains per Plant	Spikes per Plant	Grains per Spike	Splkelets per Spike	1000-grain Weight	Hecto-liter Weight	Protein Content
Leaf area	– 0.06														
Days to head.	– 0.09	0.17													
Days to mat.	– 0.12	0.18	0.22*												
Duration of g.f.	0.05	– 0.11	– 0.94**	0.14											
Biol. yield	0.32**	0.14	0.02	0.11	0.02										
Grain yield	0.28**	0.15	– 0.01	0.1	0.04	0.68**									
Harvest index	0.05	– 0.09	– 0.09	– 0.11	0.05	– 0.16	– 0.08								
Grains/plant	0.11	0.17	0.16	0.22*	– 0.8	0.61**	0.84**	– 0.09							
Spikes/plant	0.23	0.13	0.07	0.16	– 0.01	0.60**	0.61**	– 0.08	0.64**						
Grains/spike	– 0.1	0.08	0.14	0.12	– 0.1	0.18	0.45**	– 0.01	0.64**	– 0.16					
Spikelets/spike	0.02	0.14	0.29	0.14	– 0.24*	0.38**	0.35**	– 0.06	0.45**	0.16	0.41				
1000-g.w.	0.31**	– 0.09	0.29**	– 0.27**	0.20	– 0.01	0.05	0.06	– 0.39**	– 0.15	– 0.36**	– 0.31**			
Hectoliter w.	0.25**	– 0.02	– 0.17	– 0.07	0.15	– 0.01	0.02	0.14	– 0.14	0.01	– 0.18	– 0.16	0.33**		
Protein content	– 0.21*	0.01	0.07	0.21*	0.00	– 0.08	– 0.12	– 0.19	0.00	0.01	– 0.02	0.01	– 0.18	– 0.14	
Sedimen. value	– 0.05	0.12	0.05	0.01	– 0.04	0.09	0.09	0.05	0.10	0.1	0.02	0.11	– 0.04	0.32**	– 0.11

*: Significant at 5 per cent; **: Significant at 1 per cent.

Table 33.2: Phenotypic Correlation Coefficients for Different Pairs of Characters in Bread Wheat Under Heat-stress Environment (E_2)

	Plant Height	*Leaf Area*	*Days to Heading*	*Days to Maturity*	*Duration of Grain Filling*	*Bio-logical Yield*	*Grain Yield*	*Harvest Index*	*Grains per Plant*	*Spikes per Plant*	*Grains per Spike*	*Splkelets per Spike*	*1000-grain Weight*	*Hecto-liter Weight*	*Protein Content*
Leaf area	– 0.07														
Days to head.	0.08	0.22*													
Days to mat.	– 0.01	0.24*	0.48**												
Duration of g.f.	– 0.09	– 0.02	– 0.64**	0.38**											
Biol. yield	0.08	0.02	0.03	– 0.06	– 0.09										
Grain yield	0.16	– 0.04	– 0.09	– 0.07	0.02	0.64**									
Harvest index	0.11	– 0.08	– 0.14*	– 0.03	0.12	– 0.30**	0.50**								
Grains/plant	0.20*	0.04	0.05	– 0.06	– 0.1	0.61**	0.67**	0.13							
Spikes/plant	0.22*	– 0.1	– 0.10	– 0.13	– 0.01	0.58**	0.63**	0.13	0.63**						
Grains/spike	0.06	0.05	0.18	0.05	– 0.14	0.13	0.17	0.06	0.58**	– 0.24*					
Spikelets/spike	0.09	0.01	0.18	0.04	– 0.16	0.21*	0.15	– 0.03	0.28**	0.13	0.25**				
1000-g.w.	0.01	– 0.08	– 0.19	– 0.13	0.08	0.15	0.31	0.21*	– 0.13	0.15	– 0.32**	– 0.21			
Hectoliter w.	0.01	– 0.08	0.39**	– 0.19	0.24*	– 0.12	0.02	0.15	– 0.09	0.07	– 0.17	– 0.23	0.25**		
Protein content	0.00	0.20	0.41**	0.18	– 0.27**	– 0.03	0.00	0.02	0.50	– 0.11	0.18	0.11	– 0.15	– 0.28**	
Sedimen. value	– 0.09	0.09	0.08	0.04	– 0.05	0.06	0.00	– 0.06	0.03	– 0.05	0.08	– 0.10	0.03	– 0.09	0.04

*: Significant at 5 per cent; **: Significant at 1 per cent.

References

Al-Jaibouri, H.A., Miller, P.A. and H.P. Robinson (1958). Genotypic and environmental variances and covariances in an upland cotton cross of inter specific origin. *Agron. J.*, 50: 633–636.

Fischer, R.A. and F. Yates (1963). *Statistical Tables for Biological, Agricultural and Medical Research.* Oliver and Boyed, Edinburgh, London.

Nanda, G.S., Hazarika, G.N. and K.S. Gill (1980). Association analysis in wheat. *Science & Culture*, 46: 200–202.

Rehman, M.M., Hossain, A.B.S., Saha, N.K. and P.K. Malker (1997). Selection of morphological traits of heat tolerance in wheat. *Bangladesh J. Sci. Ind. Res.*, 32: 161–165.

Reynolds, M.P., Balota, M. Delgado, M.I.B., Amin, A. and R.A. Fischer (1994). Physiological and morphological traits associated with spring wheat yield under hot irrigated conditions. *Aust. J. Plant. Physiol.*, 21: 717–730.

Singh, I., Radhu, J.S. and Y. Jindal (1997). Harvest index abetter selection criterion for yield improvement in bread wheat. *Haryana Agri. Univ. J. Res.*, 7: 27–30.

Tiwari, V.N. and G.S. Rawat (1993). Variability and correlation studies between grain yield and its components in segregating generations of *aestivum* wheat. *Bhartiya Krishi Anusandhan Patrika*, 8: 19–24.

Chapter 34

Stability Analysis in Sorghum [*Sorghum bicolor* (L.) Moench]

S.P. Patil, M.R. Manjare,** S.R. Kamdi*** and A.M. Dethe****

M.Sc. (Agri.) Student, **Associate Professor of Agricultural Botany, * Ph. D. Scholar Cytogenetics and Plant Breeding, Department of Agricultural Botany, Mahatma Phule Krishi Vidyapeeth, Rahuri – 413 722*

ABSTRACT

Nineteen genotypes of sorghum comprising varieties, hybrids and elite lines were evaluated under three environments (sowing dates) for ten quantitative traits including grain yield. The significant value of G × E interactions revealed differential response of the genotypes to varying environmental conditions. Stability parameters revealed that the genotype SPV–1592 possesses average stability for grain yield, suggesting its suitability for inclusion in breeding programme for the development of stable variety.

Keywords: *Sorghum bicolor, Stability, G × E interaction.*

Introduction

The ultimate aim of any plant breeding is to develop cultivars with high-yielding potential with consistent performance over diverse environments. Productivity of population is the function of its adaptability, while later is the compromise of fitness, stability and flexibility. Thus, the predictability of performance *i.e.* stability depends upon the adaptability of the genotype.

Sorghum [*Sorghum bicolor* (L.) Moench] is one of the most important food and fodder crops of the world in general and of semi-arid tropics in particular. In the Maharashtra State as a main cereal crop, *rabi* sorghum is grown in varying conditions as residual moisture and major area under this is on medium to light soils. The present investigation was planned to identify well buffered stable genotypes among varieties, hybrids and elite populations of sorghum.

Materials and Methods

Nineteen genotypes involving the existing varieties, hybrids and newly developed lines were evaluated during *rabi* 2002 in three environments *i.e.* sowing dates, (*i*) 27th September, 2002, (*ii*) 11th October, 2002, (*iii*) 26th October, 2002 at Post Graduate Institute Farm, Mahatma Phule Krishi Vidyapeeth, Rahuri (M.S.). Each entry was sown in a 4-row plot of 4.5 m length with 45 cm interrow and 15 cm distance between the plants. All the treatments received recommended package of practices to ensure satisfactory crop growth. Ten randomly selected plants per plot in each replication were used to record data for the characters *viz.*, days to 50 per cent flowering, days to maturity, plant height, earhead length, earhead girth, number of grains per earhead, 1000 grain weight, grain yield per plant, grain yield per hectare and dry fodder yield per hectare. Data from three environments as well as pooled data were subjected to analysis of variance (Panse and Sukhatme, 1995). The traits, which showed significant genotype-environment (G × E) interaction, were subjected to stability analysis as per Eberhart and Russell (1966). The three stability parameters:

1. Overall mean performance of each genotype across the environments (X)
2. The regression of genotype on the environmental index (bi), and
3. Squared deviation from regression coefficient (S^2di) were worked out for all the genotypes.

The significance of stability parameter (bi) and its deviation from unity were tested by Students' 't' test.

Results and Discussion

The analysis of variance for phenotypic stability (Table 34.1) indicated that the mean differences due to genotypes were significant for all the characters except days to maturity when tested against G × E interaction. Significant differences were also observed for all the characters when tested against pooled deviation. Except for earhead length and earhead girth, environmental variances were found significant for all the characters. The highly significant G × E interaction observed for all the characters indicated that genotypes showed varied response to different environments. Bakeit (1990) and Reddy *et al.* (2004) also reported significant G × E interaction in most of the traits studied.

Table 34.1: Analysis of Variance for Stability with Three Environments (Eberhart and Russell, 1966)

Sl.No.	Source	d.f.	Mean Sum of Squares				
			Days to 50% Flowering	Days to Maturity	Plant Height (m)	Earhead Length (cm)	Earhead Girth (cm)
1.	Genotype	18	15.47+@	11.802@	0.208+@	3.729+@	2.785+@
2.	Environment (E)	2	197.484+@	328.656+@	0.605+@	1.423	0.554
3.	G × E	36	5.028*	6.366*	0.0261*	0.583*	0.783*
4.	E + (G × E)	38	15.157	23.328	0.0566	0.627	0.771
5.	Environment (linear)	1	394.95	657.26	1.211	2.847	1.107
6.	G × E (linear)	18	8.305@	9.023@	0.0395@	0.606	0.889
7.	Pooled deviation	19	1.659	3.517*	0.0120*	0.530*	0.641*
8.	Pooled error	54	1.017	1.0202	0.00316	0.170	0.239

Table 34.1–Contd...

Sl.No.	Source	d.f.	Mean Sum of Squares				
			Number of Grains per Earhead	1000 Grain Weight (g)	Grain Yield per Plant (g)	Grain Yield per ha (kg)	Dry Fodder Yield per ha (kg)
1.	Genotype	18	101560.90+@	31.824+@	346.63+@	878195.60+@	5058631.00+@
2.	Environment (E)	2	132128.00+@	6.531+@	15.593+@	193200.00+@	1112192.00+@
3.	G × E	36	4509.77*	0.263*	0.884*	8056.00*	46435.56*
4.	E + (G × E)	38	11226.53	0.593	1.659	1?800.42	102528.00
5.	Environment (linear)	1	26425.80	13.073	31.170	386421.40	2224072.00
6.	G × E (linear)	18	7124.46@	0.412@	1.718@	15652.31@	90158.31@
7.	Pooled deviation	19	1795.01*	0.107*	0.0516	443.475**	2612.72*
8.	Pooled error	54	177.33	0.035	0.0833	139.11	801.185

+ = Significant at 5 per cent and 1 per cent level respectively against the G × E interaction;
@ = Significant at 5 per cent and 1 per cent level respectively against the pooled deviation;
* = Significant at 5 per cent and 1 per cent level respectively against the pooled error

On partitioning the variances of G × E into linear and non-linear components, the linear component of G × E interaction was significant for all the traits except earhead length and girth suggesting that a large portion of G × E interaction was accounted for by the linear regression. Also, non-linear component (pooled deviation) was found highly significant for most of the characters except days to 50 per cent flowering and grain yield per plant, indicating that the genotypes differed considerably with respect to their stability for traits under study. The predominance of linear component noticed would help in predicting the performance of the genotypes across the environment (Table 34.1). Several workers reported importance of both linear and non-linear components of G x E interaction.

The environmental indices depicted that E_1 environment (first sowing date) was observed to be favourable for most of the traits; while E_2 was only favourable for earhead girth. However, E_3 environment was not found favourable for any character.

Different measures of stability have been used by various workers. Finlay and Wilkinson (1963) explained linear regression as a measure of stability; whereas Eberhart and Russell (1966) emphasized that both linear (bi) and non-linear (S^2di) components of G × E interaction be considered while judging the phenotype stability of a genotype; whereas deviation from regression (S^2di) should be considered as measure of stability. Accordingly, the mean (X) and deviation variance (S^2di) of each genotype were considered for stability and linear regression (bi) was used for testing varietal response. Genotypes with lowest or non-significant mean squared deviation being the most stable and *vice-versa*. The three parameters X, bi and S^2di together gave an idea of adaptability of genotypes across the environments. An ideal and stable genotype, according to Eberhart and Russell (1966) may be characterised as with high performance, unit regression and zero or minimum deviation from regression. Thus, taking into account all these points in the present study, the genotypes were categorized accordingly. The mean (X), regression coefficient (bi) and deviation from regression (S^2di) for the traits studied are presented in Table 34.2. In the present investigation, the magnitude of regression coefficient and deviation from regression varied from genotype to genotype.

Table 34.2: Stability Parameters (Eberhart and Russell, 1966)

Sl.No.	Genotype	Days of 50% Flowering			Days to Maturity			Plant Height (cm)			Earhead Length (cm)			Earhead Girth (cm)		
		Mean (X)	Bi	S^2di	Mean (X)	Bi	S^2di	Mean (X)	Bi	S^2di	Mean (X)	Bi	S^2di	Mean (X)	Bi	S^2di
1.	SPV–1411	76.16	1.23	1.87	124.16	1.20	– 0.58	2.28	– 0.01*	– 0.0031	13.67	2.15	0.008	18.50	2.43	– 0.23
2.	SPV–1504	67.33	0.31	– 0.86	120.66	0.60	– 0.47	2.05	2.18*	– 0.0015	15.11	1.67	0.94*	16.07	9.72	3.64*
3.	SPV–1546	74.00	1.31	– 0.55	123.33	1.39	3.25	2.38	1.82	– 0.0002	15.27	0.02	0.56	17.73	– 0.26	0.15
4.	SPV–1587	71.83	0.01	– 0.85	123.16	1.20	– 0.58	2.51	0.14	– 0.0032	14.68	3.63	0.10	17.34	1.00	– 0.24
5.	SPV–1588	74.66	– 0.23	– 0.01	121.83	– 0.11**	– 0.83	2.61	0.02*	– 0.0030	15.21	2.23	1.49*	16.91	– 0.32	– 0.23
6.	SPV–1589	75.33	1.08	1.46	122.66	1.27	5.18*	2.82	0.05*	– 0.0031	15.99	0.48	– 0.15	15.83	– 4.30	– 0.11
7.	SPV–1590	73.83	1.15	6.35*	122.16	0.29*	4.19	2.12	1.43	0.0436*	15.22	– 0.60	0.71*	14.78	– .76	– 0.22
8.	SPV–1591	73.50	1.86**	– 0.50	122.83	1.44	1.07	2.20	1.72	– 0.0007	16.30	0.19	0.97	16.69	5.84	0.81
9.	SPV–1592	83.50	1.78*	– 0.51	122.00	1.50	– 0.37	2.19	1.09	– 0.0052	16.36	0.27	– 0.16	16.92	– 0.70	0.54
10.	RSV–117	75.16	1.39	– 0.46	122.50	1.32	– 0.99	2.07	1.35	– 0.0004	15.31	4.40	– 0.11	16.14	1.31	1.34*
11.	RSV–143	74.83	1.46	1.23	122.83	1.19	3.80	2.16	2.13*	0.0238*	15.18	– 3.44*	0.12	16.73	1.15	– 0.16
12.	RSV–268	76.83	1.31	– 0.94	121.00	0.47	0.63	2.33	1.62	– 0.0031	17.46	– 0.79	– 0.09	17.65	3.34	– 0.19
13.	RSV–271	77.16	1.31	0.17	126.33	1.36	15.42*	2.31	1.91	0.0865*	14.54	1.88	0.12	17.19	– 2.61	– 0.17
14.	RSV–272	77.00	1.16	– 0.71	126.50	1.32	– 0.99	2.13	1.26	– 0.0016	13.37	– 0.88	0.42	16.48	– 0.13	– 0.02
15.	RSV–491	75.83	1.16	3.74	124.16	1.20	– 0.58	2.69	0.02*	– 0.0031	15.79	0.14	0.34	15.98	– 4.32	– 0.14
16.	CSV–216	72.50	0.00	– 1.01	123.00	1.39	7.29*	2.45	0.69	0.0310*	16.48	0.01	– 0.16	16.22	– 0.54	3.03*
17.	Phule Maulee	72.33	1.24	– 0.90	118.00	0.11*	0.51	2.05	0.93	– 0.0031	13.71	2.88	0.46	15.08	9.22*	0.01
18.	M–35–1	74.16	1.32	5.65*	122.33	1.33	6.19*	1.97	0.36	– 0.0028	13.61	4.56	1.43*	16.07	2.31	0.07
19.	Swati	72.66	0.07	– 0.97	121.50	0.47	5.28*	1.76	0.22	0.0041	16.17	0.15	– 0.15	15.09	– 1.24	– 0.22
	Grand Mean	74.17	–	–	122.57	–	–	0.077	–	–	15.23	–	–	16.49	–	–
	S.E. ±	0.910	0.281	–	0.319	–	–	0.442	–	–	0.515	3.003	–	0.566	3.320	–

Contd...

Table 34.2–Contd...

Sl.No.	Genotype	Number of Grains per Earhead			1000-grain Weight (g)			Grain Yield per Plant (g)			Grain Yield per ha (kg)			Grain Yield per ha (kg)		
		Mean (X)	Bi	S²di	Mean (X)	Bi	S²di	Mean (X)	Bi	S²di	Mean (X)	Bi	S²di	Mean (X)	Bi	S²di
1.	SPV–1411	1623.50	0.29	– 152.41	41.49	0.91	– 0.02	66.40	0.37**	– 0.078	2649.3	2.01**	– 139.2	6358.3	2.01**	– 695.3
2.	SPV–1504	1426.30	0.94	413.53	33.59	3.27**	0.71*	49.34	3.49**	0.163	1973.9	2.62**	682.1*	4737.5	2.62**	4050.9*
3.	SPV–1546	1840.30	2.00*	2945.3*	42.71	0.51	0.43*	78.74	0.43**	– 0.023	3365.1	0.15**	12.70	8076.2	0.15**	77.93
4.	SPV–1587	1543.60	1.13	4913.1*	40.47	– 0.17**	0.04	63.80	1.60**	– 0.003	2726.8	0.62*	– 134.5	6544.3	0.62*	– 735.11
5.	SPV–1588	1803.50	1.55	4142.7*	44.24	0.56	– 0.02	86.53	0.33**	– 0.066	3997.9	0.13**	– 135.1	8875.2	0.13**	– 780.53
6.	SPV–1589	1936.50	0.52	– 165.58	39.41	0.50	– 0.04	77.67	0.45**	– 0.073	3319.4	0.17**	– 143.8	7966.5	0.17**	– 804.83
7.	SPV–1590	1750.00	0.42	6.68	35.14	1.59	0.18*	63.94	0.95	– 0.003	2732.9	0.37**	– 99.42	6559.1	0.37**	– 566.53
8.	SPV–1591	1586.80	0.44	– 140.3	37.77	1.57	– 0.04	62.70	0.97	0.003	2561.6	1.25	6346.5*	6147.9	1.25	3688.4*
9.	SPV–1592	1576.30	2.57**	8503.5*	39.46	0.64	– 0.030	66.19	1.04	– 0.044	2640.3	0.37*	– 138.2	6336.6	0.37**	– 771.97
10.	RSV–117	1519.30	1.30	3521.3*	38.22	0.87	– 0.031	57.59	3.70**	– 0.048	2297.9	1.33*	228.7	5514.5	1.25	1419.1
11.	RSV–143	1417.20	2.03**	1606.6*	37.40	0.94	– 0.030	58.78	0.70	– 0.074	2178.6	1.89**	– 140.1	5228.5	0.37**	– 714.81
12.	RSV–268	1633.00	0.20	– 165.25	36.68	0.46	– 0.031	61.06	0.25**	– 0.073	2261.6	1.82**	– 133.1	5427.8	1.33*	– 668.02
13.	RSV–271	1833.50	0.30	– 40.51	42.83	0.30	– 0.035	79.33	0.51*	– 0.058	3390.4	0.19**	– 73.37	8137.1	1.89**	– 69.70
14.	RSV–272	1482.20	1.44	3472.1*	36.29	0.68	0.01	57.97	0.49*	– 0.077	2148.1	1.80**	– 139.6	5155.4	1.82**	– 715.56
15.	RSV–491	1883.80	0.87	– 43.84	38.00	0.68	0.07	79.39	0.21**	– 0.060	3392.9	0.08**	– 87.87	8142.9	0.08**	– 487.57
16.	CSV–216	1763.30	0.17*	93.18	32.71	0.95	0.04	59.13	0.42**	– 0.052	2527.2	0.16**	– 115.9	6065.3	0.16**	– 662.88
17.	Phule Maulee	1177.10	1.62	1223.6*	34.63	1.56	– 0.009	45.90	0.67	– 0.078	1831.9	1.53**	– 136.5	4396.8	1.53**	– 720.31
18.	M–35–1	1672.20	0.74	775.8*	34.60	2.19**	– 0.011	59.31	1.89**	0.121	2199.8	2.28**	266.1	5279.6	2.28**	1630.31
19	Swati	1632.80	0.35	– 173.46	36.44	0.91	0.17*	62.17	0.37**	– 0.079	2657.1	0.14**	– 136.5	6377.1	0.14**	– 756.20
	Gram Mean	1636.90	–	–	37.98	–	–	65.04	–	–	2660.7	–	–	6385.6	–	–
	S.E. ±	29.96	0.36	–	0.231	0.395	–	0.16	0.176	–	14.89	0.150	–	36.143	0.149	–

* and ** Significant at 5 and 1 per cent level, respectively.

The genotypes *viz.*, SPV–1546, SPV–1591 and SPV–1592 for days to 50 per cent flowering; SPV–1592 for days to maturity, SPV–1546 and RSV–268 for plant height; RSV–117 for earhead length and SPV–1591 for earhead girth exhibited below average stability indicating their suitability for favourable environment.

Whereas, the genotypes *viz.*, SPV–1504, SPV–1587, CSV–216 and Swati for days to 50 per cent flowering; SPV–1504, SPV–1588, RSV–268 and Phule Maulee for days to maturity; SPV–1411, SPV–1587, SPV–1588, SPV–1589 and RSV–491 for plant height; RSV–271 for earhead girth; SPV–1589, SPV–1590, RSV–271 and CSV–216 for number of grains per earhead; SPV–1587, SPV–1588, SPV–1589 and RSV–271 for 1000 grain weight; SPV–1411, SPV–1546, SPV–1588, SPV–1589, RSV–271 and RSV–491 for grain yield per plant; SPV–1546, SPV–1587, SPV–1588, SPV–1599, SPV–1590, RSV–491 for grain yield per hectare as well as dry fodder yield per hectare depicted above average stability suggesting their suitability for poor environment.

None of the genotype was found stable for all the characters under study. For days to 50 per cent flowering Phule Maulee; for days to maturity RSV–117; for earhead length SPV–1546, SPV–1589, SPV–1592, RSV–268, RSV–491, CSV–216 and Swati; for earhead girth SPV–1411, SPV–1546, SPV–1587, SPV–1588, SPV–1592, RSV–143 and RSV–268; for number of grains per earhead RSV–491; for 1000 grain weight SPV–1411, SPV–1592, RSV–117 and RSV–491 and for grain yield per plant SPV–1592 showed general adaptability *i.e.* average stability. No genotype was found average stable for plant height, grain yield per hectare and dry, fodder yield per hectare.

'SPV–1592' had stability (general adaptability) for major traits including grain yield per plant; while M–35–1 was found to be unstable for major characters (Table 34.3). 'Phule Maulee' and 'SPV–1504' were observed with good stability for earliness. Thus, present results are in conformity with those of Heinrich *et al.* (1983), Bakeit (1990) and Narkhede *et al.* (1997).

Table 34.3: Nature of Stability of Genotypes under Different Environment for Ten Characters in Sorghum

Sl.No.	*Character*	*Genotypes Showing Stability*			*Unstable Genotypes*
		Average	*Above Average*	*Below Average*	
1.	Days to 50% flowering	Phule Maulee	SPV–1504, SPV–1587, CSV–216, Swati	SPV–1591, SPV–1592, SPV–1546	SPV–1590, M–35–1
2.	Days to maturity	RSV–117	SPV–1504, SPV–1588, RSV–268, Phule Maulee	SPV–1590	SPV–1590, M–351, Swati
3.	Plant height	–	SPV–1411, SPV–1587, SPV–1588, SPV–1589, RSV–491	SPV–1546, RSV–268	RSV–271, CSV–216
4.	Ear head length	SPV–1546, SPV–1589, SPV–1592 RSV–268, RSV–491, CSV–216, Swati	–	RSV–117	SPV–1591
5.	Earhead girth	SPV–1411, SPV–1546, SPV–1587, SPV–1588, SPV–1592, RSV–143, RSV–268	RSV–271	SPV–1591	–

Contd...

Table 34.3–Contd...

Sl.No.	Character	Genotypes Showing Stability			Unstable Genotypes
		Average	Above Average	Below Average	
6.	Number of grains per earhead	RSV–491	SPV–1589, SPV–1590, RSV–271, CSV–216	–	SPV–1546, SPV–1588, M–35–1
7.	1000-grain weight	SPV–1411, SPV–1592, RSV–117, RSV–491	SPV–1587, SPV–1588, SPV–1589, RSV–271	–	SPV–1546
8.	Grain yield per plant	SPV–1592	SPV–1411, SPV–1546, SPV–1588, SPV–1589, RSV–271, RSV–491	–	–
9.	Grain yield per ha	–	SPV–1546, SPV–1587, SPV–1588, SPV–1589, SPV–1590, RSV–271, RSV–491	–	–
10.	Dry fodder yield per ha	–	SPV–1546, SPV–1587, SPV–1588, SPV–1589, SPV–1590, RSV–271, RSV–491	–	–

Present investigation illustrated the most stable and promising sorghum genotypes *viz.*, SPV–1592, SPV–1546, SPV–1587, SPV–1588, SPV–1504 and Phule Maulee; which can be recommended for multi-location evaluations to assess their suitability across locations or they can be used as the donor parents for generating the breeding material suited for development of new varieties with wider adaptability over seasons/locations or with specific adaptability to a particular environment.

References

Bakeit, B.R. (1990). Stability of grain yield and its components of grain sorghum genotypes [*Sorghum bicolor* (L.) Moench] as affected by different irrigation regimes. *Cereal Res. Comm.*, 18(1–2): 117–124.

Eberhart, S.A. and W.A. Russell (1966). Stability parameters for comparing varieties. *Crop Sci.*, 6: 36–40.

Finlay, K.W. and G.N. Wilkinson (1963). Analysis of adaptation in a plant breeding programme. *Aust. J Agrjc. Res.*, 14: 742–754.

Heinrich, G.M., Francis, C.A. and J.D. Eastin (1983). Stability of grain sorghum yield components across diverse environments. *Crop Sci.*, 23(2): 209–212.

Narkhede, B.N., Shinde, M.S. and S.P. Patil (1997). Stability performance of *rabi* sorghum hybrids. *J. Maharashtra Agric. Univ.*, 22(2): 174–175.

Panse, V.G. and P.V. Sukhatme (1995). *Statistical Methods for Agricultural Workers*, Revised Edn. by Sukhatme V.G. and V.N. Amble. ICAR Publication, New Delhi.

Reddy Raghu Ram, P., Maruthi Sankar, G.R. and N.D. Das (2004). Genotype X Environment interactions in *rabi* sorghum. *J. Maharashtra Agric. Univ.*, 29(1): 021–024.

Chapter 35

Seed Yield and Quality as Influenced by Sulphur Nutrition in Blackgram

S. Aruna Geetha, P.S. Senthilkumar, S. Maragatham and M. Govindaswamy

Department of Soil Science and Agricultural Chemistry,
Tamil Nadu Agricultural University, Coimbatore – 641 003

ABSTRACT

In blackgram, field experiments were conducted to study the effect of sulphur application in improving the seed yield and quality. The results revealed that application of sulphur @ 20 kg S ha^{-1} with farm yard manure @ 12.5 ha^{-1} and *Thiobacillus* increased the seed yield significantly over no sulphur application. The agronomic efficiency and the apparent S recovery worked out showed that 20 kg S ha^{-1} is sufficient for blackgram in increasing the yield and quality.

Keywords: Blackgram, Sulphur, Seed yield, Protein content, Agronomic efficiency, Apparent S recovery.

Introduction

In India, more than 85 per cent of blackgram area is under rainfed condition. Irregular and erratic rainfall as well as poor nutrition are probable reasons for low yield of blackgram (Ramamoorthy *et al.*, 1997). Among the plant nutrients, phosphorous and sulphur play significant role in the production of blackgram. Upadhyay *et al.* (1991) observed that applications of N, P and S significantly increased the grain yield. Ghosh *et al.* (1996) opined that 20 kg S ha^{-1} is sufficient for blackgram production. Also Surendra Singh *et al.* (1998) reported that the protein content of blackgram grain increased with increasing level of sulphur up to 36 kg ha^{-1} and decreased at 46 kg ha^{-1}. With this view the present study was carried out to evaluate the efficiency of elemental sulphur in improving the yield and quality of blackgram.

Materials and Methods

Field experiments were conducted at Tamil Nadu Agricultural University, Coimbatore with blackgram (Cv. CO5). The experiments were conducted in split plot design and the experimental soils come under Typic and Vertic Ustropept. The characteristics of initial soil samples are given in Table 35.1. Elemental sulphur which is obtained as a byproduct from oil refineries was used as a source of sulphur. The main plot contained absolute control, elemental sulphur + *Thiobacillus*, naturally oxidized elemental sulphur (applied one week before sowing), Elemental sulphur farm yard manure @ 12.5 t ha^{-1} and *Thiobacillus* pellets. The pellets of *Thiobacillus* were prepared by mixing the culture of *Thiobacillus* obtained from Department of Agricultural Microbiology, Tamil Nadu Agricultural University, with fine clay. In the subplots elemental sulphur was applied a five levels *viz.*, 0, 20, 40, 60 and 80 kg S ha^{-1} and replicated thrice. The seed yield parameters were calculated as follows:

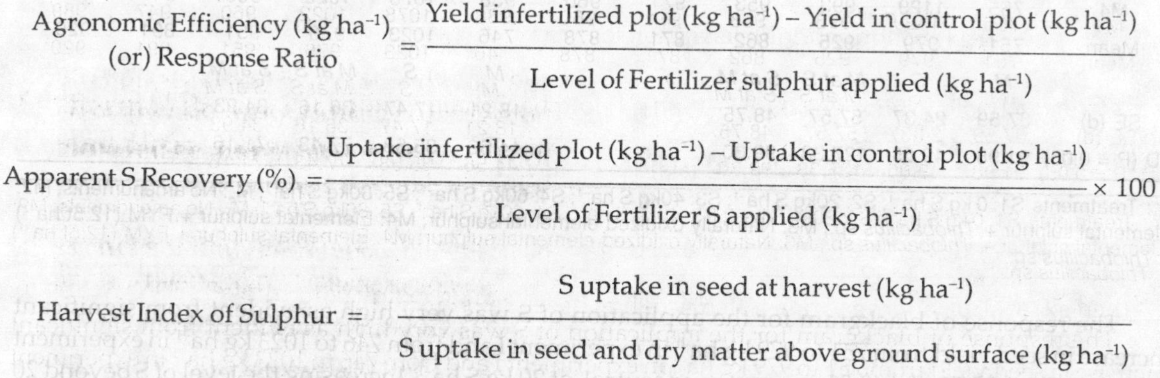

$$\text{Agronomic Efficiency (kg ha}^{-1}\text{) (or) Response Ratio} = \frac{\text{Yield infertilized plot (kg ha}^{-1}\text{)} - \text{Yield in control plot (kg ha}^{-1}\text{)}}{\text{Level of Fertilizer sulphur applied (kg ha}^{-1}\text{)}}$$

$$\text{Apparent S Recovery (\%)} = \frac{\text{Uptake infertilized plot (kg ha}^{-1}\text{)} - \text{Uptake in control plot (kg ha}^{-1}\text{)}}{\text{Level of Fertilizer S applied (kg ha}^{-1}\text{)}} \times 100$$

$$\text{Harvest Index of Sulphur} = \frac{\text{S uptake in seed at harvest (kg ha}^{-1}\text{)}}{\text{S uptake in seed and dry matter above ground surface (kg ha}^{-1}\text{)}}$$

Table 35.1: Characteristics of Experimental Soils

Characteristics	*Field Experiment I*	*Field Experiment II*
pH	8.5	8.6
EC (d sm^{-1})	0.43	0.08
CEC (c mol p(+) kg^{-1})	22.0	17.0
Organic Carbon (%)	0.38	0.30
Available N (kg ha^{-1})	180	190
Available P (kg ha^{-1})	10	10
Available K (kg ha^{-1})	400	408
Available S (ppm)	7.40	3.25

Results and Discussion

Seed Yield

The data on seed yield of blackgram indicated significant difference among the amendments, the highest influence being recorded with FYM + *Thiobacillus* spp. An increase of 166 and 150 kg ha^{-1} of seed in I and II experiments respectively were recorded for FYM addition, registering an increase of 20.91 and 13.6 per cent over no amendment (Table 35.2). The beneficial role played by FYM and S oxidizing bacteria has been clearly brought out in this study. By this way, the byproducts which contains sulphur in totally unavailable form can be profitably used in crop production with cheaper cost. Increase in seed yield with FYM addition as obtained in this study has been earlier reported by Singh and Nad (2000).

Table 35.2: Influence of Sulphur Application on Seed Yield (kg ha^{-1}) of Blackgram

Treatments	*Field Experiment I*						*Field Experiment II*					
	S1	*S2*	*S3*	*S4*	*S5*	*Mean*	*S1*	*S2*	*S3*	*S4*	*S5*	*Mean*
M1	710	803	845	813	800	794	622	1000	948	958	843	874
M2	707	942	905	832	851	847	683	1000	945	962	892	886
M3	834	1042	957	848	860	908	742	1015	988	973	882	920
M4	755	1129	993	953	971	960	938	1078	1022	960	947	989
Mean	751	979	925	862	871	878	746	1023	976	951	891	920
	M	*S*	*M at S*	*S at M*			*M*	*S*	*M at S*	*S at M*		
SE (d)	37.59	24.37	57.57	48.75			18.21	17.47	36.16	34.93		
CD (P = 0.05)	91.98	49.65	127.33	99.30			44.56	35.58	77.43	71.16		

Tr.: Treatments S1: 0 kg S ha^{-1}; S2: 20kg S ha^{-1}; S3: 40kg S ha^{-1}; S4: 60kg S ha^{-1}; S5: 80kg S ha^{-1}; M_1: No amendments; M2: Elemental sulphur + *Thiobacillus* sp., M3: Naturally oxidized elemental sulphur; M4: Elemental sulphur + FYM (12.5t ha^{-1}) + *Thiobacillus* sp.

The response of blackgram for the application of S was very high as evident from significant increase in seed yield from 751 to 979 kg ha^{-1} in experiment I and from 746 to 1023 kg ha^{-1} in experiment II representing 25.71 and 43.21 per cent over control, at 20 kg S ha^{-1}. Increasing the level of S beyond 20 kg S ha^{-1} failed to produce perceptible increase and hence it could be concluded that 20 kg S ha^{-1} is sufficient for blackgram. Similarly Ravichandran *et al.* (1995), Singh *et al.* (1996) and Surendra Singh and Sarkar (1999) reported for blackgram. Further the significant interaction existed between amendments and S levels confirmed the superiority of FYM + *Thiobacillus* with 20 kg S ha^{-1} registering 59.01 and 75.31 per cent increase in yield in the I and II crop respectively over control. Thus, the results forecast the necessity to improve the contributing factors by proper fertilization for enhancing the yield level of blackgram.

Protein Content

The data obtained for amendments and S levels in relation to protein synthesis in blackgram. It was evidenced from higher percentage of protein with sulphur addition than control and also with FYM + *Thiobacillus* over no amendment. Maximum protein content of 20.35 and 23.05 per cent was found to be associated with experiments I and II respectively at 20 kg S ha^{-1} (Table 35.3). The definite synergism between S and N as reported by Chitralekha *et al.* (1992), and Gill *et al.* (2000) might have contributed for increase in protein synthesis. The beneficial role of FYM in increasing the absorption and translocation of N and its consequent incorporation in protein molecule needs no emphasis.

Efficiency of Applied Sulphur

Agronomic Efficiency (or) Response Ratio

The response ratio computed for the levels of sulphur applied ranges from 1.50 to 11.40 in I experiment and 1.81 to 13.85 in II experiment (Table 35.4). Decrease in response ratio with increased level of sulphur addition was noticed in this study also. Based on the data, it may be concluded that 20 kg S ha^{-1} is ideal for blackgram.

Table 35.3: Influence of Sulphur Application on Protein Content (%) in Blackgram

Treatments	Field Experiment I						Field Experiment II					
	S1	S2	S3	S4	S5	Mean	S1	S2	S3	S4	S5	Mean
M1	15.40	20.03	18.67	16.13	15.40	17.13	21.46	22.75	21.98	22.38	21.73	22.06
M2	15.67	20.10	19.13	18.30	16.37	17.91	21.54	23.02	22.81	22.58	22.31	22.45
M3	15.97	20.43	20.00	19.07	17.47	18.59	21.71	23.04	22.88	22.56	22.33	22.50
M4	16.93	20.83	20.43	20.33	20.13	19.73	21.81	23.38	23.17	22.83	22.52	22.74
Mean	15.99	20.35	18.46	18.46	17.34	18.34	21.63	23.05	22.71	22.59.	22.22	22.44
	M	S	M at S	S at M			M	S	M at S	S at M		
SE (d)	0.069	0.059	0.125	0.117			0.048	0.030	0.073			
CD (P = 0.05)	0.168	0.119	0271	0.239			0.118	0.062	0.161			

Tr.: Treatments S1: 0 kg S ha^{-1}; S2: 20kg S ha^{-1}; S3: 40kg S ha^{-1}; S4: 60kg S ha^{-1}; S5: 80kg S ha^{-1}; M1: No amendments; M2: Elemental sulphur + *Thiobacillus* sp.; M3: Naturally oxidized elemental sulphur; M4: Elemental sulphur + FYM (12.5t ha^{-1}) + *Thiobacillus* sp.

Table 35.4: Response Ratio (kg kg^{-1}) and Apparent S Recovery of S Applied in Blackgram

Levels of S Applied (kg ha^{-1})	Response Ratio (kg kg^{-1})		Apparent S Recovery (%)	
	Field Experiment I	Field Experiment II	Field Experiment I	Field Experiment II
20	11.4	13.85	8.70	14.85
40	4.35	5.75	3.25	6.40
60	1.85	3.42	1.38	3.12
80	1.50	1.81	0.64	1.36
Mean	4.78	6.21	3.49	6.43

Apparent S Recovery

As that of response ratio, more recovery of 8.70 and 14.85 per cent in experiments I and II was reported at 20 kg S ha^{-1} compared to 0.64 and 1.36 per cent with 80 kg S ha^{-1} (Table 35.4). Thus it was proved that the efficiency of any added nutrient as that of sulphur will be more at low level of application at sufficiency with a decrease at higher levels.

Harvest Index of Sulphur

The harvest index computed for the amendments did not show perceptible variation, while addition of sulphur increased it over control (Table 35.5). This brought out the beneficial role of sulphur in increasing the productivity of blackgram.

Table 35.5: Harvest Index of Applied Sulphur in Blackgram

Treatments	Field Experiment I	Field Experiment II
M1S1	0.47	0.39
M1S2	0.48	0.47
M1S3	0.49	0.52
M1S4	0.50	0.52
M1S5	0.50	0.52
M2S1	0.48	0.38
M2S2	0.52	0.42
M2S3	0.53	0.45
M2S4	0.52	0.40
M2S5	0.53	0.42
M3S1	0.54	0.39
M3S2	0.54	0.43
M3S3	0.53	0.47
M3S4	0.55	0.45
M3S5	0.52	0.46
M4S1	0.49	0.43
M4S2	0.54	0.43
M4S3	0.52	0.44
M4S4	0.52	0.43
M4S5	0.57	0.45

S1: 0 kg S ha^{-1}; S2: 20kg S ha^{-1}; S3: 40kg S ha^{-1}; S4: 60kg S ha^{-1}; S5: 80kg S ha^{-1}; M1: Absolute Control; M2: Elemental sulphur + *Thiobacillus* sp.; M3: Naturally oxidized elemental sulphur; M4: Elemental sulphur + FYM (12.5t ha^{-1}) + *Thiobacillus* sp.

Acknowledgements

The financial support and supply of elemental sulphur rendered by M/s. Kochi Refineries Limited, Cochin is highly acknowledged by the authors.

References

Chatterjee, Chitralekha, Saukla, Saubra and Neena Khurana (1992). Effect of sulphur deficiency on metabolism of greengram (*Phaseolus radiatus*). *Indian J. Agric. Soc.*, 62(7): 445–449.

Ghosh, T.K., Maitra, S., Saren, B.K., Roy, D.K. and S.S. Singh (1996). Effect of different levels and methods of sulphur application on growth and yield of blackgram. *Environ. Ecol.*, 14(4): 765–768.

Gill, M.S., Mankotia, B.S. and S.S. Walia (2000). Production technology for sustaining pulses productivity. *Fert. News*, 45(3): 33–38, 41–43.

Ramamoorthy, K., Balasubramanian, A. and A. Arokiaraj (1997). Response of rainfed blackgram (*Phaseolus mungo*) to phophorous and sulphur nutrition in red lateritic soils. *Indian J. Agron.*, 42(1): 191–193.

Ravichandran, V.K. Singaravelu and N. Balasubramanian (1995). Effect of sources and levels of sulphur application on yield and economics of blackgram. *Madras Agric. J.*, 82(3): 224–225.

Singh, D.V. and B.K. Nad (2000). N–S inter-relationships as affecting yield and nutrient uptake in mustard–moong cropping sequence under various nutrient combinations. *Crop Res.*, (Hisar), 19(3): 403–408.

Singh, K.P., Singh, Surendra, Sarkar, A.K., Singh, R.P. and Arvind Kumar (1996). Studies and responses to sulphur and micro nutrients in soils of Bihar plateau for higher crop productivity. *Fert. News*, 41(8): 41–47.

Singh, Surendra and A.K. Sarkar (1999). Responses of oil seeds and pulses to indigenous sulphur sources in Bihar plateau. *Fert. News*, 44(7): 23–26.

Singh, Surendra, Singh, K.P., Singh, S.K. and Gautam Kumar (1998). Response of blackgram (*Phaseolus mungo*) to sulphur on acid alfisol of Bihar plateau. *J. Indian Soc. Soil Sci.*, 46(2): 257–260.

Upadhyay, R.M., Singh, Bharat and S.K. Katiyar (1991). Effect of nitrogen, phosphorous and sulphur application to blackgram on yield and fate of P in Inceptisol. *J. Indian Soc. Soil Sci.*, 39(2): 298–301.

Chapter 36

Effect of Conditioning Treatments on Physiological Attributes of *Acacia catechu* Willd. Transplants

A. Vasishth, P. Kaushal,** A.N. Kaushal* and B. Dutt**

**Department of Forest Products,
Dr. Y.S. Parmar University of Horticulture and Forestry, Nauni, Solan – 173 230, H.P.
**Coordinator, Regional Centre, National Afforestation and Eco-Development Board,
Dr. Y.S. Parmar University of Horticulture and Forestry, Nauni, Solan – 173 230, H.P.*

ABSTRACT

One year old seedlings of *Acacia catechu* Willd. were root pruned at 0, 5 and 10 cm length from collar region before planting in the field and were analyzed for predawn xylem water potential, white root regeneration, total soluble sugar, starch, total carbohydrates content and survival per cent. Root pruning decreased xylem water potential and white root regeneration. In monsoon planting, seedlings root pruned for 5 cm showed increase in total soluble sugars, starch and carbohydrates in comparison to 10 cm and unpruned seedlings. In winter planting, these reserves increased in 10 cm pruned seedlings than 5 cm pruned ones. Survival per cent also decreased with increase in root pruning severity in both planting seasons. Winter transplants had showed lower survival per cent than monsoon.

Keywords: *Root pruning, Pre-dawn xylem water potential, Total soluble sugars, Starch, Total carbohydrates and Survival per cent.*

Introduction

Acacia catechu Willd. is one of the most important species found in sub-Himalayan tract and ascends up to 900 metres and sometimes as high as 1200 metres above mean sea level (Anonymous,

1983). It is a moderate sized deciduous tree and grows well in favourable localities. The most common method for raising successful plantation of this species is by transplanting the container grown seedlings, however, this method of planting is labour intensive and costly. Even root system strangulate in the polythene bag and get modified. To overcome this problem, the nursery stock is root pruned which enhances the fibrous root system and hardens off the shoot. For improving field performance root disturbance treatments is emphasized and has been found effective, (Van Dorsser and Rook 1972 and Tanaka *et al.*, 1976). Until recently, seedlings quality had been defined by the morphological characteristics of the seedlings. There has been increasing trend towards physiological characterization of the planting stock for better establishment and growth (Sutton, 1979 and Ritchi, 1984). Root pruning also remained a recommended practice for both ornamental and forestry nursery to develop a compact, dense root system which facilitates transplanting and subsequent field establishment (Mullin, 1966 and Comeron and Rook, 1969). The objective of present study was to study the effect of root pruning on physiological attributes and survival per cent of the Khair (*Acacia catechu* Willd.) seedlings.

Materials and Methods

Acacia catechu Willd. (Khair) nursery growing stock (1+0) was procured from State Forest Department nursery at Deli, Parwanoo District Solan (HP). The experiment was laid out in a randomized block design at the experiment field of Department of Silviculture and Agroforestry, Dr. YS Parmar University of Horticulture and Forestry, Nauni–Solan (HP) having 30°51'N latitude and 76°11'E longitude with elevation of 1150 m above mean sea level. The climate of area is subtropical to sub-temperate. The soil was removed from the polybags raised seedlings and from collar region of the plants roots were pruned at 0, 5 and 10 cm length. These conditioned seedlings were planted on 17–08–1998 (60 per replication) for monsoon and 20–02–1999 for winterplanting at a spacing of 60 × 40 cm. These seedlings were analyzed for pre-dawn xylem water potential, white root regeneration, total sugar content, starch, total carbohydrates and survival per cent after six months of planting.

Predawn Xylem Water Potential

The seedlings pre-dawn xylem water potential was determined by pressure chamber technique.

White Root Regeneration

White root regeneration was determined by counting the number of new/white roots on the transplants and was expressed as number of white roots/plant.

Total Sugars and Starch

Total sugars in plant samples were estimated by phenol-sulphuric acid method given by Dubois *et al.* (1951).

Total Carbohydrates

Total carbohydrates were obtained adding total sugars and starch and were expressed as mg/g dry weight.

Survival (per cent)

$$\text{Survival (per cent)} = \frac{\text{No. of plants at the time of observation}}{\text{Total number of plants}} \times 100$$

Results and Discussion

Predawn Xylem Water Potential

In monsoon planting, root pruned seedlings showed less water potential as compared to unpruned or control seedlings (L_1). Minimum water potential (–0.49 MPa) was observed in the seedlings where roots were pruned for 10 cm length (L_3) Maximum water potential (–0.23 MPa) was recorded for control seedlings (L_1). In winter planting, the minimum value of –1.08 Mpa was found in 10 cm pruned seedlings, which was statistically at. par with 5 cm pruned seedlings. The maximum value was observed (–0.60 MPa) in case of unpruned seedlings (Table 36.1).

Table 36.1: Effect of Root Pruning on Physiological Attributes of *Acacia catechu*. Wild Seedlings

Sl.No.	Attributes	Planting Season	Root Pruning Length (cm)			$CD_{0.05}$
			$L_1(0)$	$L_2(5)$	$L_3(10)$	
1.	Pre-dawn xylem water potential (–) MPa	(M)	0.23	0.41	0.49	0.02.
		(W)	0.60	1.07	1.08	0.06
2.	White root regeneration	(M)	8.60	6.45	8.12	1.21
		(W)	4.65	1.92	3.03	0.57
3.	Total soluble sugar content (mg/g dry matter)	(M)	22.23	37.22	33.84	0.90
		(W)	29.99	36.16	39.59	0.38
4.	Starch content mg/g dry matter	(M)	23.35	32.92	30.01	0.64
		(W)	24.11	28.39	28.96	0.39
5.	Total carbohydrates (mg/g dry matter)	(M)	45.70	70.15	63.85	1.27
		(W)	54.22	67.56	68.59	0.52
6.	Survival per cent	(M)	68.66 (56.02)	26.73 (30.91)	47.31 (43.33)	0.66
		(W)	55.91 (53.36)	11.53 (23.52)	19.14 (26.99)	1.59

Figures in parenthesis are square root transformed values.

M: Monsoon planting; W: Winter planting

Seedlings pruned to 5 and 10 cm in monsoon as well as in winter planting have remained under water stress which lowered the xylem water potential. However, the effect is more pronounced in winter planting. This confirms the earlier view of Kramer and Kozlowski (1979) that root pruning decreased the xylem water potential which resulted in less absorption, more transpiration and water deficit. Similar results were also observed in cotton (Stansell *et al.*, 1974) and chirpine (Dhiman, 1991).

White Root Regeneration

In case of root pruned seedlings, white root regeneration increased with increase in length of root pruning and was found in higher number (8.12) in L_3 which was at par with L_1 (8.60) and the lowest (6.45) in L_2. In winter planting, maximum of 4.65 new roots were recorded in unpruned (L_1) followed by 3.03 in 10 cm pruned seedlings (L_3) the minimum of 1.92 in case of seedlings pruned for 5 cm (L_2). The seedlings planted in winter showed lesser number of roots as compared to monsoon planting (Table 36.1). Root pruning has affected the white roots in both seasons. It reduced the shoot growth, resulting in increased translocation of photosynthate to the remaining roots and caused the compensatory root growth (Randolph and Weist, 1981).

Total Carbohydrates

Pruning length showed significant affect on total sugar content. Persual of the data shows that sugar content was significantly higher in 5 cm (L_2) and 10 cm pruned seedlings (L_3) in comparison to unpruned seedlings (L_1). The maximum sugar content (37.27 mg/g) was recorded in L_2 and minimum (22.33 mg/g) in L_1. The sugar content in winter planting was significantly higher in light pruned (L_3) than severely pruned (L_2) and lowest for the unpruned seedlings (Table 36.1). The highest starch content was observed (32.92 mg/g) in severely pruned seedlings (5 cm from collar region). Whereas lowest content (23.35 mg/g) was registered in unpruned seedlings (L_1). Starch content was more pronounced under L_2 and L_3 in comparison to unpruned seedlings. However, starch content in L_2 and L_3 were found to be statistically at par. The present investigation demonstrates increase in total soluble sugar, starch and total carbohydrate in root pruned seedlings in both planting seasons. In monsoon planting, severely pruned seedlings (5 cm from collar region) exhibited higher content of sugar, starch and total carbohydrates than the light pruned seedlings (10 cm from collar region) and reverse is the case in winter planting.

Root pruning of seedlings have resulted in reduction of photosynthetic rate which attributes to accumulation of carbohydrates (Burt, 1966) or to a decrease in translocation rates (Humphrics and Thorne 1964). Ghobrial (1983) was of the view that removal of 50 per cent roots in red kidney bean plants significantly increased the concentrations of soluble sugar contents in various plant parts. He observed that root pruning might have reduced transport of growth hormones to shoot sink thus checking shoot growth. Consequently the ability of shoots to utilize available sugars was lost resulting in their accumulation in plant parts. These findings are also in agreement with Van Dorsser and Rook (1971) who observed that starch and sugar content in pine seedlings increases following root pruning at 4 successive monthly intervals. Abod and Sandi (1983) confirmed the observation that starch content would increase with increase in severity of treatment. Higher carbohydrate content in seedling may be due to an increase in auxin and cytokinin levels in the root xylem sap which promote the accumulation of carbohydrates (Carlson and Larson, 1977).

In winter transplanting, higher concentration of carbohydrates were observed in light pruned seedlings than severely pruned ones which may be due to long dry spell of drought that has resulted into no plant growth and ultimately increased the accumulation of reserves in light pruned seedlings.

Survival Per Cent

The survival per cent of the stock planted during monsoon season 1998 (survival assessed during March 1999) had significant effect in response to root pruning. Increase in root pruning length exercised significant effect on survival. The lowest survival percentage of 26.73 per cent was observed for 5 cm pruned seedlings (L_2) in comparison to seedlings pruned for 10 cm (L_3) in which the survival per cent was recorded 47.31 per cent. However, the highest outplanting performance of 68.66 per cent was recorded for the unpruned seedlings. However, unpruned seedlings have shown higher per cent survival and increase in severity (from 10 cm to 5 cm from collar region) has reduced the survival per cent. In comparison to monsoon planting, winter transplants have given lower survival per cent in all the treatments. The survival per cent is positively related with white root regeneration, higher root regeneration better is the survival. This is related with better root soil contact in transplants. Root pruning reduced the survival per cent, this has been confirmed by the findings of Brown, 1969 and

Mullin, 1976 on coniferous species. Even Sushil (1990) did not observe significant improvement in the outplanting survival of *Acacia catechu* after undercutting and wrenching. Yadav (1992) also observed non-significant differences in the survival of 15 cm table pruned seedlings of Kachnar and Khirak.

The possible reason for low field survival in heavy pruned seedlings of *Acacia catechu* can be that seedlings have only tap root with very few laterals, the most important genetic character of the species. On root pruning, tap root has been removed, the root regeneration capacity is low and the seedlings have not developed proper fibrous root system.

Conclusion

Root pruning decreased the xylem water potential, which was –0.41 MPa for severely pruned and –0.24 MPa in unpruned seedlings in monsoon planting. Root regeneration significantly increased with increase in root pruning length in both planting seasons. The maximum number of roots (8.12) was recorded in light pruned and minimum (6.45) in severely pruned seedlings in monsoon planting. Similarly, in winter planting also, the highest (4.65) was observed in unpruned and lowest (1.92) in severely pruned seedlings. Total soluble sugars, starch and carbohydrates increased with increase in severity of root pruning. In monsoon planting, the maximum carbohydrates (70.15 mg/g) was found in severely pruned seedlings and minimum (45.70 mg/g) in unpruned seedlings. Whereas, in winter planting these reserves increased with increase in root pruning lengths. Survival per cent decreased with increase in root pruning severity. In comparison to monsoon winter transplanting had lower survival per cent in all the treatments.

Acknowledgements

Authors are highly thankful to ICFRE for providing financial support during the course of these investigations.

References

Anonymous (1983). *Troup's–The Silviculture of Indian Trees*. 4: 7–22.

Abod, S.A. and A. Sandi (1983). Effect of restricted watering and its combination with root pruning on root growth capacity. Water status and food reserves of *Pinus caribaea* var. *hondurensis* seedlings. *Pl. Soil*, 71: 123–129.

Brown, J.H. (1969). Effect of root pruning and provenance on shoot and root growth of scotch pine seedlings. *Bull. W. Virginia Agri. Expt. Sta. Virginia*, 584T.

Burt, R.L. (1966). Some effects of temperature on carbohydrate utilization and plant growth. *Aust. J. Bioi. Sci.*, 19: 711–714.

Carlson, W.C. and M.M. Larson (1977). Changes in auxin and cytokinin activity in roots for red oaks (*Quercus rubra*) seedlings during lateral root formation. *Physiol. Plant*, 41: 162–166.

Dhiman, R.C. (1991). Conditioning of *Pinus roxburghii* Sargent seedlings for bare root planting. *Ph.D. Thesis*. Dr. Y.S. Parmar University of Horticulture and Forestry, Nauni–Solan, pp. 169.

Dubois, M., Gilles, K., Hamilton, J.K., Rebers, P.A. and F. Smith (1951). A colorimetric method for the determination of sugars. *Nature*, 168: 167.

Ghobrial, G.I. (1983). Effects of root pruning on translocation of photosynthates in *Phaseolus vulgaris. L.J. Exp. Bot.*, 34(138): 20–26.

Humphrics, E.E. and G.N. Thorne (1964). The effect of root formation on photosynthesis of detached leaves. *Ann. Bot.*, 28: 391–400.

Kramer, P.J. and T.T. Kozlowski (1979). *Physiology of Woody Plants*. Academic Press, New York, pp. 811.

Mullin, R.E. (1966). Root pruning of nursery stock. *For. Chron.*, 42(3): 256–264.

Mullin, R.E. (1976). Practical guidelines to nursery stock quality. *Plantation Symposium*, Kirklnd Lake. *Can. For. Serv. Great Lakes for Res. Centre, Sualt Ste. Marie, Ont.*, p. 1–11

Randolph, W.S. and C. Wiest (1981). Relative importance of tractable factors affecting: The establishment of transplanted holly (*Ilex crenata*). *J. Amer. Soc. Sort. Sci.*, 106: 207–210.

Ritchie, G.A. (1984). Assessing seedlings quality. In: *Forest Nursery Manual: Production of Bare-root Seedlings*, (Eds.) Duryea, M.L. and J.D. Landis. Martinus Nijhoff/Junk. Publ., Forest Res. Lab., Oregon State University, p. 234–259.

Stansell, J.R., Kleppe, B., Browning, V., and H.M. Taylor (1974). Effects of root pruning on water relations and growth of cotton. *Agron. J.*, 66: 591–592.

Sushil, Kumar (1990). Conditioning studies in *Acacia catechu* Wild. *M.Sc. Thesis*, Dr. Y.S. Parmar University of Horticulture and Forestry, Nauni–Solan, H.P., pp. 68.

Sutton, R.F. (1979). Planting stock quality and grading. *For. Ecol. and Management*, 2: 123–132.

Tanaka, Y., Walstad, J.D. and J.E. Borrecco (1976). The effect of wrenching on, morphology and field performance of Douglas fir and loblolly pine seedling. *Can. J. For. Res.*, 6: 453–458.

van Dorsser, J.C. and D.A. Rook (1972). Conditioning of radiata pine seedlings by undercutting and wrenching: description of methods, equipment and seedling response. *NZ J. For.*, 17: 61–73.

Yadav, V. (1992). Effect of planting stock on the performance of *Bauhinia variegata* Linn. and *Celtis australis* Linn. *M.Sc. Thesis*, Dr. Y.S. Parmar University of Horticulture and Forestry, Nauni–Solan, H.P., pp. 79.

Chapter 37

Eco-Crop Planning with Reference to Cereal Crops in West Bengal

Gunadhar Dey

Department of Agricultural Economics, Bidhan Chandra Krishi Viswavidyalaya, Mohanpur, Nadia, West Bengal

ABSTRACT

The study has been conducted in view of reckless use of ground water for irrigation in crop production particularly cereal crops in West Bengal. The objective of the study is to bring down the harvest of ground water through reallocation of agricultural land to cereal crops. Based on water requirement of different crops the study has suggested for reallocation of agricultural land to the crops which require relatively low quantity of water. In the study boro paddy has been observed to require the highest quantity of water among the cereal crops using groundwater. Land under boro paddy has been suggested for allocation to wheat with a view to abating the use of the groundwater in crop production. To maintain the existing level of cereal an additional area of monocropped land needs to be converted into double cropped land. These efforts entails a partial change in food habits of the people of West Bengal in favour of wheatmeal-foods.

Introduction

Long days ago rural people in different regions of India largely depended on ponds, lakes, fountains and rivers for their drinking water. Present days people in some pockets uses these water bodies as sources of drinking water. In reality in many parts of India hygienic drinking water is scarce. Scarcity of this essential thing is also reported in summer season in some parts of West Bengal. At present almost all the people in West Bengal use groundwater for drinking purpose. Groundwater also plays an important role in the context of irrigation status in agriculture of West Bengal. Though percentage of area irrigated by groundwater is declining in West Bengal it is reported that 46 per cent of the total irrigated area receives irrigation from groundwater resources. Harvest of groundwater resources for crop production will result in depletion of stock and ultimately will cause exhaustion if the harvest rate exceeds recharge rate. This depletion of groundwater carries a long-term negative effect on sustainable development defined as that which meets the basic needs of present without compromising the ability of future generation to meet their own needs.

Unsustainable use of groundwater in agriculture by the present generation will leave less for the future generations. In this paper an attempt has been made to reallocate agricultural land to cereal crops with a view to abating the use of groundwater.

Methodology

The study is based on secondary data obtained from different sources. In almost all the cases data have been compiled from original sources. Tabular method of analysis has been used in the study.

Results and Discussion

Before going to make the attempt to reallocate agricultural land to different cereal crops it is worthwhile to deal with land-use pattern, area, production and yield of different cereals in West Bengal. In this connection land-use pattern in West Bengal is furnished below.

A small part of cultivable agricultural land remains fallow every year in West Bengal. It is observed from the Table 37.1 that fallow land constitutes above 3 per cent of the total geographical area in West Bengal. Untimely rainfall in kharif season and scarcity of resources are considered to be major causes of fallow land in West Bengal.

Table 37.1: Land-use Pattern in West Bengal **('000ha)**

Land Utilisation	*1999–2000*	*2000–2001*	*2001–2002*	*Average Area of Triennium Ending 2001–2002*	*Percentage Coverage*
Net area sown	5471.48	5417.66	5221.99	5470.37	62.95
Current fallow	208.54	357.93	289.53	285.33	3.28
Forest	1192.13	1190.22	1184.21	1188.85	13.68
Area not available for cultivation	1658.74	1598.54	1572.87	1610.05	18.53
Other uncultivated land	158.14	123.36	126.07	135.85	1.56
	8689.03	**8687.71**	**8694.67**	**8690.45**	**100.00**

Source: Economic Review 2000–2001, 2001–2002 and 2002–2003, Statistical Appendix, Government of West Bengal.

Among cereals rice and wheat are important crops grown in West Bengal. Other cereals like barley, maize, etc. are also cultivated marginally. It is noted from the Table 37.2 that area devoted to rice accounts for about 93 per cent of the total gross cropped area under cereals. Rice is cultivated in three seasons in West Bengal. Seed sowing of aus paddy is generally done in second or third week of May and crop is harvested during period extending from last week of September to 1st week of October. The duration of aman paddy after transplanting extends from 2nd fortnight of July to 1st fortnight of November. Seeds of boro paddy are sown in 2nd fortnight of December and transplanted in 2nd fortnight of January. The crop is harvested during 1st fortnight of May. Duration in each case depends on varieties of rice selected for cultivation. Area under aus paddy accounts for 6 per cent of the total cropped area under cereals. The corresponding percentage figures for aman and boro paddy are 63.52 and 22.73 respectively. Land distributed to aus and wheat crops are more or less equal in terms of percentage though these are not grown in same season. Wheat is generally sown during the period from 2nd fortnight of November to 1st fortnight of December in plains of West Bengal. In hill areas seeds are generally sown during the period from 2nd fortnight of October to 1st fortnight of November. In both the areas late sowing after 15th November results in decrease in yield. Duration of the crop differs from variety to variety. Some high yielding varieties of wheat which are suitable for plains of West Bengal are Sonalika, Janak, Sarbati Sanora, U.P.–262, U.P.–115, etc. Varieties suitable for hill areas are Sonalika, Sailaja, Girija, etc. Sonalika possesses a wide range of adaptability from hills to plains. Other cereals like barley, maize, etc. are cultivated in an area constituting only 0.90 per cent of the total gross cropped area under cereals. Most of these cereals are grown as rainfed crops in kharif season in West Bengal.

Table 37.2: Area, Production and Yields of Cereals in West Bengal

Crops	*Average Area of Triennium Ending 2001–2002 ('000 ha)*	*Average Production of Triennium Ending 2001–2002 ('000 tonnes)*	*Average Yield of Triennium Ending 2001–2002 (kg/ha)*
Rice	5884.94 (92.67)	13814.79 (92.87)	2347.48
(*a*) Aus	407.88 (6.42)	784.58 (5.27)	1923.55
(*b*) Aman	4033.34 (63.52)	8555.34 (57.51)	2121.14
(*c*) Boro	1443.72 (22.73)	4474.87 (30.08)	3099.54
Wheat	408.05 (6.43)	956.97 (6.43)	2345.22
Other cereals	57.05 (0.90)	103.48 (0.69)	1813.84
Total cereal	**6350.04**	**14875.24**	

Notes: Data are compiled from Economic Review, 2000–2001, 2001–2002 and 2002–2003, Statistical Appendix, Government of West Bengal.

Figures in parentheses indicate percentages to total in respective columns.

It is also exhibited from the Table 37.2 that rice production constitutes about 93 per cent of the total production of cereals. Among the three types of rice aman is observed to occupy the highest position in terms of percentage followed by boro rice. Wheat is noted to constitute 6.43 per cent of the total production of cereals. Productivity of boro rice is observed to be highest among the cereals grown in West Bengal. An almost same productivity has been recorded for rice and wheat. Productivity of boro rice is substantially higher than those of other cereals.

In the context of reallocation of agricultural land in favour of cereal crops which require low quantity of irrigation water it is worthwhile to present information on water requirement of different cereal and non-cereal crops in the study.

Aus paddy is generally grown as rainfed crop in pre-kharif season. Water requirement of aman paddy is highest among the cereals. This crop is grown as rainfed crop in kharif season. Rarely this crop requires irrigation in the situation of low or untimely rainfall. Boro paddy is cultivated in irrigated land requiring 175 ha cm of water and more than 80 per cent of the cropped area is reportedly irrigated by groundwater through deep tubewells, shallow tube wells, etc. Wheat is also cultivated in irrigated land requiring 25 ha cm of water. With an objective of reducing the use of groundwater as irrigation in cereal crops attention has to be paid to the crop which require relatively low quantity of water. At the same time it must be taken into consideration that the level of production of cereals should not go down. In the production of cereals huge quantity of groundwater is used for irrigation purpose. It is noted from the Table 37.3 that water requirement of born paddy is as much as five times of wheat crop. These two crops have been brought into the floor of comparison because of their closeness to each other in respect of some criteria which are discussed below.

Table 37.3: Water Requirement of Cereal Crops in West Bengal

Crops	*Water Requirement*	*Crops*	*Water Requirement*
Autumn (aus) paddy	88 ha cm	Potato	50 ha cm
Winter (aman) paddy	200 ha cm	Mustard	32 ha cm
Summer (boro) paddy	175 ha cm	Lentil	24 ha cm
Wheat	35 ha cm	Green gram	24 ha cm
Maize	50 ha cm	Groundnut	28 ha cm
Barley	50 ha cm		

For bringing down the use of groundwater boro paddy has to be replaced by other crop. Among the crops wheat requires least quantity of water. Both these crops are cultivated in irrigated condition. In this study rice and wheat have been considered substitute of each other from the point of view of similarity of food value obtained in more or less same quantity from these two cereals. Wheat may be placed in superior position in respect of percentage composition of protein, minerals and some vitamins. Another fact which attracts attention to the matter of substituteness is the preference of people of West Bengal in accepting wheatmeal-food to any other cereal excepting rice. In the list of preference of food wheatmeal-food occupies the second position. Water requirement of boro paddy is much more higher than that of wheat crop. Penetration of wheat seeds in agricultural field by switching off land from boro paddy will result in decrease in harvest of groundwater. To minimise the use of groundwater production of wheat has to be emphasised by allocating land in favour of the crop. It is noted from Table 37.2 that a quantity of 4474.87 thousand tonnes of boro rice is produced in the state. To maintain the existing level of production same quantity of wheat must be produced alongwith the production of other cereals. But as the yield of boro rice is much more higher than the yield of wheat a land area already devoted to boro rice, if allocated to wheat, will not suffice for producing the required quantity of cereals. It is to be observed how much deficit will arise in this respect. An area of 1443.72 thousand hectare of land is capable of producing 3385.84 thousand tonnes of wheat. It is observed that total production of cereals will decrease by 1089.03 thousand tonnes if land is allocated in favour of wheat by switching off land from boro rice. To compensate this deficit of production of cereals an additional area of 464.36 thousand hectare of land is estimated to be required for the purpose. It is exhibited from Table 37.1 that an area of 285.33 thousand hectare remains as fallow land. This fallow land that can produces 669.16 thousand tonnes of wheat may be brought under cultivation. An area of 179.03 thousand hectare of land that is required for producing 419.87 thousand tonnes of wheat may be used for second time after growing aman paddy in kharif season.

Table 37.4: Allocation of Land to Wheat Crop and its Production

Sl.No.	*Source of Land*	*Area of Land in Thousand Hectares*	*Production of Wheat in Thousand Tonnes*
1.	Land switched from boro paddy	1443.72	3385.84
2.	Current fallow land	285.33	669.16
3.	Land under aman paddy	179.03	419.87
4.	Existing land under wheat crop	408.05	956.97
	Total land	**2316.13**	**5431.84**

Before reallocating agricultural land to different cereal crops for attaining specific objectives like minimisation of the harvest of groundwater a sound knowledge about cropping pattern is necessary. But no information on cropping pattern of monocropped, double cropped and triple cropped land in West Bengal has been obtained so far separately. Practically there is no such state level information in details in the province. One thing has to be considered here. Land allocation in favour of wheat crop has an impact on non-cereal crops. This attempt, if made, will limit the possibility of growing non-cereal crops in the state. It has been observed that area under aus and aman rice accounts for 81 per cent of the total net cropped area. The rest area is devoted to other crops which may be cereal or noncereal in type. A part of area under aus rice is used for second time in growing crops other than wheat in winter season. A similar picture has also been reported in case of land area under aman rice which is preceded by vegetables grown in winter season or boro rice in summer season. A part of land under aman rice is devoted to potato and other non-cereal crops and then to boro rice in some districts. So allocation of this land in favour of wheat to be grown in next season will adversely affect the production of potato and other non-cereal crops by reducing area under these crops. It is may be mentioned that potato and wheat are contemporary to each other. Cultivation of wheat after aman rice in the same pieces of land calls for harvest of rice before time when seed sowing of wheat is started in the state. Keeping this in mind variety of proper duration of aman rice has to be selected, sown and transplanted in due time. It is explicit from the above discussion that reallocation of land in favour of wheat will bring about a change not only in cropping pattern of cereal crops but also a change in cropping pattern of non-cereal crops. It has implicitly been mentioned earlier that a portion of land under boro paddy gets irrigated by surface water from canals, rivers, ponds, etc. Allocation of this amount of land to wheat crop will save huge quantity of surface water. This water can successfully be used in growing non-cereal crops like potato, vegetables, pulses, oil seeds, etc. For this purpose an additional amount of land over the requirement of land for growing wheat after aman paddy needs to be brought under cultivation in next season.

For decreasing the use of groundwater boro rice which requires huge quantity of this resource must be traded off for wheat. Agricultural land has to be switched from boro rice to wheat. To bridge the gap in production of cereal an additional area of monocropped land growing aman rice in kharif season has to be designed for bringing under cultivation of wheat in winter season. This effort would reduce the use of groundwater. Trade-Off of boro rice for wheat calls for partial change of food habit of the people in West Bengal in favour of wheat-based food. Bengalies generally takes rice as staple food. Change of food habit is sometimes associated with sacrifice of taste and preference. But this sacrifice of present generation is less than the gain which would be achieved by the future generation owing to the relatively less exhaustion of an important natural resource like ground water.

References

Boyle, Alan (1994). Economic growth and protection of the environment: The impact of international law and policy. In: *Environmental Regulation and Economic Growth,* (Ed.) Alan Boyle. Clarendon Press, Oxford.

Chakraborty, Ramanda and Bijonkumar Mondal (1995). *Principles of Crop Production.* West Bengal State Book Board.

Economic Review, Government of West Bengal, 2000–2001, 2001–2002 and 2002–2003.

Handbook of Agriculture. (1992). Indian Council of Agricultural Research, New Delhi.

Chapter 38
Botanical Derivative in Mosquitoes Control Programme to Minimize Pesticides Pollution Hazards

R.K. Tenguria, Versha Rai, P.K. Mishra and Sapan Patel

Department of Botany, Government M.V.M., Bhopal, M.P.

Department of Botany, Government J.H.P.G. College, Betul, M.P.

ABSTRACT

Alkaloid compound has been isolated from the P, ether chloroform, methanol and water extract of the aerial part of the *Lantana camara* L. The compound demonstrated, strong insecticidal activity against *Culex quinaquifasciatus.* The chemical derived from plants have been projected as weapons in future mosquito control programme as they shown to toxic growth and reproductive inhibitor of laboratory.

Introduction

Long before the invent of synthetic insecticides plants and their derivatives were used to kill pest of agriculture veterinary and public health importance. The chemical derived from plants have been projected as weapons in future mosquito control programme as they are shown to function as general toxic growth and reproductive inhibitor of laboratory test and field trials of a series of a plant extract as well as purified phyto-chemical as mosquito larvicidal concentration have shown promising result. The present project plants species against the early fourth instars larvae of *culex quinquefasciatus.* The present study is a part of our continuous effort for the last ten years or more to investigate the phytochemical against vector control. It is reported by Saxena and Sharma (1993). The isolation of alkaloid compound from *Lantana camara* against fourth instars larvae of *culex quinquefasciatus.*

Material and Method

Lantana camara L. (Verbenaceae) grows wildly shrub. It was identified by Dr. R.K. Tenguria (Department of Botany, M.V.M., Bhopal). It was collected from Vidisha and Bhopal voucher specimen plant is procured herbarium maintained at pest control research laboratory. The collected plant material leaf of *Lantana camara* was washed with tap water and air dried material was extracted in soxhlet apparatus. Rearing of laboratory test mosquito for bioassay.

Culture of Test Insects

Culex quinfasciatus, larvae were collected from cesspools and ditches. It was then cultured in laboratory. The larvae were red on yeast tablet and dog biscuits (3 : 1) powdered material. The culture was maintained in insectary at controlled temperature 27 ± 2C, RH 75 ± 5 per cent and L : D 14 : 10 photoperiod.

Extraction and Purification

Extraction was done in soxhlet apparatus using following solvent reported by Ram P. Rastogi (1960) n-haxane, Benzene, Chloroform, P. ether, Methenol water.

Purification was done by column chromatography using solvent system Alkaloid is reported by Harnbone (1984).

The biological active component where reported form the three crude extracts by column chromatography.

1. Different concentration of purified fraction applied on larvae.
2. WHO (1971) Methods for bioassay adopted.

Bioassay Procedure

The larvicidal activity of the extracts was evaluated as per the method recommended by WHO. the stock solution of the plant extract was volumetrically diluted to 250 ml with filtered tab water to obtain the test solution of 10, 20, 40, 60, 80 mg the test solution for assaying the larvae of *culex quinquefasciatus*.

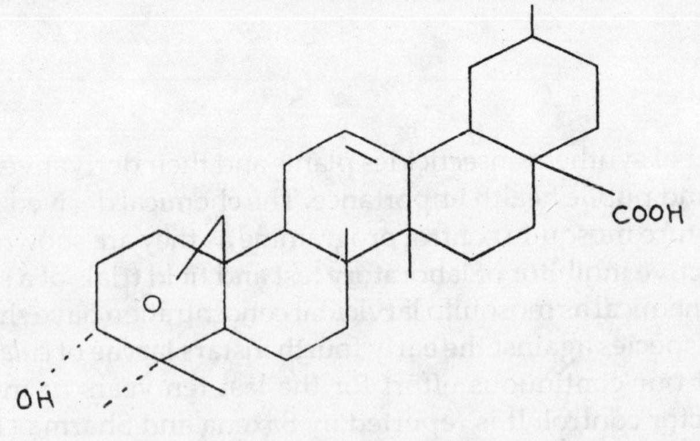

Figure 38.1: Lactic Acid

It was prepared in saline water of salinity 15 10.3 was used emulsified at a concentration of 0.001 per cent in these test solution two control were maintained at a time. One consisted of acetone and the other tap water only early fourth instars larvae (25) were introduced to each of the test solution as well as control for each dose four replicates were run at time. The larval mortality was recorded in each set of stock solution and with three different batches of mosquito larvae.

Botanical evaluation by Finney (1971) ANOVA was carried out.

Result and Discussion

The result maintained on the effect of Lactic acid of *culex quinquejasciatus* larvae. The larvae used for experimental work were laboratory cultured, second and fourth instars larvae calculated LC_{50} values to determine larvicidal potential for further experimentation of larvicidal LC_{50} observed in *culex quinquejasciatus*. Fecundity and fertility of larvae treated adults showed significant difference (PL 0.05) than control and loss was markedly observed in *Lantana camara.* Fertility experiments produced shorter egg rates in *Lantana camara* and decrease in hatching per cent noticed *Lantana camara* than control some developmental difference and some morphological abbreviation were observed.

1. Botanical derivatives and eco-compatible and do not cause any pollution hazards.
2. They are quite so be to the non target organisms including human beings. Hence use of natural products for vector control program is quite promising.

Acknowledgement

The authors acknowledged with thanks for the financial support from MAPCOST project No. 2–15/93.

References

Finney (1971). *Probit Analysis: A Book,* Revised edn. Cambridge University Press, London, pp. 318.

Harnbone, T.B. (1984). *Phytochemical Methods: A Guide to Modern Techniques of Plant Analysis.* Chapman and Hall, p. 228.

Rastogi, Ram P. and R.N. Malhotra (1960-69). *Indian Medicinal Plants.*

Saxena, R.C. Sharma, M.C., Dixit, O.P., Jacob K. Babey and Laxman Kumar (1983). Larvicidal and growth discrupting activity of some indigeneous plant extract to *Culex quinquefasciatus. J. Appl. Zool. Res.,* (1993)4(1): 79–82.

Chapter 39
Bioefficacy of Conventional and Neem Insecticides Against Insect Pests of Okra

Rabindra Prasad

Department of Entomology, Birsa Agricultural University, Kanke, Ranchi – 834 006

ABSTRACT

Experimental findings of the field experiments conducted during two consecutive seasons, spring-summer and kharif of 1996 for the evaluation of bio-efficacy of five insecticides used in two different dosages against pests of okra *viz.* jassid (*Amrasca bigutulla bigutulla* Ishida) and fruit borer (*Earias vitella* Fab.) revealed that their effectiveness were found to be almost in order of monocrotophos> chlorpyriphos> quinalphos> neem oil> fenvalevate in reducing the incidence of both of the pest species. The insecticides hold also almost the same rank in terms of enhancement of yield of healthy and marketable fruits of okra and cost benefit ratio. As such, yield of marketable fruits *viz.* 90.2, 73.8, 69.5, 76.1 and 67.7 q/ha and CBR of 1 : 16.4, 1 : 12.8, 1 : 10.6, 1 : 9.6 and 1 : 7.6 were obtained due to reduction in the incidence of jassids to the tune of 88.9, 80.9, 72.4, 72.3 and 59.9 per cent and that of fruit borer up to 90.8, 86.6, 74.5, 73.9 and 67.7 per cent respectively over the unprotected crop by the foliar spray of the respective chemical insecticides @ 500g a.i/ha, and neem oil @ 3.5 per cent. The expected avoidable loss in yield was found up to 42.0 per cent over the unprotected crop of okra.

Keywords: Okra pests, Bioefficacy, Insecticides, Neem pesticide.

Introduction

Okra (*Abelmoschus esculentus* L. Moench) an important vegetable crop grown in summer and kharif seasons is usually endangered by the damage caused by jassid (*Amrasca bigutulla bigutulla* Ishida) and fruit (*Earias vitella* Fab.). Loss in yield of marketable fruits of okra were estimated to be 35 per cent (Krishnaiah, 1980) and even up to 40 to 53 per cent caused by fruit borer (Prasad, 2000). Various formulations of conventional insecticides from different commercial firms are available in the market, hence, the periodic evaluation of their effectiveness is necessary. Keeping these views and objectives, in mind, the present experiment was planned for bio-efficacy evaluation of some insecticides against the prevailing pest problem of okra.

Materials and Methods

Field experiments were conducted at Birsa Agricultural University Campus, Kanke, Ranchi during two consecutive seasons, spring-summer and Kharif, 1996.

Four conventional insecticides and one botanical insecticide (neem oil) were evaluated in their two dosages against the pest complex of okra. The experiments were executed with eleven treatments including untreated check (Table 39.1) and three replications. The plot size was 4.5 × 2.7 m. Distance of 45 and 30cm between rows and plants were maintained. Ad hoc Economic threshold level (ETL) of 2 jassids per leaf and 4 per cent fruit damage caused by *E. vitella* were considered for initiating foliar spraying. Three spraying were made at 14 days interval on attainment of ETL during both of the cropping seasons. Jassids incidence were recorded by counting the number of nymphs and adults on three leaves per plants from five randomly selected plants in each plot a day prior and 5 and 10 days after insecticidal application. Fruit damage (by weight) was recorded at the time of each picking by weighing healthy and damaged fruits per treatment per replications and converted into per cent fruit damage.

Table 39.1: Bioefficacy of Insecticides Against Jassids (*Amrasca bigutulla bigutulla Ishida*) Infesting Okra

Sl.No.	*Treatment*	*Dose g ai/ha*	*Mean Number of Jassids per 15 Leaves Day After Application of Foliar Spary*			*Over All Mean*
			1	*5*	*10*	
1.	Chlorphyriphos 20 EC	500	5.1 (87.4)*	8.3 (80.5)*	11.3 (75.6)*	8.2 (80.9)*
2.	Chlorphyriphos 20 EC	300	19.6 (51.5)	20.2 (52.5)	23.2 (43.5)	21 (51.3)
3.	Monocrotophos 36 EC	500	3.7 (90.8)	3.2 (92.5)	7.5 (83.8)	4.8 (88.9)
4.	Monocrotophos 36 EC	300	15.7 (61.1)	16.0 (62.3)	19.1 (58.8)	16.3 (62.2)
5.	Quinalphos 25 EC	500	12.2 (69.8)	10.3 (75.7)	13.3 (71.3)	11.9 (72.4)
6.	Quinalphos 25 EC	300	22.6 (44.1)	22.4 (47.3)	25.6 (44.8)	25.5 (40.8)
7.	Fenvalerate 20 EC	50	16.3 (59.6)	16.5 (61.2)	19.0 (59.0)	17.3 (59.9)
8.	Fenvalerate 20 EC	30	28.2 (30.2)	27.7 (34.8)	29.5 (36.4)	28.5 (33.9)
9.	Neem oil**	3.50%	12.9 (68.1)	11.8 (72.2)	13.8 (70.2)	12.8 (70.3)
10.	Neem oil**	1.50%	26.9 (33.4)	24.7 (41.8)	28.8 (37.9)	26.8 (37.8)
11.	Untreated check	–	40.4	42.5	46.4	43.1
	S.E. (m)	–	1.2	1.6	1.4	1.6
	C.D. (P = 0.05)	–	3.6	4.7	4.2	4.9
	C.V. (%)	–	8.2	10.3	9.4	11.4

*: Figures in the parentheses are percentage reduction in the population of jassids due to the insecticidal treatment over untreated check.

**: Dose in percentage of neem oil in the spray solution of water.

Yield of healthy and marketable fruits was also recorded during each picking. Data of the three sprays in terms of jassid and fruit borer incidence were pooled together. Ultimately data in terms of the pest incidence and yield of fruits were pooled (Table 39.1) to draw the final conclusion. Percentage or yield gain due to the respective insecticidal treatments and loss in yield (per cent) due to damage caused by the pest fauna in the absence of the corresponding insecticidal protection measures were calculated by the following formula:

$$\textbf{Yield Gain (per cent)} = \frac{\text{Yield of Protected Crop} - \text{Yield of Control Plot}}{\text{Yield of Control Plot}} \times 100$$

$$\textbf{Yield Loss (per cent)} = \frac{\text{Yield of Protected Crop} - \text{Yield of Control Plot}}{\text{Yield of Protected Crop}} \times 100$$

The data were subjected to analysis of variance.

Results and Discussion

Jassid

The pooled results of two cropping seasons (Table 39.1) revealed the effectiveness of all the insecticides used in two different dosages against the incidence of jassid proved superior over control. It is obvious that the higher dosage of all the insecticides proved superior to their own lower dosage. The bioefficacy of almost all the treatments were found to be decreased drastically on 10 days after their application against the incidence of leaf hopper (*Amrasca bigutulla bigutulla* Ishida).

Minimum jassid incidence was recorded throughout the period when the crop was treated with monocrotophos 36 SL @ 500g ai/ha. followed by chlorpyriphos 20 EC @ 500 g a.i./ha. and quinalphos 25 EC @ 500g a.i./ha Neem oil used @ 3.5 per cent gave rise to reduction in the population of jassids from 68.1 to 72 per cent. Earlier, Easwaramoorthi *et al.* (1976), Dhawan *et al.* (1988) found that monocrotophos provided effective control of sucking pests *e.g.* jassid and others of okra.

All the five tested insecticides in their higher dosages showed their maximum effectiveness against the incidence of A. *bigutulla bigutulla* upto 5 days after application (DM) in order of monocrotophos > chlorpyriphos > quinalphos > neem oil > fenvalerate, though they were able to reduce the pest intensity upto 10 DM from 59 to 83.8 per cent.

Sidhu *et al.* (1979) reported that monocrotophos and quinalphos applied @ 0.3 kg a.i./ha were found to be highly effective against jassids.

Patel *et al.* (1997) also found chlorpyriphos and quinalphos to be very effective against *A. b. bigutulla* up to 7 days after application.

Fruit Borer (*E. vitella*)

Results (Table 39.1) reveals that all the five tested insecticides proved significantly superior in reducing the incidence of fruit borer (*Earias vitetta* Fab.) at their higher dosage as compared to their lower dosage. Monocrotophos proved most effective followed by chlorpyriphos against fruit borer when they were used @ 500 g a.i./ha. Quinalphos @ 500 g a.i./ha and neem oil used @ 3.5 per cent remained at par in this respect. Okra fruit damage was significantly lower on insecticide sprayed crop than unsprayed crop. The effectiveness of monocrotophas against the fruit borer of okra have also been reported by Krishnaiah *et al.* (1976) and that of chlorpyriphos and quinalphos by Patel *et al.* (1997).

Fruits Yield

The highest fruit yield of okra (90.2 q/ha) was obtained when the crop was sprayed with monocrotophos @ 500 g a.i./ha followed by neem oil applied @ 3.5 per cent (76.1 q/ha) which remained at par with that of chlorpyriphos applied @ 500 g a.i./ha. Patel *et al.* (1984) and Patel *et al.* (1997) revealed that the insecticidal protection resulted into considerable enhancement in yield over untreated control (Table 39.2).

Table 39.2: Effect of Various Insecticidal Treatments on Fruit Borer Incidence and Yield of Healthy Fruits of Okra, Its Increase and Cost Benefit Ratio

Treatment	*Percentage of Fruit Damage Due to Fruit Borer Incidence*		*Yield of Healthy Fruits*				*Cost Benefit Ratio (CBR)*
	Fruit Damage (%)	*Reduction in Fruit Damage Over Control (%)*	*Yield (q/ha)*	*Increased Yield Over Control (q/ha)*	*Gain in Yield Over Control (%)*	*Expected Yield Loss in absence of the treatment over control (%)*	
1	5.3 (13.3)	86.0	73.8	21.5	41.2	29.1	1 : 12.8
2	18.6 (25.5)	51.0	59.9	11.4	21.8	12.7	1 : 7.6
3	3.5 (10.7)	90.8	90.2	37.8	72.3	42.0	1 : 16.4
4	14.3 (22.2)	62.4	64.8	12.4	23.7	29.3	1 : 8.9
5	9.7 (18.1)	74.5	69.5	17.2	32.7	24.7	1 : 10.6
6	20.9 (27.2)	45.0	66.2	13.8	26.3	21.0	1 : 6.2
7	14.3 (22.2)	62.4	67.7	15.4	29.9	22.7	1 : 7.6
8	25.1 (30.1)	33.9	60.8	8.5	16.3	13.9	1 : 5.0
9	9.9 (18.3)	73.9	76.1	23.7	45.4	31.3	1 : 9.6
10	33.9 (35.6)	10.8	59.0	6.6	12.7	11.0	1 : 3.2
11	38.0 (38.1)	–	52.3	–	–	–	–
S.E.(m) ±	(1.4)	–	2.2	–	–	–	–
C.D. (P = 0.05)	(4.2)	–	6.4	–	–	–	–
C.V. (%)	(9.2)	–	8.5	–	–	–	–

Figures in the parentheses are angular transformed values.

On keeping in view the difference between the yield of healthy marketable fruit obtained with the help of the treatments from that of unprotected (control) plants, it is evident that the yield loss caused by jassid and fruit borer was found to be up to 37.8 q/ha which was equivalent to 42.0 per cent over control (Table 39.2).

On considering per cent increase in yield over control, the highest rank was occupied by monocrophos applied @ 500 g ai./ha. which gave rise to 72.3 per cent increase in yield (Table 39.2).

Cost Benefit Ratio

Data presented in Table 39.2 reveals that relatively higher benefit cost ratio, ranging from 1 : 7.6 to 1 : 16.4, were obtained with the higher dosage of the insecticides. The highest benefit cost ratio was obtained with monocrotophos (1 : 16.4) followed by chlorphyriphos (1 : 12.8), quinalphos (1 : 10.6), neem oil (1 : 9.6) and fenvalerate (1 : 7.76) at their higher concentration in the spray fluid.

Conclusively, on the basis of the two seasons experimentation, it may be concluded that the tested insecticides remained in order of monocrotophos > chlorpyriphos > quinalphos> neem oil > fenvalerate in reducing the incidence of jassid and fruit borer and in enhancing the yield of marketable fruit of okra as well as cost benefit ratio.

References

Dhawan, A.K., Simwat, G.S. and A.S. Sidhu (1998). Field evaluation of monocrotophos for the control of sucking pest of cotton. *Pesticides*, 22(6): 25–28.

Easwaramoorthy, S, Cheliah, S. and S. Uthamaswamy (1976). Efficacy of certain insecticides against the pests of bhindi: *Abelmoschus esculentus* (I) Moench. *Madras Agric. J.*, 6(4): 254–256.

Krishaniah, K., Tondon, P.L. and A.C. Mathur (1976) Evaluation of insecticides for the control of major insect pests of okra. *Indian J. Agric. Sci.*, 46(4): 178–186.

Krishnaiah, K. (1980). Methodology for assessing crop loss due to pests of vegetables. In: *Assessment of Crop Loss due to Pest and Diseases*, (Ed.) H.C. Govindu *et al.*, USA Tech Series, 33: 259–267.

Patel, A.R., Chairi, M.S. and D.R. Vanshi (1984). Control of okra fruit borer, (*E. vitella*) with synthetic pyriethroids. *Pestology*, 8: 22–24.

Patel, N.C., Patel, J.J., Jayani, D.B., Patel, J.R. and B.D. Patel (1997). Bioefficacy of conventional insecticides against pests of okra. *Indian J. Ent.*, 59(1) 51–53.

Prasad, Rabindra (2000). Integrated pest management for sustainable production of okra. *The Bihar J. Agric. Mktg.*, 8(4): 466–470.

Sidhu, A.S., Dhawan, A.K.G. and K. Singh (1979). Testing new chemicals for control of cotton pests. *Pesticides*, 13(11): 7–11.

Chapter 40

Management of *Earias vitella* Fabricious Infesting Okra through Companion Croppings and Soil Application of Insecticides

Rabindra Prasad

Department of Entomology, Birsa Agricultural University Kanke, Ranchi – 834 006

ABSTRACT

The findings of the field experiment conducted during the two consecutive years, 1996 and 1997 in *kharif* revealed that the independent effects of insecticides, *viz.* aldicarb, cartap hydrochloride and carbofuran, applied in the soil were found to be more or less comparable with that of the intercrops, *viz.* marigold, french bean and brinjal in reducing the incidence of shoot and fruit borer (*Earias vitella* Fab.) of okra and enhancing the yield of healthy marketable fruits of okra. Moreover, pest management through the influence of crop association is safer and eco-friendly. Marigold took the lead among the companion crop(s) in reducing the incidence of *Earias vitella* Fab., while French bean as companion crop was found responsible for the highest yield of marketable fruits of okra (91.9 q/ha), among all the intercrops as well as all the three insecticides. Untreated sole crop of okra resulted to the lowest yield of healthy fruits, ranging from 46.0 to 67.8 q/ha. The findings of the present study suggests that use of impact of companion crops *e.g.* marigold or French bean with okra should be included in IPM programmes in vegetable crops in general and okra in particular keeping in view their role in eco-friendly management of the pest fauna of vegetable agro-ecosystem in order to minimise the excess use of chemical insecticides in vegetable cultivation. Among the insecticides, soil application of carbofuran @ 1.25 kg a.i./ha at the sowing time may also be included in the IPM of okra as suggested by the present study as soil application of granular insecticides is usually less hazardous than foliar spray of insecticides particularly for vegetable crops in general.

Keywords: *Companion crop, Insecticides, Fruit borer, Okra.*

Introduction

Okra (*Abelmoschus esculentus* L. Moench) an important vegetable crop is usually endangered by shoot and fruit borer, *Earias vitella* Fab. The insect pest usually causes severe damage to shoot (upto 18.6 per cent) and fruit (25.4 to 28.8 per cent) of okra (Prasad, 2002). There is considerable loss in yield of healthy marketable fruits of okra amounting to 35 per cent (Krishnaiah, 1980) and even upto 40 to 53 per cent (Prasad, 2001). Despite the well known ill after-effects of the chemical insecticides, farmers are still solely dependent on them for protecting their vegetable crops, including okra. Frequent foliar spraying of the insecticides are applied by the farmers to combat the pest problems for achieving the desirable fruits production. The use of crop association (inter-cropping) supplemented with soil application of granular insecticides may be not only much safer but also highly effective both from ecological and economic terms as compared with foliar application of insecticides. With this objective, the present experiment was undertaken to assess the relative efficacy of crop association (companion crops) and soil application of granular insecticides for management of *E. vitella* infesting okra, for realizing higher yield of healthy fruits.

Materials and Methods

A field experiment was conducted in the research farm of Birsa Agricultural University, Kanke, Ranchi during *kharif* season of 1996 and 1997. The experiment was laid out in FRBD with 16 treatment and three replications having plot size 4.0 × 2.5m. Recommended agronomical practices were applied for raising the crops excepting pest control measures. Okra var. Pusa Sawani was sown in intercropping system with French bean c. v. S–9 (*Phaseolus vulgaris* L.) african tall marigold (*Tagetus erecta* L.) and brinjal (*Solanum melongena* L.) var. pusa purple in control plots as well as in treated plots with granular insecticides either with aldicarb 100, cartap hydrochloride 4G and carbofuran 3G applied each @ 1.25 kg a.i./ha. at the time of seed sowing. The main crop (*i.e.* okra) was grown with the inter crops in the combination of 3 : 1 and as sole crop. The sowing of seeds of okra and French bean and transplanting of seedling of marigold were done during 3rd week of May 1996–97. Harvesting of okra fruits was done periodically. Pest incidence in terms of total number of shoots and damaged shoots and total weight of fruits and weight of damaged fruits, due to *E. vitella*, were recorded periodically in order to work out shoot damage and fruit damage and to calculate yield of healthy fruits. Year-wise data and pooled data on the pest incidence and yield of both of the years were statistically analysed by employing factorial randomized block design (FRBD), after suitable transformation by bifurcating the data into inter-cropping and insecticides.

Results and Discussion

Results are presented in the form of effect of crop association (Table 40.1) and insecticides (Table 40.2) separately on the pest incidence and on yield as well.

Effect of Companion Cropping(s)

The incidence of shoot borer and fruit borer (*E. vitella*) differed significantly in different crop associations. Okra grown in association with marigold suffered from the least incidence of shoot borer 9.4 and 11.5 per cent during 1996 and 1997. Fruit borer incidence showed almost similar trend. Effect of crop association were found to be in order of : okra grown with marigold > okra grown with French bean > okra grown with brinjal > okra grown alone in terms of suppression of incidence of both shoot and fruit borer (Table 40.1). The companion crops, *viz.* marigold and French bean probably created more obstruction against the entry and incidence of *E. vitella* on okra as compared to brinjal by creating bio-physical barrier for the pest, hence lesser infestation of shoot and fruit borer (*E. vitella*)

Table 40.1: Effect of Crop Association on the Incidence of Shoot and Fruit Borer (*Earias vitella* Fab.) and Yield of Healthy Fruits of Okra

Treatment	Shoot Damage (%)				Fruit Damage (%)				Yield of Healthy Fruits of Okra (g/ha)		
	1996	1997	Pooled Mean	Reduction Over Control (%)	1996	1997	Pooled Mean	Reduction Over Control (%)	1996	1997	Pooled Mean
Sole crop of okra	17.8 (29.9)	20.3 (26.8)	19.1 (28.3)	–	25.5 (30.3)	27.6 (31.7)	26.5 (31.0)	–	66.3	69.4	67.8
Okra + marigold (3 : 1)	9.4 (17.8)	11.5 (19.8)	10.4 (18.8)	45.5	10.2 (18.6)	13.3 (21.4)	11.7 (20.0)	55.8	85.1	88.5	86.8
Okra + brinjal (3 : 1)	11.7 (20.0)	13.7 (21.7)	12.7 (20.8)	35.5	17.4 (24.6)	21.0 (27.3)	19.2 (25.9)	27.5	76.4	79.8	78.1
Okra + French bean (3 : 1)	10.2 (18.6)	12.5 (20.7)	11.3 (19.6)	40.8	14.3 (22.2	16.8 (24.2)	15.5 (23.2)	41.5	90.3	93.5	91.9
S.E.(m) (±)	0.2	0.2	0.2	–	0.9	0.8	1.1	–	2.3	2.4	1.4
C.D. (P = 0.05)	0.6	0.7	0.5	–	2.8	2.6	3.0	–	6.6	7.4	4.0
C.V. (%)	8.6	7.2	9.6	–	6.6	8.4	10.6	–	9.6	8.5	10.2

Figures in parentheses are angular transformed values.

Table 40.2: Effect of Soil Application of Granular Insecticides on Shoot and Fruit Borer (*Earias vitella* Fab.) Incidence and Yield of Healthy Fruits of Okra

Treatment	Shoot Damage (%)				Fruit Damage (%)				Yield of Healthy Fruits of Okra (g/ha)		
	1996	1997	Pooled Mean	Reduction Over Control (%)	1996	1997	Pooled Mean	Reduction Over Control (%)	1996	1997	Pooled Mean
Control	22.4 (28.2)	25.4 (30.3)	23.9 (29.2)	–	30.7 (33.6)	33.2 (35.2)	31.9 (34.4)	–	47.7	44.4	46.0
Aldicarb	8.8 (17.3)	10.6 (19.0)	9.7 (18.1)	59.4	9.3 (17.8)	11.8 (20.1)	10.5 (18.9)	67.1	71.5	74.3	72.9
Cartap hydrochloride	10.5 (18.9)	13.2 (21.3)	11.8 (20.1)	50.6	12.6 (20.8)	14.9 (22.7)	13.7 (21.7)	57.0	74.3	77.2	75.7
Carbofaran	12.4 (18.8)	11.7 (20.0)	12 (19.4)	49.8	11.4 (19.7)	13.9 (21.9)	12.6 (20.8)	60.5	77.4	79.6	78.5
S.E.(m) (±)	(0.5)	(0.7)	(0.4)	–	(0.6)	(0.5)	(0.6)	–	(0.8)	(0.7)	(0.8)
C.D. (P = 0.05)	(1.4)	(1.8)	(1.2)	–	(1.8)	(1.6)	(1.8)	–	(2.5)	(2.3)	(2.6)
C.V. (%)	(%)	(8.6	(2.6)	–	(6.6)	(8.4)	(10.6)	–	(9.6)	(8.5)	– 10.2

Figures in parentheses are angular transformed values.

was occurred. Some undesirable aroma emitted by the inter-crops might also be responsible for the lesser incidence of the pest of okra. Prasad (2001) and Prasad and Prasad (2002) obtained almost similar results pertaining to the reduction of shoot and fruit borer incidence of okra through the inter-cropping system.

On considering the yield of healthy and marketable fruits of okra, the effect of companion crops were found to be in order of okra + French bean > okra + marigold > okra + brinjal > okra alone.

Effect of Insecticide(s)

A perusal of Table 40.2 reveals that the independent efficacy of insecticides were found to be in order of : aldicarb > cartap hydrochloride > carbofuran > control; in terms of reduction in the shoot damage, where as they were noticed to be as aldicarb > carbofuran > cartap–hydrochloride in reducing fruit damage during both of the years.

Almost, similar trends were recorded in case of pooled results of two years too. Prasad and Prasad (2002) revealed on the basis of two years experimentation that soil application of granular insecticides were found to be in order of aldicarb > carbofuran > ethoprophos in reducing incidence of shoot and fruit, borer of okra. Sole crop of okra grown in the untreated soil was found to be suffered from maximum damage of shoot, 22.4 and 25.4 per cent and the highest damage of fruit 30.7 and 33.2 per cent during 1996 and 1997 respectively in the present study.

Overall Pooled Result of 1996 and 1997

On the basis of pooled results of two years of experimentation, it is concluded that the two inter-crop, *viz.* French bean and marigold were responsible for reduction in shoot damage to the extent of 40.8 to 45.5 per cent and fruit damage upto 55.8 to 41.5 per cent as compared to sole crop of okra. This finding clearly revealed that efficacy of the crop associations and insecticides applied around the root zone in the soil remained more or less comparable in suppressing the incidence of shoot and fruit borer (*E. vitella*) infesting okra. Moreover, pest management through the impact of crop association is safer and eco-friendly as compared to that with pest control made by insecticides. Pooled results of the two years showed that highest yield of marketable fruits of okra (91.9 q/ha) was obtained when okra was grown in association with French bean which was so, probably, being a leguminous crop, French bean spared some nitrogenous nutrient for okra, on the one side, and it suppressed the incidence of shoot borer upto 40.8 per cent and fruit borer upto 41.5 per cent. Prasad (2001) obtained almost similar results in the pest suppression and enhancement in yield of fruits of okra through the inter-cropping of French bean with okra almost by the similar mechanism of allelopathic pest control.

Among the three insecticides, highest reduction in shoot damage 59.4 per cent and that of fruit damage 67.1 per cent was found over untreated crop of okra by the action of soil treatment with aldicarb; while carbofuran caused 49.8 and 60.5 per cent reduction in shoot and fruit damage respectively. However, carbofuran treated plants resulted to the highest yield of marketable fruits of okra (78.5 q/ha) probably due to the phytotonic effect of carbofuran (Table 40.2). Earlier, Stanley and Russel (1967) and Srivastava (1975) reported the phytotonic effect of carbofuran. Rao (1981) also obtained similar result with carbofuran and found 3 fold increase in seed yield of green gram with basal application of carbofuran at the rate of 0.5 kg a.i./ha.

Thus, findings of the present study suggests that use of impact of companion crops *e.g.* marigold or French bean with okra should be included in IPM programmes in vegetable crops in general and okra in particular keeping in view their role in eco-friendly management of the pest fauna of vegetable agro-ecosystem in order to minimise the excess use of chemical insecticides in vegetable cultivation.

Among the insecticides, soil application of carbofuran @ 1.25 kg a.i./ha at the sowing time may also be included in the IPM of okra as suggested by the present study as soil application of granular insecticides is usually less hazardous than foliar spray of insecticides particularly for vegetable crops in general.

References

Krishnaiah, K. (1980). Methodology for assessing crop loss due to pests of vegetables. In: *Assessment of Crop Loss due to Pest and Diseases*, (Ed.) H.C. Govind *et al.* USA Tech. Series, 33: 259–267.

Prasad, Rabindra (2001). Management of okra fruit borer (*Earias vitella* Fab.) through the interaction of intercropping and insecticides. *Indian J. Environ & Ecoplan.*, 5(1): 125–129.

Prasad, Rabindra and Devendra Prasad (2002). Effect of crop association and insecticides on the incidence of shoot and fruit borer of okra (*Abelmoschus esculentus*). *Birsa Agril Univ. Res. J.*, 14(2): 255–258.

Rao, P.S. (1981). Eco-toxicological studies with carbofuran granules on the pest complex of green gram, *vigna radiata* Wilczek. *Ph.D. Thesis*, IARI, New Delhi.

Srivastava, K.P. (1975). Chemical Control of sorghum pests and studies on persistence of insecticides used on sorghum crop. *Ph.D. Thesis*, IARI, New Delhi.

Stanley, W.W. and W.G. Russel (1967). Growth response of flowering crop apple and control of seed banded leaf roller through flowing application of systemic insecticides. *Agron. J.*, 60: 306–308.

Chapter 41

Effect of Micronutrients on Growth, Yield Attributes and Yield of Garden Pea (*Pisum sativum* var. *Hortense*) Cultivars

Subhendu Mandal, Subimal Mondal,** Subhadeep Nath**, V.B. Yadav*** and Barun Singh****

**Department of Agronomy, ** Department of Vegetable Crops, ***Department of Fruits and Orchard Management*
Bidhan Chandra Krishi Viswavidyalaya, Mohanpur, Nadia, West Bengal – 741 252

ABSTRACT

A field experiment was conducted during winter season of 2001–02 and 2002–03 to study the effect of micronutrients on growth, yield attributes and yield of garden pea cultivars. Growth, yield and yield attributes and seed qualities of garden pea were significantly influenced by micronutrients application. Among the micronutrients, application of Mo recorded significantly higher growth, yield attributes, yield and qualities over control and zinc application but it was at par with boron application. Plant height and pod length, pod weight and number of seeds/pod were significantly higher with KS–168 but it was at par with VL–7 in case of pod weight. Maximum number of pods/plant and ascorbic acid content in seed were with VRP–2. However, highest pod diameter, crude protein content and yield were significantly with VL–7.

Keywords: *Garden pea, Micronutrients, Cultivars.*

Introduction

Realizing the role of micronutrients in the present day intensive agriculture, three micronutrients *viz.* zinc, boron and molybdenum have been identified to be applied to commercial garden pea cultivation. These micronutrients are applied to improve the yield. Although yield improvement by micronutrient application is well known, but information regarding growth, yield attributes and yield of pea cultivars is meager. Cultivation of garden pea as a cash crop in Lower Gangatic Zone of West Bengal during winter season after harvest of *kharif* rice is gaining popularity. Besides this, selection of suitable cultivars for specific agro-climatic region is also important for increasing the production of this crop. Therefore, an attempt was made to study the effect of Zn, B and Mo on growth, yield attributes and yield of garden pea cultivars.

Materials and Methods

The present investigation was conducted during winter season of 2001–02 and repeated in 2002–03 at Central Research Farm, Gayeshpur, Bidhan Chandra Krishi Viswavidyalaya. The soil of experiment field was sandy loam in texture, containing 0.06 per cent total N, available phosphorous 17.8 kg/ha and potash 192.4 kg/ha with pH 7.4. Zinc (0.005 M Diethylene triamine penta acetic acid), boron (barium chloride 10 per cent) and molybdenum contents in soil were 0.70, 0.03 and 0.04 mg/kg of soil, respectively. The experiment was laid out in split plot design replicated thrice with five cultivars (VRP–1, VRP–2, VL–7, KS–168 and Arkel) as main plot and four treatments of micronutrierits (C = Control, Zn = Zinc sulphate @ 25 kg/ha, B = Boron @ 10 kg/ha and Mo = Ammonium molybdate @ 1.5kg/ha) as sub plot treatment. Cop was sown on 2nd week of September in both the years with spacing 30 cm × 10 cm. The crop was fertilized with 50 kg N, 80 kg/ha P_2O_5, 50 kg K_2O/ha and micronutrients were applied as per treatment. Micronutrients were broadcasted at final land preparation. Nitrogen, phosphorous and potash were applied as ammonium sulphate, single super phosphate and muiate of potash respectively. Half dose of N and full dose of phosphorous and potash were applied as basal. The remaining N was top dressed at flowering stage. Observations on growth, yield and yield attributes were recorded at harvesting in both the years.

Results and Discussion

Growth and Yield Attributes

Garden pea cultivars KS–168 produced the maximum plant height as compared to other cultivars. Among the micronutrients molybdenum recorded significantly higher plant height, which was at par with boron application. The number of pods/plant, pod length, pod diameter, pod weight and number of seeds per pod were increased significantly by application of micronutrients over control. Among the micronutrients molybdenum produced significantly higher all the above yield attributes as compared to control and zinc application. Though zinc application improved all the yield attributes and yield significantly over control but it was significantly lower than boron and, molybdenum application. These results are in close conformity with the findings of Ibrahim *et al.* (1989) and Dwivedi *et al.* (1992). Among the cultivars VRP–2 recorded significantly highest number of pods/plant whereas, Arkel VL–7 and KS–168 produced significantly lower number of pods per plant than other cultivars but were at par among them. Pod length, pod weight and number of seeds per pod were significantly higher with KS–168 but were at par with VL–7 and lowest were with VRP–1, except for number of seeds per pod. It was lowest in Arkel. Significantly highest pod diameter was recorded with VL–7 and lower was found in VRP–1 and Arkel, which were at par among themselves. Shukla and Kohli (1991) also reported similar observation on garden pea.

Table 41.1: Growth and Yield Components of Garden Pea Cultivars as Influenced by Micronutrients Application

Treatment	Plant Height (cm)			No. of Pods/Plant			Pod Length (cm)			Diameter of Pods (cm)			Pod Weight (g)		
	2001-02	2002-03	Pooled	2001-02	2002-03	Pooled	2001-02	2002-03	Pooled	2001-02	2002-03	Pooled	2001-02	2002-03	Poolea
Cultivars															
VRP–1	74.1	74.5	74.3	42.3	42.2	42.3	7.3	7.2	7.3	1.9	1.8	1.9	4.5	4.4	4.5
VRP–2	71.9	73.5	72.7	56.4	55.7	56.1	8.1	8.2	8.2	2.1	2.2	2.2	5.7	5.5	5.6
VL–7	75.8	76.2	76.0	24.3	25.0	24.7	9.7	9.4	9.6	3.3	3.1	3.2	8.2	8.0	8.1
KS–168	92.1	93.4	92.8	25.9	26.1	26.0	10.5	10.7	10.6	2.6	2.5	2.6	9.0	9.4	9.2
Arkel	76.0	75.7	75.9	34.6	35.8	35.2	8.5	8.7	8.6	19	1.8	1.91	6.6	6.3	6.5
C.D (0.05)	9.1	8.9	7.2	12.5	12.3	10.2	0.6	0.5	0.4	0.3	0.2	0.1	1.8	1.9	1.4
Micronutrients															
Control	65.8	65.5	65.5	29.1	29.3	29.2	6.8	6.5	6.7	1.8	1.9	1.9	5.8	5.7	5.8
Zn	70.5	70.7	70.7	35.6	3-6.1	35.9	8.4	8.5	8.5	2.4	2.6	2.5	6.9	7.0	7.0
B	75.3	75.7	75.7	42.5	43	42.8	9.3	9.5	9.4	2.9	3.1	3.0	8.6	8.4	8.5
Mo	78.3	78.0	78.2	44.7	45.8	45.3	9.8	9.4	9.6	3.1	3.2	3.2	8.8	9.0	8.9
CD (0.05)	4.5	4.3	3.6	5.7	5.2	4.2	0.5	0.5	0.4	0.4	0.4	0.3	1.1	1.2	1.0

Table 41.2: Yield Components, Yield and Quality of Garden Pea Cultivars as Influenced by Micronutrients Application

Treatment	No. of Seed per Pod			Pod Yield (kg/ha)			Ascrobic and Content (mg/100 g)			Crude Protein Content (%)		
	2001–02	2002–03	Pooled	2001–02	2002–03	Pooled	2001–02	2002–03	Pooled	2001–02	2002–03	Pooled
Cultivars												
VRP–1	6.5	6.5	6.5	3416	3455	3436	22.2	21.8	22.0	23.4	22.4	22.9
VRP–2	4.7	4.5	4.6	5689	5567	5628	47.8	47.0	47.4	28.1	27.6	27.9
VL–7	7.3	7.2	7.3	8671	8576	8624	25.9	25.4	25.7	30.3	31.2	30.8
KS–168	9.5	9.1	9.3	6483	6547	6515	37.1	37	36.8	28.7	27.6	28.2
Arkel	5.5	5.7	5.6	4867	4808	4838	32.2	32.3	32.3	25.0	25.3	25.2
CD (0.05)	0.4	0.4	0.3	582	592	407	4.2	4.4	3.4	1.1	1.2	0.8
Micronutrients												
Control	5.5	5.6	5.6	4135	4109	4122	22.1	22.7	22.4	21.9	22.1	22.0
Zn	6.2	6.3	6.3	4866	4896	4881	24.8	25.2	25.0	24.8	24.2	24.5
B	7.7	7.8	7.8	6138	6168	6153	28.4	29.3	28.9	27.1	26.8	27.0
Mo	8.1	8.0	8.1	6518	6493	6506	30.2	30.7	30.5	28.3	27.6	28.0
CD (0.05)	0.5	0.5	0.4	517	531.0	467	2.5	2.4	2.1	1.3	1.2	1.1

Yield

The maximum pod yield was registered with VL–7, being significantly superior to other cultivars. Whereas, VRP–1 recorded significantly lowest pod yield. Shridhar *et al.* (1997) also suggest that growing of VL–7 is suitable for increasing the yield of garden pea. Application of micronutrients significantly increased the pod yield over control. Among the micronutrients molybdenum produced significantly higher pod yield as compared to control and zinc application. But, it was at par with boron application. Jasinka and Kotecki (1991) Dwivedi *et al.* (1992) reported similar results.

Quality

Ascorbic acid and crude protein content in seed among the cultivars and micronutrients differ significantly. Cultivar VRP–2 recorded highest ascorbic acid content and lowest was with VRP–1. Application of molybdenum significantly increased the ascorbic acid content but it was at par with boron application. Jasinka and Kotecki (1991) and Vinay Singh (1993) also obtained similar results. Amino acid content was highest in VL–7 and lowest was with VRP–1. Application of molybdenum recorded highest crude protein content and lowest was in control. These findings are in conformity with Shukla and Kohli (1991) and Sridhar *et al.* (1997).

References

Dwivedi, B.S., Ram, M., Das, B.P., Das, M. and R.N. Prasad (1992). Effect of liming on boron nutrition of pea and corn grown in sequence an acid Alfisol. *Fert. Res.*, 31(3): 257–262.

Ibrahim, S.A. (1989). Growth, yield and nutrient uptake response of pea plant to phosphorous and micronutrients. *Egyption J. Soil. Sci.*, 29(3): 251–259.

Jasinka, Z. and A. Kotecki (1991). Effect of molybdenum on development and yield of pea. *Produkeja Roslinna*, 108(3): 163–172.

Shridhar, V.P., Nani, S.D., Dube, Shankar Lal and Ved Prakash (1997). VL–7 and Matar–7 new garden pea. *Indian. Hort.*, 41(4): 47–49.

Shukla, Y.R. and U.K. Kohli (1991). Influence of varieties and phosphorous fertilization on seed vigour of garden pea. *Ann. Agric. Res.*, 12(3): 284–287.

Singh, Vinay (1993). Iron-zinc relation in pea. *Ann. Agric. Res.*, 14(4): 497–499.

Chapter 42

Study on Social Aspects of Sustainable Dryland Agriculture as Perceived by the Farmers

Sube Singh, P.S. Shehrawat, Milakh Raj and R.C. Hasija

College of Agriculture, CCSHAU, Hisar – 125 004

ABSTRACT

The study was conducted in the Bhiwani district of Haryana state, which was selected purposively on the basis of maximum area under dryland agriculture. From the four blocks in the Bhiwani district 200 dryland farmers (50 farmers from each block, taking two villages from each block) were selected randomly. The study revealed that majority of the farmers perceived minimum tillage, crop diversification, soil fertilization application, integrated nutrient management, weed control, integrated pest management, maintaining plant population, drought resistant varieties, moisture and water conservation practices, agro–forestry and subsidiary occupation like livestock, poultry, horticulture, vegetable etc., as socially acceptable and socially sustainable for dry land agriculture.

Introduction

The concept of sustainable agriculture involves the evolution of a new type of agriculture rich in technology and information, with much less than intensive energy use and market purchased inputs. Thus, sustainability is the successful management of resources to satisfy the challenging human needs, while maintaining or enhancing the quality of environment and conserving natural resources.

Rainfed/Dry land farming has a distinct place in Indian agriculture occupying 67 per cent of the cultivated areas, contributing 44 per cent of the production. Two-third of the livestock lives in these regions. The resource poor infrastructure and low investment in technology and inputs characterize it. Recently, government has given high priority for the development of dry land areas for which several programmes and projects have been launched for the utilization of potential of these areas.

Effective technology transfer is one of the systems, which may ensure this great task of dry land agriculture development. Unless the needs for extension machinery is geared to the maximum, it will not be possible to transfer the proven and viable technologies to the ultimate users *i.e.* farmers.

Farmers all over the world are working as managers of their farm, irrespective of the economic, social, cultural, physical, technological and environmental; the farmers manage the production system to get returns from it. "Think nationally but plan and act locally" is the only relevant method of promoting sustainability in farming. Agriculture extension is a part of technology transfer system, which is primarily concerned, with the transmission of technology from the point of generation to its utilization.

Keeping in view the ever-increasing population, development of dry land agriculture, the depletion of natural resources, environmental pollution and limitations of sustainable agriculture this study was conducted.

Research Methodology

The study was conducted in the Bhiwani district of Haryana state. From the Bhiwani district all four blocks were selected and from these blocks two villages were selected. To make the study more comprehensive, from these villages 200 dryland farmers 25 (farmers from each village) were selected to collect the information. A schedule was developed and data were collected at farmer's fields.

For the measurement of perception of the respondents about the social aspects of sustainability, the dry land technology was divided into ten (10) broad components, which includes land preparation, crop diversification, soil fertility, weed control, integrated pest management, plant population adjustment, varieties of crops, soil and water conservation measures, agro–forestry, and subsidiary occupations. These broad components were sub-divided into specific technological components. The respondents were asked to indicate perception of each component in terms of social acceptability (SA), and social sustainability (SS). Each category of social dimensions were rated on the basis of dichotomous reply, *i.e.* Yes or No. The score of 1 (one) was allotted to yes reply and the score of 0 (zero) was allotted to the negative answer. The data were analyzed using percentage technique.

Results and Discussion

Land Preparation

The data in Table 42.1 speaks that 60.00 per cent of the dry land farmers perceived minimum tillage as socially acceptable and 67.50 per cent of them perceived the minimum tillage as socially sustainable. However, 31 per cent and 32.50 per cent of the farmers perceived that minimum tillage was not socially acceptable and sustainable, respectively.

Crop Diversification

It is observed from the Table 42.1 that 100.00 per cent of the farmers perceived crop rotation as socially acceptable and socially sustainable. Further it was resulted that 87.50 per cent farmers, followed by 75.00 per cent of them perceived mixed cropping as socially acceptable and sustainable, respectively. As majority, 84.00 per cent of the farmers, followed by 85.00 per cent of farmers perceived intercropping as socially acceptable and sustainable, respectively. In case of vegetable production an equal percentage (60.00 per cent) of them perceived vegetable production as socially acceptable and sustainable. The 90.00 per cent and 85.00 per cent of the farmers perceived the dairy farming as socially acceptable and sustainable, respectively. As for as bee-keeping is concerned, majority of farmers (77.00 per cent and 74.00 per cent, respectively) perceived as socially acceptable and sustainable, respectively. In case of poultry, 45.00 per cent and 39.00 per cent farmers respectively perceived it as socially acceptable and sustainable.

Table 42.1: Perception of Farmers about Social Aspects of Sustainable Dryland Agriculture (N = 200)

Sl.No.	Items	Social Aspects			
		Socially Acceptable		Socially Sustainable	
		Yes	No	Yes	No
1.	**Land Preparation**				
	Minimum tillage	138 (69.00)	62 (31.00)	135 (67.50)	65 (32.50)
2.	**Crop Diversification**				
	(*a*) Crop rotation	200 (0.00)	–	200 (100.00)	–
	(*b*) Mixed cropping	175 (87.50)	25 (12.50)	150 (75.00)	50 (25.00)
	(*c*) Inter cropping	168 (84.00)	32 (16.00)	170 (85.00)	30 (15.00)
	(*d*) Vegetable production	120 (60.00)	80 (40.00)	120 (60.00)	80 (40.00)
	(*e*) Dairy farming	180 (90.00)	20 (10.00)	170 (85.00)	30 (15.00)
	(*f*) Bee keeping	154 (77.00)	46 (23.00)	148 (74.00)	52 (26.00)
	(*g*) Poultry	90 (45.00)	110 (55.00)	78 (39.00)	122 (61.00)
3.	**Soil Fertility**				
	(*a*) *Soil fertilization*				
	(*i*) Broadcasting of fertilizer	156 (78.00)	44 (22.00)	166 (83.00)	34 (17.00)
	(*ii*) Banding use of fertilizer	166 (83.00)	34 (17.00)	158 (79.00)	52 (26.00)
	(*b*) *Integrated Nutrient Management*				
	(*i*) Fertilizer	124 (62.00)	76 (38.00)	128 (64.00)	72 (36.00)
	(*ii*) FYM	200 (100.00)	–	200 (100.00)	–
	(*iii*) Biofertilizers	20 (10.00)	180 (90.00)	22 (11.00)	178 (89.00)
	(*iv*) Crop residue	146 (73.00)	54 (27.00)	164 (82.00)	36 (18.00)
	(*v*) Green manure	160 (80.00)	40 (20.00)	172 (86.00)	28 (14.00)
	(*vi*) Legume cropping	135 (67.50)	65 (32.50)	130 (65.00)	70 (35.00)
4.	**Weed Control**				
	(*a*) Cultural method	200 (100.00)	–	200 (100.00)	–
	(*b*) Mechanical method	200 (100.00)	–	200 (100.00)	–
	(*c*) Crop rotation	200 (100.00)	00 (0.00)	200 (100.00)	
	(*d*) Chemical method	110 (55.00)	90 (45.00)	120 (60.00)	80 (40.00)
5.	**Integrated Pest Management**				
	(*a*) Cultural method	159 (79.50)	41 (20.50)	105 (52.50)	95 (47.50)
	(*b*) Summer ploughing	200 (100.00)	–	200 (100.00)	–
	(*c*) Hand collection and destruction	120 (60.00)	80 (40.00)	120 (60.00)	80 (40.00)
	(*d*) Biological control	200 (100.00)	–	200 (100.00)	–
	(*e*) Use of insecticides/pesticides	180 (90.00)	20 (10.00)	175 (87.50)	25 (12.50)

Contd...

Table 42.1–Contd...

Sl.No.	Items	Social Aspects			
		Socially Acceptable		Socially Sustainable	
		Yes	No	Yes	No
6.	**Maintaining Plant Population**	195 (97.50)	5 (2.50)	175 (87.50)	25 (12.50)
7.	**Drought Resistant Variety**	200 (100.00)	–	200 (100.00)	–
8.	**Moisture and Water Conservation Practices**				
	(*a*) Farm ponds	200 (10.00)	–	200 (100.00)	–
	(*b*) Vegetative barrier	200 (100.00)	–	200 (100.00)	–
	(*c*) Mulching	120 (60.00)	80 (40.00)	104 (52.00)	96 (48.00)
	(*d*) Summer ploughing	200 (100.00)	–	200 (100.00)	–
9.	**Agro Forestry**	200 (100.00)	–	200 (100.00)	–
10.	**Subsidiary Occupation**				
	(*a*) Live stock	200 (100.00)	–	200 (100.00)	–
	(*b*) Dairying	200 (100.00)	–	200 (100.00)	–
	(*c*) Poultry	180 (90.00)	20 (10.00)	178 (89.00)	22 (11.00)
	(*d*) Horticulture	170 (85.00)	30 (15.00)	165 (82.50)	35 (17.50)
	(*e*) Vegetable	90 (45.00)	110 (55.00)	100 (50.00)	100 (50.00)
	(*f*) Bee keeping	100 (50.00)	100 (50.00)	125 (62.50)	75 (37.50)

Figures in parentheses indicate percentage.

Soil Fertility

Soil Fertilization

The data from the Table 42.1 reveals that majority (78.00 per cent) of the farmers perceived broadcasting of fertilizers as socially acceptable while as 83.00 per cent of the farmers perceived it as socially sustainable, respectively. The banding use of fertilizer placement was perceived by majority (83.00 per cent and 79.00 per cent) of them as socially acceptable and sustainable.

Integrated Nutrient Management

The perusal of data from the Table 42.1 indicates that 62.00 per cent and 64.00 per cent of farmers perceived fertilizer application as socially acceptable and socially sustainable. On the other hand, 38.00 per cent and 36.00 per cent of the farmers not perceived this practice as socially acceptable and sustainable, respectively. As for as biofertilizer under integrated nutrient management is concerned, 10.00 per cent and 11.00 per cent of the farmers perceived this as socially acceptable and sustainable. In case of crop residue, 73.00 per cent, followed by 82.00 per cent of the farmers perceived it as socially acceptable and sustainable, respectively. The use of green manure was perceived by majority (80.50 per cent and 86.00 per cent) as socially acceptable and sustainable, respectively. As for as legume cropping is concerned, it was found that the majority of the farmers (67.50 per cent and 65.00 per cent) perceived this operation as socially acceptable and sustainable, respectively.

Weed Control

The Table 42.1 indicates that an equal 100.00 of them perceived cultural method of weed control as socially acceptable and sustainable, respectively. The same was the trend in case of mechanical method of weed control to the extent of an equal 100.00 per cent of the farmers. In case of crop rotation an equal 100.00 per cent perceived this technique of weed control as socially acceptable and sustainable, respectively. The use of chemicals for weed control perceived by 55.00 per cent and 60.00 per cent of the farmers as socially acceptable and sustainable.

Integrated Pest Management

The perusal of the data from the Table 42.1 reveals that 52.50 per cent of farmers perceived cultural method of pest control as socially acceptable and sustainable, respectively. Summer ploughing was perceived by an equal 100.00 per cent of them as socially acceptable and sustainable as a method of pest control. In case of biological method of pest control an equal 100.00 per cent perceived as socially acceptable and sustainable. The use of insecticides/pesticides to control insect pest it was found that 90.00 per cent of farmers followed by 85.00 per cent of them perceived it as socially acceptable and sustainable, respectively.

Maintaining Plant Population

The Table 42.1 reveals that a majority (97.50 per cent and 87.50 per cent) of the farmers perceived this technique as socially acceptable and sustainable, respectively.

Drought Resistant Variety (DRY)

The Table 42.1 indicates that an equal 100.00 of them perceived introduction to drought resistant variety as socially acceptable and sustainable, respectively. Thus, it was concluded that the dry land farmers had better perception about drought resistant variety because it is essential for dry land areas.

Moisture and Water Conservation Practices

It is concluded from the data in Table 42.1 that an equal 100.00 of them perceived farm ponds as socially acceptable and sustainable, respectively. Regarding mulching, 60.00 per cent, followed by 52.00 per cent of farmers who perceived it as socially acceptable and sustainable, respectively. Summer ploughing was perceived by an equal 100.00 per cent to be socially acceptable and sustainable. Thus, it was concluded that farmers had very good perception of farm ponds, vegetative barrier and summer ploughing as well as mulching regarding moisture and water conservation practices.

Agro-forestry

The data in Table 42.1 indicates that all 100.00 per cent of the farmers perceived agro-forestry as socially acceptable and sustainable. This practice helps to maintain environment and gives extra income other than field crops to the farmers.

Subsidiary Occupation

It reveals from the Table 42.1 that all 100.00 per cent of the dry land farmers perceived livestock and dairying as socially acceptable and sustainable. Regarding poultry as a subsidiary occupation, a majority (90.00 per cent and 89.00 per cent) of the farmers perceived it as socially acceptable and sustainable, respectively. In case of horticulture, 85.00 per cent, followed by 82.50 per cent of the farmers perceived this practice as socially acceptable and sustainable. Regarding vegetable as subsidiary occupation of the farmers 45.00 per cent, 50.00 per cent of them perceived as socially acceptable and sustainable, respectively. As regards to bee keeping, 50.00 per cent and 62.50 per cent of farmers perceived it as socially acceptable and sustainable as a subsidiary occupation. Thus, the conclusion indicates that farmers had better perception for livestock, dairying poultry, horticulture as well as vegetable.

In general, the conclusion regarding farmers' perception about technological components was that they had very good perception about tillage, crop diversification, soil fertility, integrated nutrient management, weed control, integrated pest management, maintaining plant population, moisture conservation measures, agro-forestry, livestock, dairying and poultry. This may be attributed to their better financial position and social status, etc.

References

Singh, Sube (2002). Extension Management for Sustainable Dryland Agriculture in Haryana. *Ph.D. Thesis* (Unpublished), CCSHAU, Hisar.

Vijayaragwan, K., Ramesh Babu, A. and Y.P. Singh (1990). Input management in dryland agriculture: A case study. *Indian J. Ext. Edu.*, 26(3&4): 30–37.

Chapter 43
Green Manure Crop for Nutrient Management

N.K. Bohra

Arid Forest Research Institute, Jodhpur, Rajasthan

ABSTRACT

Green manures both non-grain legumes and grain legumes improve the productivity especially in rice field. Their value in substituting inorganic N and also improving soil fertility are now well established. It is the need of the present era to overcome some of the constraints in its use. There is a scope for land improvement and nutrient management by using green manure.

Keywords: Legumes, Green manure and Fertilizer.

Introduction

Fertilizers and manures are used in agriculture to supplement the nutrients that the plants can obtain from the soil alone. The result is usually an increase in yield. The role of green manure in improving soil fertility and supplying a part of the nutrient requirement of crops is well known.

Green manure refers to fresh plant matter that is added to the soil largely for supplying the nutrients contained in its biomass. Such biomass can both be grown *in situ* and incorporated or grown elsewhere and brought in for incorporation in the field to be manured. Their use in crop production is recorded to have been practised in China as early as 1134 B.C. These are one of the main components of integrated nutrient supply system along with inorganic fertilizers and biofertilizers. Green manures can meet a part of the nutrient needs (Particularly N) of crops for optimum production and to that extent can result in fertilizers saving. In India, an estimated 7 million ha. were green manured during 1990–91 (nearly 4.5 per cent of the net sown area). Andhra Pradesh and Uttar Pradesh accounted for 60 per cent of green-manured area. The six states of A.P., Karnataka, Madhya Pradesh, Orissa, Punjab and U.P. comprised 88 per cent of total area (FAI, 1990).

Plants for Green Manure

An ideal green manure should possess early establishment and high seedling vigour, tolerance to drought, shade, flood and adverse temperature, possess early onset of N fixation and its efficient sustenance. It must have an ability to accumulate large biomass and N in 4–6 weeks, easy to incorporate, quickly decomposable, and tolerant to pests and diseases. (FAG, 1977; IRRI, 1988; Cosico, 1990)

Green manures may be either non-grain legumes or leguminous plants. Non-grain legumes include Crotolaria, *Sesbania, Centrosema, Stylosanthes* and *Desmodium* while leguminous green manure plants include pigeon pea, green gram, cowpea, Soyabean and perrenial woody legumes such as *Leucaena leucocephala* (Subabul), *Giricidia sepium* and *Cassia siamea* etc.

Green manure crops suitable for some field crops were listed by Krishnamurthy (1978) and Sharma *et al.* (1991).

Table 43.1: Green Manure Crops

Field Crop	*Recommended Green Manure*
Rice	Sun hemp (*Crotolaria juncea*)
	Sesbania and Wild indigo (*Indigofera tinctoria*)
Sugar cane	Sun hemp (*Crotolaria juncea*)
Finger millet	Sun hemp (*Crotolaria juncea*)
Wheat	Sun hemp (*Crotolaria juncea*)
Sorghum	Sun hemp (*Crotolaria juncea*)
	Subabul (*Laucaena leucocephala*)
Banana	Leaves of *Gliricidia sepium*
Potato	Lupin (*Lupinus albus*)
	Sun hemp (*Crotolaria juncea*)
	Cowpea (*Vigna sinensis*)
	Guar (*Cyamopsis tetragonoloba*)
	Buckwheat (*Fagopyrum* sp.)

Leguminous Green Manures

These differ widely in nitrogen concentration and yield. Among 86 species of leguminous green manure crops used in India their nitrogen content was found from 2 to 4.9 per cent (Vachhani and Murthy, 1964; Ghai *et al.*, 1985). Their utility in rice field was studied by several workers. Some common green manure crops for rice fields and their potential nitrogen contribution is given in Table 43.2.

Non-grain Legume

Evaluation of some of the promising green manure crops at Coimbatore indicated the potential of Dhaincha and the newly introduced stem nodulating *S. rostrate* (Table 43.3) The exotic stem nodulating *S. rostrate* of Senegalese origin has much promise for low land rice especially with adequate irrigation. Another promising stem nodulating introduction from Medagascar is *Aeschynomene afraspera*, which is capable of withstanding water stress to some extent. Its potential under Indian conditions is under study *Tephrosia purpurea* and *Phaseolus trilobus* grow rather slowly and accumulate much less nitrogen than sesbania and are more adopt to drought (Palaniappan *et al.*, 1990). Among them *Tephrosia purpurea* has the additional advantage of self-sowing and is not browsed by cattle. Wild Indigo is another drought tolerant green manure crop suitable for rainfed rice, and it can contribute 62–182 Kg N/ha. to the succeeding rice crop depending on soil, climatic and cropping conditions (Garrity and Flinn, 1988).

Table 43.2: Potential of Various Green Manure Crops in Rice

Local Name	*Botanical Name*	*Growing Season*	*Output in 45–60 Days Green Matter (t/ha)*	*Nitrogen Contribution (Kg/ha.)*
Sun hemp	*Crotolariajuncea*	Wet	21.2	91
Dhaincha	*Sesbania aculeate*	Wet	20.2	86
Pillipesara	*Phaseolus trilobus*	Wet	18.3	201
Green Gram	*Vigna radiata*	Wet	8.0	42
Cowpea	*Vigna sinensis*	Wet	15.0	74
Guar	*Cyamopsis tetragonoloba*	Wet	20.0	68
Senji	*Melilotus alba*	Dry	28.6	163
Barseem	*Trifolium alexandium*	Dry	15.5	67
Khesari	*Lanthyrus sativus*	Dry	12.3	66

(Abrol and Palaniappan, 1988).

Table 43.3: Evaluation of Green Manures for Rice Based Cropping System at Coimbatore

Green Manure	*Output in 60 Days*			*Special Feature*
	Biomass	*N Accumulation*		
		Kg/ha	*Kg/ha/day*	
Crotolaria juncea (Sun hemp)	16.8	159	2.65	Quick growing, easy seed production
Sesbania aculeate (Dhaincha)	26.3	185	3.08	High biomass accumulation, wide adaptability
Sesbania rostrata	24.9	219	3.65	Stem nodulating tolerant to water logging
Tephrosia purpurea (Kolinji)	16.8	115	1.92	Drought tolerant self seedling
Phaseolus trilobus	17.6	126	2.10	Green manure-cum-fodder-cum-cover crop

(Palanippan *et al.*, 1990).

Grain Legumes

Some annual grain legume crops are also used as green manure after all or part of the grain is harvested. Green gram Stover incorporated in the soil after pod harvest contributed about 60 Kg. N/ha to the succeeding rice crop (Rakhi and Meelu, 1983) The green gram yielding about 1 ton grain and 2.5 ton dry matter/ha equivalent to 50 Kg.N/ha (Meelu *et al.*, 1986). It is concluded that green gram, black gram (Phaseolus mungol and cowpea could provide about 50-60 kg N/ha for the succeeding rice crop. (Kulkarni and Pandey, 1988). Green manuring is a very old established practice, known to Romans in the early centuries A.D. using leguminous crops such as vetches and lupins. The essential characteristics of green manure crops are rapid growth, vigorous root development and abundant tops. Cheap, reliable and rapidly germinating seed is essential and the crop must not present difficult husbandry problems. The main functions of green manures include addition to the soil of readily mineralizable plant nutrients particularly nitrogen, supply of organic matter to the soil, provision of soil cover in areas where erosion is a problem, prevention of leaching of nutrients by recycling them, control of weeds by smothering them. Green manures used properly should either increase humus content or increase the immediate supply of available nitrogen and other nutrients, but cannot effectively do both at the same time. If the main aim of growing a green manure crop is to increase the supply of nutrients, with emphasis on nitrogen, a legume should be grown and ploughed in while immature. These, partly because of the N_2 fixing symbiotic bacteria in the root nodules, give material with a very low carbon-nitrogen ratio. After ploughing in, rapid decomposition takes place with release of available nitrogen for the following commercial crop. Much of the organic matter is lost to the atmosphere as carbon dioxide, so that the humus content of the soil is not greatly increased, For this suitable leguminous crops are red clover (*Trifolium pratense*), sweet lovers (*Melilotus* spp.), Common vetch (*Vicia sativa*), trefoil (*Medicago lupulina*) and yellow lupin (*Lupinus luteus*).

If the main aim of growing the green manure crop is to maintain or increase soil organic matter, it is preferable to grow a non-leguminous crop and allow it to grow further, towards maturity, though still green, before ploughing in. This material will have a higher carbon-nitrogen ratio and will contribute more persistent humus to the soil. Suitable species are mustard (*Sinapis alba*), rye (*secale cereale*), rape (*Brassica napus*) turnip rapes (*Brassica rapa* and crosses) and ryegrass (*Lolium* spp.), Sowing leguminous green manure crops by, for example, direct drilling into stubble after early-harvested cereals followed by autumn ploughing–in, which can supply 40-60 kg/ha of available nitrogen for the following crop, Benefits of green manuring are generally interpreted in terms of their capacity to provide N or substitute fertilizer N. The role of green manures in enhancing the availability of other macro and micronutrients has been under study. Several workers reported increase in N and P contents in agricultural crops by using green manures, Enhanced availability of iron upon green manuring has been reported (Takkar and Nayyar, 1986).

Economics of Green manuring

Green manuring increased rice yield by 78 per cent in low fertile compared to 22 per cent in high fertile soils (Gu and Wen, 1981). In a rice–rice-mung bean system, inclusion of black gram or cow pea in the pre-rice summer season increased the productivity level of the system over the conventional rice–rice-mung bean system (Siddeswaram, 1992). Besides these the residual effect of green manuring in cropping system improve nutrient status in the soil. However it depends on soil types, environment condition, quality and quantity of green manure application etc. (Meelu and Morris, 1984). It is now well established fact that incorporatation of grain legume haulms as organic source gave net returns comparable to that with green manuring. Combined application of green manure and recommended fertilizer N to rice produced high net returns (Srinivasaulu Reddy, 1988; Sivabal, 1989; Padmavathy, 1992; Siddeswaram, 1992).

Wider adoption of green manure technology at the farm level and its practical and economic feasibility are critical issues. Tillage and seeding operations are major components of the labour and cash outlay for farmers who grow green manure crops (Garrity and Flinn, 1988). The financial analysis seeks to compare costs and benefits, between the combined use of green manure and inorganic N and inorganic N alone. Overall an added cost to produce a green manure crop is estimated to be about

Rs. 700/ha to 1000/ha that is equivalent to nearly 100 kg. urea and without any harmful effects as a plus point.

Constraints

Besides many benefits green manure crops have several constraints hampering its wide spread adoption. They are:

1. In intensively cropped areas, farmers do not wish to set apart 6-8 weeks exclusively for growing a green manure crop with no direct cash benefit.
2. When rice is grown after wheat, farmers find it difficult to do the farm operations in the intense heat of May and June.
3. The most costly item in green manuring is the seed. Inadequate availability of quality seeds of desired species at reduced cost is one of the major problems in the adoption of green manuring practice.
4. Benefits of green manuring are not perceptible to the farms because these are not directly visible as for fertiliser N.
5. The inter-cropped green manures may compete for resources with the main crop. The present day investigation should be focused to develop viable solution to overcome these constraints.

Further Research Needs

The cost of fertilizers and their harmful effects on environment, growing awareness of ecological impact of agrochemicals, integrated nutrient management with organic, inorganic and biofertilizer sources have become imperative for sustainable crop production. To make green manure technology more viable and feasible following efforts are required in future:

1. Collection, conservation, cataloguing and evaluation of green manure species that show promise for different ecosystems and their screening for N accumulation patterns under controlled and field situations.
2. Identification of the optimum combination of fast and slow N-release green manure species. A combination of two types of legumes may provide the best agro ecosystem management.
3. Research to improve the establishment and stand of green manures particularly on the aspects of seed treatments for early and uniform germination.
4. Contribution of nutrients other than N in different soils and environments with long-term economic advantages including sustainability facts.
5. Multipurpose trees and short season legumes with pests and disease tolerance that can be grown on bunds or wastelands. Legumes that provide both forage and green manure or fuel would be more appropriate. Research on legume species with efficient rhizobium strains for symbiotic N fixation even in adverse soil environments.
6. Design of cropping systems including green manure crops without displacement of any economic or crop in the system.

Conclusion

Although green manuring is a 3000 years old practice it is still relevant in today's context and must be made use of wherever feasible and economical. For the proper rotting of the green manure, it is necessary that the green material should be succulent and there should be adequate moisture to the soil. Plants at the flowering stage contain the greatest bulk of succulent organic matter with a low carbon/nitrogen ratio. The incorporation of the green manure crop into the soil could enforce fertility of the soil by 30 to 50 per cent.

References

Abrol, I.P. and S.P. Palaniappan (1988). Green manure crops in irrigated and rainfed lowland rice based cropping system in south Asia. *Proc. Symp. on Sustainable Agriculture: The Role of Green Manure Crop in Rice Farming Systems*, IRRI, Philippines, p. 71–82.

Cosico, W.C. (1990). Studies on green manuring in the Philippines. *Extension Bulletin* FFTC, Taiwan, 314: 14.

FAI (1990). *Fertilizer Statistics.*

FAO (1997). Recycling of organic waste in agriculture, Report on FAO/UNDP tour to Republic of China. *Soils Bulletin*, FAO, Rome

Garity, D.P. and J.C. Flinn (1988). Farm level management system for green manure crops in Asian rice environments *Proc. Symp. on Sustainable Agriculture: The Role of Green Manure Crop in Rice Farming Systems*, IRRI, Philippines, p. 111–129.

Ghai, S.K. *et al.* (1985). Comparative study of the potential of *Sesbanias* for green manuring. *Tropical Agric.*, (Trinidad), 62: 52–56.

Gu, R. and Q. Wen (1981). Cultivation and application of green manure in paddy fields of China. *Proc. of Symp. on Paddy Soil*, Beijing, China, Institute of Soil Science, Academia, Sinica and Science Press, p. 207–219.

IRRI (International Rice Research Institute) (1988). Symposium recommendations: Green manure in rice farming. *Proceeding of a Symposium on Sustainable Agriculture*, 25–29th May 1987, Los Banos, Philippines, IRRI, p. 1–9.

Krishnamurthy, K.K. (1989). Plantation crops: Research and development in the 21st century. *J. Plantation Crops* (supplement), p. 1–6.

Krishnamurthy, R. (1978). *A Manual on Compost and other Organic Manures.* Today and Tomorrow publishers, New Delhi, p. 85–119.

Kulkarni, K.R. and R.K. Pandey (1988). Annual legumes for food and as green manure in a rice based cropping system. *Proc. Symp. on Sustainable Agriculture: The Role of Green Manure Crop in Rice Farming Systems*, IRRI, Philippines, p. 289–299.

Meelu, O.P. (1986). Green manuring in low land rice. Paper at planning meeting workshop of international network on soil fertility and evaluation research. Hangzhon, China.

Meelu, O.P. and R.A. Morris (1984). Integrated management of plant nutrients rice and rice based cropping system. *Fertilizer News*, 29(2): 65–75.

Padmavathy, V.K. (1992). Effect of green manure and green gram on the economy of nitrogen management in rice based cropping system. *M.Sc. (Ag.) Thesis*, Tamil Nadu Agricultural University, Coimbatore, India, pp. 133.

Palaniappan, S.P. (1990). Green manure evaluation for rice farming system. *Proc. Int. Symposium on Notational Resources Management for Sustainable Agriculture.*

Rekhi, R.S. and O.P. Meelu (1983). Effect of complementary use of mung straw and inorganic fertilizer N on the nitrogen availability and yield of rice. *Oryza*, 20: 125–129.

Sharma, R.C. *et al.* (1991). Nitrogen management in potato. *Tech. Bull.*, CPRI, Shimla, 32: 74.

Siddeswaran, K. (1992). Integrated nitrogen management with green manure and grain legumes in rice based Cropping systems. *Ph.D. Thesis,* Tamilnadu Agricultural University, Coimbatore, India, pp. 204.

Sivabal, M.K. (1989). Integrated nitrogen management practices with green manures for low land rice. *M.Sc. (Ag.) Ph.D. Thesis,* Tamilnadu Agricultural University Coimbatore, India.

Srinivas Reddy, P. (1988). Integrated nitrogen management in rice-based cropping system. *Ph.D. Thesis,* Tamilnadu Agricultural University, Coimbatore, India, pp. 327.

Takkar, P.W. and V.K. Nayyar (1986). Integrated approach to combat micronutrient deficiency. *Proc. FAI seminar on Growth and Modernization of Fertilizer Industry*, New Delhi, PS III/22/1–16.

Vachhani, M.V. and K.S. Murthy (1964). *Green Manuring for Rice*. ICAR Research report serial No. 17, New Delhi, India, Indian Council of Agricultural Research, pp. 50.

Chapter 44

Yield and Quality of Soybean as Influenced by the Application of Sulphur as Elemental Sulphur

S. Aruna Geetha, P.S. Senthil Kumar, S. Maragatham and M. Govindaswamy

Department of Soil Science and Agricultural Chemistry,
Tamil Nadu Agricultural University, Coimbatore – 641 003

ABSTRACT

To study the influence of application of elemental sulphur on yield and quality of soybean field experiments were conducted. The results of the study revealed that the elemental sulphur @ 20kg S ha^{-1} increased the seed yield, protein and oil content of soybean. Application of elemental sulphur beyond 20 kg S ha^{-1} did not produce appreciable increase over control. Elemental sulphur @ 20 kg S ha^{-1} applied along with farm yard manure @ 12.5 t ha^{-1} + Thiobacillus pellets increased the yield and quality of soybean than when applied alone. The agronomic efficiency and apparent sulphur recovery were also high when elemental sulphur was applied @ 20 kg S ha^{-1} in soybean.

Keywords: Soybean, Sulphur, Seed yield, Protein content oil content, Agronomic efficiency, Apparent S recovery.

Introduction

Sulphur deficiency is common under intensive cropping systems due to heavy use of high analysis sulphur free fertilizers for getting the maximum potential of crop plants. The deficiency of sulphur is more pronounced in pulses than in cereals due to comparatively high sulphur need of pulses for producing grain. Bansal (1991) reported that seed yield of soybean increased with increasing S rate from 10 to 80 kg S ha^{-1}. Bansal and Motiramani (1993) also reported increased plant dry matter yield with increasing S rate. Teotia *et al.* (2000) observed significant response to sulphur doses with increase in grain and straw yield over no sulphur application.

Apart from its effect on yield of pulses, sulphur plays an important role in improving the quality and marketability of crop produce. Sulphur is a part of aminoacids cysterna and methionine, hence, essential for protein production. The effect on sulphur on yield, oil and protein content was significant in two sites with soybean as reported by Nagar *et al.* (1993). Seed yield, protein and oil content significantly increased with S rates up to 50 kg ha^{-1} (Dwivedi and Bapat, 1998).

The major source of sulphur in the fertilizers available in India is the elemental sulphur which is imported and supplied to fertilizers manufactures as a raw material. Elemental sulphur used in this study is a by-product from oil refineries whose effect was studied with the combination of organic amendment and biofertilizer in soybean.

Materials and Methods

Field experiments were conducted at Tamil Nadu Agricultural University, Coimbatore in two locations with soybean (Cv. CO_2). The experimental soils come under Typic and Vertic Ustropept with low soil sulphur status. The initial soil characteristics are given in Table 44.1. The experiment was conducted with split plot design. The main plot contained absolute control, elemental sulphur which is a by-product from oil refineries with Thiobacillus pellets. Naturally oxidized elemental sulphur (applied one week before sowing) and combination on farm yard manure @ 12.5 t ha^{-1} with elemental sulphur and Thiobacillus pellets. The pellets were prepared with Thiobacillus culture obtained from department of agricultural microbiology, Tamil Nadu Agricultural University, with fine clay. The sub plot contained five levels of sulphur *viz.*, 0, 20, 40, 60 and 80 kg S ha^{-1}. The treatments were replicated thrice. The grain yield, oil and protein content, agronomic efficiency, Harvest index and apparent. S recovery were recorded and calculated as follows:

$$\text{Agronomic Efficiency (kg ha}^{-1}\text{) (or) Response Ratio} = \frac{\text{Yield in fertilized plot (kg ha}^{-1}\text{)} - \text{Yield in control plot (kg ha}^{-1}\text{)}}{\text{Level of Fertilizer sulphur applied (kg ha}^{-1}\text{)}}$$

$$\text{Apparent S Recovery (\%)} = \frac{\text{Uptake in fertilized plot (kg ha}^{-1}\text{)} - \text{Uptake in control plot (kg ha}^{-1}\text{)}}{\text{Level of Fertilizer S applied (kg ha}^{-1}\text{)}} \times 100$$

$$\text{Harvest Index of Sulphur} = \frac{\text{S uptake in seed at harvest (kg ha}^{-1}\text{)}}{\text{S uptake in seed and dry matter above ground surface (kg ha}^{-1}\text{)}}$$

Table 44.1: Chemical Characteristics of Experimental Soils

Characteristics	*Field Experiment I*	*Field Experiment II*
pH	8.6	8.5
EC (d sm^{-1})	0.43	0.08
CEC (c mol p(+) kg^{-1})	24.0	25.1
Organic Carbon (%)	0.330	0.333
Available N (kg ha^{-1})	174	185
Available P (kg ha^{-1})	16	17
Available K (kg ha^{-1})	405	410
Available S (ppm)	7.5	6.5

Results and Discussion

Seed Yield

The seed yield ranged from 1611 to 2061 and from 1506 to 1856 kg ha (kg ha^{-1}) for the different amendments and from 1445 to 2142 and 1412 to 1792 kg ha (kg ha^{-1}) for the levels of S in the I and II crops respectively in the field experiments (Table 44.2). From the range of yield obtained for the different amendments and levels of S, it can be stated that the average yield of soybean is appreciably higher as variations in treatments is highly significant. Application of FYM along with Thiobacillus recorded 27.93 and 23.24 per cent increased yield over no amendments in the I and II experiments respectively, indicating a superiority among the amendments. Application of Thiobacillus alone with sulphur could produce only marginal increase in yield compared to the combination with FYM. The increased dry matter production with this combination might have contributed for higher seed yield. Soil fertility management involving organic manures and biofertilizers has a direct effect on available S status of soil as well as productivity of crops (Gupta and Dubey, 1998). The better performance of soybean under combined application of FYM and Thiobacillus may be reasoned to the fact that FYM could have been the base and carbon substrate for Thiobacillus for the growth, multiplication and spread. Due to this, quick and unfailing colonization might have been occurred leading to effective solubilization of S. Cengal and Okur (1999) reported the same under Thiobacillus inoculation.

Table 44.2: Effect of Application of Sulphur and Amendments on Seed Yield in Soybean

Treatments	*Seed Yield (kg ha^{-1})–Field Experiment I*						*Seed Yield (kg ha^{-1})–Field Experiment II*					
	S1	*S2*	*S3*	*S4*	*S5*	*Mean*	*S1*	*S2*	*S3*	*S4*	*S5*	*Mean*
M1	11.13	19.83	17.10	16.33	16.16	16.11	13.16	17.83	15.16	14.66	14.50	15.06
M2	15.33	20.33	18.46	16.33	15.50	17.19	15.16	17.33	15.33	15.83	16.33	16.00
M3	15.83	21.83	21.33	18.33	17.16	18.90	13.66	14.16	17.16	16.33	17.33	15.73
M4	15.50	23.66	23.50	21.56	18.83	20.61	14.50	22.33	20.63	18.50	16.83	18.56
Mean	14.45	21.48	20.10	18.14	16.92	18.20	14.12	17.91	17.07	16.33	16.24	16.33
	M	*S*	*M at S*	*S at M*			*M*	*S*	*M at S*	*S at M*		
SE (d)	60.96	36.42	89.22	72.83			25.85	45.81	85.93	91.63		
CD (P = 0.05)	149.16	74.18	198.81	148.36			63.26	93.32	178.24	186.64		

Tr.: Treatments S1: 0 kg S ha^{-1}; S2: 20kg S ha^{-1}; S3: 40kg S ha^{-1}; S4: 60kg S ha^{-1}; S5: 80kg S ha^{-1}; M_1: No amendments; M2: Elemental sulphur + *Thiobacillus* sp.; M3: Naturally oxidized elemental sulphur; M4: Elemental sulphur + FYM (12.5t ha^{-1}) + *Thiobacillus* sp.

The magnitude of response by soybean for the application of S was very high as evident from the significant increase in seed yield for the addition of sulphur. The yield increase was in the order of 48.24 and 26.85 per cent over control in the I and II experiments respectively for the addition of 20 kg S ha^{-1}. Increase in yield with addition of sulphur in soybean has been reported by Bansal (1991), Dubey and Bellore (1995) and Teotial *et al.* (2000).

Increase in the dose of sulphur beyond 20 kg S ha^{-1} did not produce any significant increment in yield. Joshi and Bellore (1998) expressed the same, emphasizing the superiority of 20 kg S ha^{-1}. They also observed a negative response beyond 40 kg S ha^{-1} as obtained in the present study. Ravankar *et al.* (1999) also suggested the application of 20 kg S ha^{-1} along with the recommended N, P, K for higher seed yield of soybean.

Protein and Oil Content

In the evaluation of the quality of pulses, protein content occupies a prime position since many of other parameters are influenced by this. Increase in protein content is of paramount importance as it is considered as the building block of the living system. In this study significant differences were noticed due to different amendments and levels of S. The results showed that there was better assimilation of N in the presence of FYM with 20 kg S ha^{-1} (Table 44.3). Although amendments and S levels exhibited a smaller magnitude of influence on the protein synthesis, their combination exhibited a marked effect. Increased absorption of nitrogen as a result of the combined influence of these factors might have contributed for increased incorporation of N into aminoacids leading to improved protein synthesis as reported by Chitralekha *et al.* (1992), Ganeshamoorthy (1996) and Dwivedi and Bapat (1998).

Table 44.3: Application of Sulphur and Amendments on Protein Content (per cent) of Soybean

Treatments	*Seed Yield (kg ha^{-1})–Field Experiment I*						*Seed Yield (kg ha^{-1})–Field Experiment II*					
	S1	*S2*	*S3*	*S4*	*S5*	*Mean*	*S1*	*S2*	*S3*	*S4*	*S5*	*Mean*
M1	39.43	40.57	40.53	40.47	39.97	40.19	38.48	39.27	40.17	36.73	38.61	38.65
M2	39.23	41.30	40.87	40.40	40.00	40.37	38.33	40.63	39.65	39.32	39.21	39.43
M3	38.77	40.17	39.87	39.07	39.40	39.45	38.39	40.59	40.21	39.98	39.88	39.81
M4	38.83	41.33	40.97	39.07	39.00	39.84	38.52	42.73	42.15	41.75	38.89	40.81
Mean	39.07	40.85	40.56	39.75	39.59	39.96	38.43	40.80	40.54	39.44	39.15	39.67
	M	*S*	*M at S*	*S at M*			*M*	*S*	*M at S*	*S at M*		
SE (d)	0.121	0.115	0.239	0.231			0.279	0.339	0.669	0.679		
CD (P = 0.05)	0.297	0.235	0.513	0.469			0685	0.692	0.411	1.385		

Tr.: Treatments S1: 0 kg S ha^{-1}; S2: 20kg S ha^{-1}; S3: 40kg S ha^{-1}; S4: 60kg S ha^{-1}; S5: 80kg S ha^{-1}; M_1: No amendments; M2: Elemental sulphur + *Thiobacillus* sp.; M3: Naturally oxidized elemental sulphur; M4: Elemental sulphur + FYM (12.5t ha^{-1}) + *Thiobacillus* sp.

Table 44.4: Application of Sulphur and Amendments on Oil Content (%) of Soybean

Treatments	*Oil Content (%)–Field Experiment I*						*Oil Content (%)–Field Experiment II*					
	S1	*S2*	*S3*	*S4*	*S5*	*Mean*	*S1*	*S2*	*S3*	*S4*	*S5*	*Mean*
M1	19.3	20.3	19.5	19.6	19.5	19.6	18.95	19.65	19.51	19.46	19.39	19.39
M2	20.6	20.8	19.7	20.4	20.5	20.4	19.37	20.42	20.35	20.25	20.01	20.08
M3	18.6	20.6	20.4	18.8	19.7	19.6	19.06	20.53	20.29	20.26	20.03	20.05
M4	19.4	20.4	18.9	20.4	20.7	19.8	20.06	20.64	20.52	20.51	20.31	20.41
Mean	19.5	20.5	19.6	19.8	20.1	19.9	19.36	20.31	20.17	20.12	19.96	19.98
	M	*S*	*M at S*	*S at M*			*M*	*S*	*M at S*	*S at M*		
SE (d)	0.083	0.088	0.177	0.175			0.019	0.015	0.032	0.029		
CD (P = 0.05)	0.204	0.178	0.378	0.356			0.046	0.029	0.069	0.059		

Tr.: Treatments S1: 0 kg S ha^{-1}; S2: 20kg S ha^{-1}; S3: 40kg S ha^{-1}; S4: 60kg S ha^{-1}; S5: 80kg S ha^{-1}; M_1: No amendments; M2: Elemental sulphur + *Thiobacillus* sp.; M3: Naturally oxidized elemental sulphur; M4: Elemental sulphur + FYM (12.5t ha^{-1}) + *Thiobacillus* sp.

Soybean occupies an unique position based on its utility both as pulse and oil seed. Results of past work have established the importance of sulphur for the synthesis of oil and protein (Nagar *et al.*, 1993, Khajanchilal *et al.* (1996) and Dwivedi and Bapat, 1998). Both the content and yield of oil were favourably increased by amendments and S levels (Table 44.4). Application of FYM + Thiobacillus and S at 20 kg ha^{-1} either individually or in combination highly favoured the oil content and yield in both the crops.

Table 44.5: Response Ratio and Apparent S Recovery or S Applied in Soybean

Levels of S Applied (kg ha^{-1})	*Response Ratio (kg kg^{-1})*		*Apparent S Recovery (%)*	
	Field Experiment I	*Field Experiment II*	*Field Experiment I*	*Field Experiment II*
20	35.14	18.95	19.70	27.40
40	14.13	7.38	7.45	10.50
60	6.15	3.68	3.68	5.32
80	3.08	2.65	1.83	2.54
Mean	14.63	8.17	15.88	11.44

Efficiency Parameters

The response ratio which is an indicator of the seed produced per kg of applied S showed wide variation and the highest being associated with 20 kg S ha^{-1} (Table 44.5).

Table 44.6: Harvest Index of Applied Sulphur in Soybean

Treatments	*Field Experiment I*	*Field Experiment II*
M1S1	0.44	0.44
M1S2	0.47	0.46
M1S3	0.46	0.45
M1S4	0.50	0.50
M1S5	0.49	0.48
M2S1	0.49	0.45
M2S2	0.48	0.43
M2S3	0.46	0.42
M2S4	0.45	0.45
M2S5	0.49	0.47
M3S1	0.44	0.42
M3S2	0.46	0.38
M3S3	0.51	0.42
M3S4	0.49	0.44
M3S5	0.48	0.49
M4S1	0.50	0.43
M4S2	0.54	0.48
M4S3	0.65	0.50
M4S4	0.63	0.48
M4S5	0.62	0.46

S1: 0 kg S ha^{-1}; S2: 20kg S ha^{-1}; S3: 40kg S ha^{-1}; S4: 60kg S ha^{-1}; S5: 80kg S ha^{-1}; M1: Absolute Control; M2: Elemental sulphur + *Thiobacillus* sp.; M3: Naturally oxidized elemental sulphur; M4: Elemental sulphur + FYM (12.5t ha^{-1}) + *Thiobacillus* sp.

This indicated that the response is high at low level compared to higher levels. Therefore, a careful management is needed to derive maximum benefit from sulphur addition. The ideal level may be fixed in conjuction with higher response ratio.

The apparent S recovery computed for different S levels exhibited considerable variation from a low value at 80 kg S ha^{-1} to a high recovery at 20 kg S ha^{-1} in both experiments (Table 44.5). Decrease in the recovery of nutrients with the increase in levels of addition is a common phenomenon and the recovery of S may not be an excemption to it.

The quality of sulphur transported to the economic part, the seed, out of the total S absorbed by the crop is computed as harvest index which decides the nutritional value of protein supply. The data obtained broughtout a favourable effect for the addition of FYM + Thiobacillus in increasing the harvest index from 0.49 to 0.62 in I experiment and from 0.38 to 0.50 in the II experiment (Table 44.6). Addition of sulphur also improved the harvest index when integrated with FYM + Thiobacillus. Thus indicates that S is best utilized in the presence of organic manure and biofertilizer.

Acknowledgements

The supply of elemental sulphur and the financial support rendered by M/s. Kochi refineries limited, Cochin during the study is highly acknowledged by the authors.

References

Bansal, K.N. (1991). Effect of levels of sulphur on the yield and composition of soybean [*Glycine Max* (L.) Merr.] Greengram (*Vigna radiata L.*), Blackgram (*Vigna mungo*) and cowpea (*Vigna sinensis* L.). *Madras Agric. J.*, 78(5–8): 188–190.

Bansal, K.N. and D.P. Motiramani (1993). Uptake of native and applied sulphur by soybean in vertisols of Madhya Pradesh. *J. Nuclear Agri. Biol.*, 22(1): 42–46.

Cengel, M. and N. Okur (1999). Effect of S bacteria obtained from thermal springs in Aegean Region on oxidation of elemental sulphur, pH and H_2SO_4 formation as influenced by some trace elements. *Ege Universitesi Ziraal-Fakultesi, Dergisi*, 36(1/2/3): 73–80.

Chitralekha Chatterjee, Seiubra Saukla and Neena Khurana (1992). Effect of sulphur deficiency on metabolism of greengram (*Phaseolus hadiatus*). *Indian J. Agric. Soc.*, 62(7): 445–449.

Dubey, S.K. and S.D. Bellore (1995). Effect of level and source of sulphur on symbiotic and biometrical parameters of soybean (*Glycine max*). *Indian J. Agric. Soc.*, 65(2): 140–144.

Dwivedi, A.K. and P.N. Bapat (1998). Sulphur-phosphorus interactions on the synthesis of nitrogenous fractions and oil in soybean. *J. Indian Soc. Soil Sci.*, 46(2): 254–257.

Geneshamoorthy, A.N. (1996). Critical plant sulphur content and effect of S application on grain and oil yield of rainfed soybean in Vertic Ustropept. *J. Indian Soc. Soil Sci.*, 44(2): 290–294.

Gupta, R.K. and S.K. Dubey (1998). Sulphur management in rainfed cropping system in Madhya Pradesh. *Fert. News*, 42(7): 57–60 & 63–67.

Joshi, O.P. and S.D. Bellore (1998). Economic optima of sulphur fertilizer for soybean (*Glycine max*). *Indian J. Agri. Sci.*, 68(5): 244–246.

Khajanchilal, D., L. Deb, M. and S. Sachdev (1996). Phosphorus-sulphur inter-relationship in soybean using P32 and S35. *J. Nuclear Agric. Biol.*, 25(4): 196–204.

Nagar, R.P., G.C. Mali and P. Lal (1993). Effect of phosphorus and sulphur on yield and chemical composition of soybean in Vertisols. *J. Indian Soc. Soil Sci.*, 41(2): 385–386.

Ravankar, H.N., R.S. Deore and P.W. Deshmukh (1999). Effect of nutrient management through organic and inorganic sources of soybean yield and soil fertility. *PKV Res. J.*, 23(1): 47–50.

Teotia, U.S., D. Ghosh and P.C. Srivastava (2000). Influence of sulphur on yield and S uptake in soybean-wheat cropping system in Mollisols of Namital Javai. *Feril. News*, 45(8): 65–68.

Chapter 45
Delayed Effect of Neem Extracts on the Fitness Parameters of *Aedes aegypti*

R.S. Mohanraj and B. Dhanakkodi

Department of Zoology, Kongunadu Arts and Science College, Coimbatore – 29, Tamil Nadu, India

ABSTRACT

Crude neem seed kernel water extract and neem oil were found to reduce fecundity, fertility and longevity of adult *Aedes aegypti*, resulting from larvae reared through sublethal concentrations of the neem products.

Keywords: Aedes aegypti, Neem products, Fecundity, Fertility, Longevity.

Introduction

Aedes aegypti is the only known potential vector of dengue and urban yellow fever (Mekuria, 1991; Mazzarri and Georghiou, 1995). This species of mosquito was shown to be a competent laboratory vector of Chikungunya (CHIK) virus (Jupp and McIntosh, 1990). *Aedes aegypti* has also been noted to transmit filariasis and encephalitis (Dorland's illustrated Medical Dictionary, 1982).

Reports on changes in fitness parameters such as fecundity, fertility and longevity of adult mosquitoes survived through larval treatment with chemical and microbial insecticides, chemosterilants, insect growth regulators are available (Judson, 1967; Saleh *et al.*, 1987; Sahgal and Pillai, 1993). Neem products are relatively advantageous over synthetic insecticides in being eco-friendly and non-toxic to non-target organisms including humans. Purohit *et al.* (1989) demonstrated that plant products are easily biodegradable. It was also emphasized that they do not leave poisonous residues and insects do not develop resistance. However except the findings of Su and Mulla (1999) in

Culex tarsalis and *Culex quinquefasciatus* which indicated the delayed effects of larval treatment with neem formulation on reproductive capacity of resultant adult, studies of this kind are scanty. The delayed effects of crude neem seed kemel water extract and neem oil on the fitness parameters of *Aedes aegypti* are presented in this article.

Materials and Methods

First instar larvae of *Aedes aegypti,* obtained from a stock of mosquitoes maintained since 1995, were exposed to any of the sub-lethal concentrations of test compounds *viz.*, crude neem seed kernel water extract (50, 75 and 100 ppm), and neem oil (50, 75 and 100 ppm). They were maintained with fish food till pupation. From those surviving adults emerging within the same 24 hours period, 25 pairs were chosen randomly and kept in cages in respect to each concentration of a particular test compound.

The adult mosquitoes were provided with glucose solution (10 per cent) in wicks, and the females were blood fed on human arm on day 1 (2 days after emergence) and alternate days. An ovitrap was placed at a corner of every cage. Each morning, the number of eggs laid during the previous day and live females as well as males was counted. Subsequently, the eggs were kept in unchlorinated water for hatching and the number of hatched and unhatched eggs were recorded at 96 hours. Three replicates for each concentration were conducted.

The daily data of egg yield and the number of live females for each concentration and each replicate were appropriately pooled. The total reproductive period, based on the average reproductive period of control group, was divided into three time periods: P1, days 1 (the day on which first batch of eggs obtained) through 10; P2, days 11 through 20; P3, days 21 through 28 in order to conveniently carry out statistical analysis as suggested by Ferrari and Georghiou (1981). The average eggs per female per day (fecundity) produced during each time period and average fertility (percent hatch of eggs) were calculated according to Ferrari and Georghiou (1981), only when the reproductive period of mosquitoes developed in any concentration of a neem product extended for the whole period of a time period. For example, the reproductive period of mosquitoes survived through 100 ppm of crude extract, and neem oil covered 10 days but did not extend upto day 20, therefore average fecundity and average fertility were calculated only for the first time period (P1) and the remaining data were not taken into account. Fecundity and fertility were expressed in terms of mean eggs per female per day and mean per cent hatch, respectively according to Ferrari and Georghiou (1981).

In order to know the survivability of adults emerged from treated larvae, mortality of adult female and male mosquitoes was recorded daily till the last one died for each concentration and each replicate of every neem product tested. Similar counting was also made for control. Average longevity was expressed in terms of L_{50} (longevity 50, days when 50 per cent of adult mosquitoes survived through larval treatment were dead) as suggested by Dorn (1986).

Results and Discussion

Control female mosquitoes laid a total number of 3008 eggs during the I reproductive period. The average number of eggs per female per day was 4.43. The mosquitoes survived through larvae treated with different concentrations of neem derivatives laid remarkably less number of eggs. During the time period P1, the total number of eggs laid by mosquitoes emerged out of 100 ppm media of crude extract and neem oil was drastically reduced to 231 and 175, respectively. Mean number of eggs per female per day for mosquitoes developed in 100 ppm of neem crude extract, and neem oil was 0.489 and 0.082, respectively which were significantly ($P < 0.001$) different from that of control (Table 45.1).

Table 45.1: Changes in Fecundity and Fertility of *Aedes aegypti* Survived through Larvae Treated with Different Sublethal Concentrations of Neem Products and Control

Neem Products	*Concentrations (ppm)*	*Fecundity*			*Fertility*		
		P1	*P2*	*P3*	*P1*	*P2*	*P3*
Control		4.43±1.914	5.52±0.473	5.22±0.592	79.55±0.717	78.68±0.146	77.77±1.158
CNSKWE	50	1.051*±0.206	–	–	49.54*±2.643	–	–
	75	0.618*±0.066	–	–	35.23*±0.25	–	–
	100	0.489*±1.150	–	–	5.5*±1.167	–	–
NO	50	1.53*±0.175	1.108± 0.298	–	49.98*±1.331	45.79±3.569	–
	75	1.005*±0.314	–	–	30.25*±1.364	–	–
	100	0.082 *±0.05	–	–	3.47*±0.043	–	–

Values are mean (± SE) of 3 replicates; P1, P2 and P3 are time periods 10, 10 and 9 days, respectively; Values of treated significantly different from control *: P < 0.001; CNSKWE: Crude neem seed kernel water extract; NO: Neem oil.

Out of the total number of eggs laid by control female *Aedes aegypti*, 77.77 to 79.56 per cent successfully hatched at different time periods. However, the per cent hatching (fertility) of eggs obtained from mosquitoes survived through neem. treated larvae was greatly reduced in comparison with control. The delayed effect of even the lowest concentration (50 ppm) of crude extract and neem oil on the fertility of surviving mosquitoes was conspicuously exhibited from the significantly (P < 0.001) low hatchability (49.54 to 49.98) of eggs laid during the time period P1. Remarkably, still less hatching (3.47 to 5.5 per cent) occurred in eggs obtained from mosquitoes developed in 100 ppm of neem extracts and neem oil.

Mosquitoes emerged out of control medium survived for a total period of 35.22 days (female) and 15.2 days (male) with L50 (longevity 50) values of 24.8 days for females and 10.3 days for males, whereas those developed in neem products, crude extract and neem oil lived relatively short duration (Table 45.2). Post-emergence L_{50} of *Aedes aegypti* developed in 100 ppm of crude extract and neem oil, was 9.0 and 4.4 days as well as 10 and 4.2 days for female and male, respectively.

Table 452: Changes in Longevity and Reproductive Period (RP) of Adult *Aedes aegypti* Developed in Sublethal Concentrations of Neem Products and Control

Neem Products	*Concentration (ppm)*	*Total Period of Survival (Days)*		*L_{50} (Days)*		*RP*
		Male	*Female*	*Male*	*Female*	*Female*
Control		15.2 ± 0.133	35.22 ± 0.133	10.3± 0.353	24.8 ± 0.072	28.2 ± 0.304
CNSKWE	50	11.0* ± 0.577	24.3* ± 0.333	5.6 ± 0.663	15.03 ± 0.554	17.3 ± 0.33
	75	10.0* ± 0.577	21.16* ± 0.333	5.2 ± 0.491	13.23 ± 0.673	14.66 ± 0.333
	100	6.0* ± 0.577	18.5* ± 0.577	4.4 ± 0.207	9.0 ± 0.463	11.0 ± 0.577
NO	50	10.0** ± 0.577	27.16** ± 0.333	6.9 ± 0.121	17.83 ± 0.440	20.66 0.333
	75	9.0** ± 0.577	23.0** ± 0.577	6.8 ± 0.435	14.2 ± 0.901	17.0 ± 0.577
	100	6.0** ± 0.577	20.5** ± 0.577	4.2 ± 0.317	10.0± 0.990	14.0 ± 0.796

Values are mean (± SE) of three replicates; Significantly different from control *: p < 0.05; **: p < 0.001; CNSKWE: Crude neem seed kernel water extract; NO: Neem oil.

In consequence, among the treated mosquitoes reproductive period of females decreased in proportion with L_{50}. The reproductive period of mosquitoes survived larval treatment with neem products was ranging between 11 and 20.66 days, for the highest and lowest doses, respectively in contrast to that of control (28.2 days).

Published data of this kind, the delayed effect of neem on reproductive., potential and longevity of surviving adult *Aedes aegypti* are not available for comparison. However, the present observation could be referred to the report of Su and Mulla (1999) that when 3rd or early 4th instar larvae of *Culex tarsalis* and *Culex quinquefasciatus* were treated with sub-lethal dosages of neem insecticide, until pupation, oviposition rate of the resulting adults was reduced. These fitness characters were found to decrease in adult *Aedes aegypti* derived from, larval treatment with sub-lethal concentrations of IGR, methoprene (Sawby *et al.*, 1992) as well as OMS 2017 and diflubenzuron (Fournet *et al.*, 1993). Mosquito, *Culex quinquefasciatus* selected through organophosphate insecticide, temphose laid less number of eggs which contained more poor viable eggs (Ferrari and Georghiou, 1981).

Microbial preparations, *Bacillus thuringiensis* and *B. sphaericus* caused such a reduction in fecundity and fertility of surviving adult mosquito, *Culex pipiens* (Saleh *et al.*, 1987). Similar reports on delayed effect of several forms of neem products on the reproductive potential of insects of different orders are available (Gaaboub and Hayes, 1984b; Parkman and Pienkowski, 1990; Richter *et al.*, 1997; Kaur *et al.*, 2001). Reduced longevity and reproductive period in adult insects formed from neem treated larvae was reported (Dorn, 1986).

It was suggested that reduced fecundity in mosquitoes may be due to ingestion of neem ingredient, azadirachtin which inhibits the previtellogenic development of the ovaries (Su and Mulla, 1999) and yolk incorporation (Sieber and Rembold, 1983).

Delayed effect of neem products on longevity of surviving *Aedes aegypti* has several advantages. Firstly, reduced longevity simultaneously bring down reproductive period available for both female and male. Secondly, as a consequence it arrests the progeny and subsequent increase in population of vectors. Furthermore, short life-span lessens chances of host-vector contact and thus vector potential of the mosquito is decreased (De Moor and Steffens, 1970).

References

De Moor, P.P. and Steffens, F.E. (1970). A computer-stimulated model of an arthropod-borne virus transmission cycle, with special reference to Chikungunya virus. *Transactions of the Royal Society of Tropical Medicine and Hygiene*, 64: 977–937.

Dorland's Illustrated Medical Dictionary (1982). 26th Edition, W.B. Saunder Company. Igaku-Shoin/ Saunders, London, pp. 35.

Dorn, A. (1986). Effects of azadirachtin on reproduction and egg development of the heteropteran *Oncopeltus fasciatus* Dallas. *J. Appl. Ent.*, 102: 313–319.

Ferrari, J.A. and G.P. Georghiou (1981). Effects of insecticidal selection and treatment on reproductive potential of resistant, susceptible and heterozygous strains of the southern house mosquito. *J. Econ. Entomol.*, 74: 323–327.

Fournet, F., Sannier, C. and N. Monteny (1993). Effects of the insect growth regulators OMS 2017 and diflubenzuron on the reproductive potential of *Aedes aegypti*. *J. Am. Mosq. Control Assoc.*, 9: 426–430.

Gaaboub, I.A. and D.K. Hayes (1984b). Effect of larval treatment with azadirachtin, a molting inhibitory component of the neem tree, on reproductive capacity of the face fly, *Musca autumnalis* De Geer (Diptera: Muscidae). *Environ. Entomol.*, 13: 1639–1643.

Judson, C.L. (1967). Alternation of feeding behaviour and fertility in *Aedes aegypti* by the chemosterilant apholate. *Ent. Exp. Appl.*, 10: 387–394.

Jupp. P.G. and B.M. McIntosh (1990). *Aedes furcifer* and other mosquitoes as vectors of Chikungunya virus at mica. northeastern transvaal, South America. *J. Am. Mosq. Control Assoc.*, 12: 415–419.

Kaur, J.J., Rao, D.K., Sehgal, S.S. and R.K. Seth (2001). Effect of hexane extract of neem seed kernel on development and reproductive behaviour of *Spodoptera litura* Fabr. *Annals of Plant Protection Science*, 9: 171–178.

Mazzarri, M.B. and P. Georghiou (1995). Characterization of resistance to organophosphate carbomate and pyrethroid insecticides in field populations of *Aedes aegypti* from Venezuela. *J. Am. Mosq. Control Assoc.*, 11: 315–322.

Mekuria, Y., Gwinn, T.A., Williams, D.C. and M.A. Tidwell (1991). Insecticide susceptibility of *Aedes aegypti* from Santo Domingo, Dominican Republic. *J. Am. Mosq. Control Assoc.*, 7: 69–72.

Parkman, P. and R.L. Pienkowski (1990). Sub-lethal effects of neem seed extract on adults of *Liriomyza trifolii* (Diptera: Agromyzidae). *J. Econ. Entomol.*, 83: 1246–1249.

Purohit, P., Jyotsna, D. and G. Srimannarayana (1989). Antifeedant activity of indigenous plant extracts against larvae of castor semi-looper. *Pesticides*, 25: 23–26.

Richter, K., Bohm, G.A. and H. Kleeberg (1997). Effect of neem Azal: A natural azadirachtin-containing preparation on *Periplaneta americana* L. (Orthopt., Blattidae). *J. Appl. Ent.*, 121: 59–64.

Sahgal, A. and M.K.K. Pillai (1993). Ovicidal activity of permethrin and deltamethrin on mosquitoes. *Entomon.*, 17: 149–154.

Saleh, M.S., Kelada, N.L. and A.A. Al-Fazairy (1987). Toxicity of two bacterial insecticides against the larvae of mosquito *Culex pipiens* and their action on the reproductive potential and morphometric characteristics of surviving females. *Insect Science and its Application*, 8: 107–110.

Sawby, R., Klowden, M.J. and D. Sjogren (1992). Sub-lethal effects of larval methoprene exposure on adult mosquito longevity. *J. Am. Mosq. Control Assoc.*, 8: 290–292.

Sieber, K.P. and H. Rembold (1983). The effects of azadirachtin on the endocrine control of moulting in *Locusta migratoria*. *J. Insect Physiol.*, 29: 523–527.

Su, T.Y. and M.S. Mulla (1999). Effects of neem products containing azadirachtin on blood feeding, fecundity and survivorship of *Culex tarsalis* and *Culex quinquefasciatus* (Diptera: Culicidae). *J. Vec. Eco.*, 24: 202–215.

Chapter 46
Prevalence of Methicillin Resistance *Staphylococcus aureus* from Clinical Samples in Kanchipuram Town, Tamil Nadu, India

M. Prakash, V. Karthikeyan and N. Karmegam

Department of Microbiology, Kanchi Shri Krishna College of Arts and Science, Kilambi – 631 551, Kanchipuram District, Tamil Nadu, India

ABSTRACT

A total of 60 clinical samples were collected from medical laboratories in Kanchipuram town, Tamil Nadu, South India to find out the prevalence of methicillin resistant *Staphylococcus aureus* (MRSA). Among the different age groups of sample donors categorized, a maximum of 18 samples were collected from the age group 21-30 years. Out of a total of 60 samples, 29 samples showed coagulase positive *Staphylococcus aureus* isolates. A percentage of 35 isolates of *Staphylococcus aureus* showed methicillin resistance from a total of 60 samples analysed. The antibiotic sensitivity of MRSA isolates showed a maximum of 90.4 per cent to vancomycin which was followed by other antibiotics in the following order: Erythromycin > ciprofloxacin > streptomycin > levofloxacin. The MRSA isolates also showed resistance to vancomycin (9.6 per cent).

Keywords: Methicillin resistant, Staphylococcus aureus, MRSA, Antibiotic sensitivity, Clinical samples.

Introduction

The emergence of antibiotic resistance in microorganisms and their spread is threatening the medical community. This is particularly true in case of *Staphylococcus aureus. S. aureus* is the most common cause of nosocomial infection and is of increasing concern because of their tendency to multiple antibiotic resistance which often complicates treatment. Many isolates of *S. aureus* have been found to be resistant to new semi-synthetic β-lactums (methicillin, oxacillin and flucloxacillin) which become known as Methicillin Resistant *Staphylococcus aureus* (MRSA). The resistance to antibiotics in MRSA is due to the presence of plasmid DNA and those separate plasmids encodes resistance to gentamycin and chloramphenicol, also revealed that determines for penicillinase production, resistance to heavy metals and tetracycline were plasmid borne (Gillespie *et al.*, 1985). During the period between 1910s and 1980s MRSA strains emerged and now have been distributed throughout the world (Coia, 1990; Zaman and Dibb, 1994).

The antibiotic resistance mechanism and the prevalence of MRSA isolates from clinical samples has been studied by Cookson (1990), Bhat *et al.*, 1990 and 1991, Coimbra De Silva *et al.* (2000), Humphreys (2002) and others. Since it is very essential to know the prevalence of MRSA in any environment, because of public health importance and the threat posed by MRSA infection. Hence the present study has been undertaken to find out the prevalence of MRSA from clinical samples in Kanchipuram town, Tamil Nadu, South India.

Materials and Methods

A total of 60 samples were collected from the patients of different age groups in the laboratory of hospitals in Kanchipuram town, Tamil Nadu. Sterile cotton swabs were used for collecting samples from pus, sputum, ear and eye. The obtained swabs were put in to a sterile test tube containing Amines transport media or Nutrient broth to prevent drying of swab. Each specimen was labeled with needed particulars such as patients name, age, date and time of collection, type of specimen and chemotherapy if undertaken and the specimen was brought to the laboratory within six hours. The obtained clinical specimens were plated on three different media namely Nutrient agar, Blood agar and Mannitol Salt agar and were incubated at 37°C in an incubator.

The isolated *S. aureus* were then subjected to morphological and biochemical studies such as microscopic examination, Gram's staining, motility, catalase, urease, coagulase, oxidase, indole, methyl red, Voges-Proskauer, citrate utilization test, carbohydrate fermentation test, phosphatase and DNase activity to confirm the identification by using standard methods (Collee *et al.*, 1989). Then the isolates of *S. aureus* were subjected to antibiotic susceptibility test by adopting Kirby-Bauer disc diffusion method with standard antibiotic discs of methicillin (5 μg) (Bauer *et al.*, 1966). The *S. aureus* isolates which showed resistance to methicillin (MRSA isolates) were tested for antibiotic sensitivity pattern with ciprofloxacin, erythromycin, levofloxacin, streptomycin and vancomycin. The zone of inhibition was measured and the results were interpreted.

Results

A total of 60 clinical samples were collected for the study from different medical laboratories of Kanchipuram town. A maximum of 18 samples were obtained from the age group of 21–30 years followed by the age group 31–40 years (15 samples). From the age groups 1–10 and 61–70 years, 2 and 3 samples were obtained respectively.

The phenotypic characteristics of bacterial isolates collected from clinical samples are indicated in Table 46.1. A percentage of 58.3 isolates were Gram positive cocci in clusters. From this, 29 isolates showed coagulase actlvity and 21 isolates with methicillin resistance (Figure 46.1).

Table 46.1: Phenotypic Characteristics of Bacterial Isolates, Collected from Clinical Samples (n = 60)

Sl.No.	*Phenotypic Characters*	*Percentage Activity*
1.	Gram positive cocci in clusters	58.3
2.	Motility	33.3
3.	Growth on nutrient agar (Golden yellow colonies)	58.3
4.	Growth on blood agar	
	(*a*) α–hemolysis	13.3
	(*b*) β–hemolysis	48.3
5.	Growth on Mannitol Salt Agar (yellow colour colonies)	58.3
6.	Growth on DNase agar (DNase activity)	58.3
7.	Growth on Phenolphthalein phosphatase agar (phosphatase activity)	58 3
8.	IMViC	
	(*a*) Indole	41.6
	(*b*) Methyl red	65.0
	(*c*) Voges-Proskauer	68.3
	(*d*) Citrate	36.6
9.	Coagulase	48.3
10.	Catalase	65.0
11.	Oxidase	53.3
12.	Urease	61.6
13.	Gelatin liquefaction	63.3
14.	Sugar fermentation	
	(*a*) Glucose	68.3
	(*b*) Lactose	65.0
	(*c*) Mannitol	58.3
	(*d*) Sucrose	18.3
15.	Methicillin resistance	35.0

Prevalence of methicillin resistant *S. aureus* (MRSA) isolated among different age groups of sample donors is depicted in Figure 46.2. A higher percentage of 38.1 isolates were obtained from the age group of 21–30 years, and was followed by 3–40 years of age group. There were no MRSA isolates found among the age group of 1–10 and 61–70 years.

The results on the antibiotic sensitivity of MRSA isolates of the present study showed a higher percentage sensitivity of 90.40 to vancomycin. All the other antibiotics which were tested in this study, *i.e.*, ciprofloxacin, erythromycin, streptomycin and levofloxacin exhibited a sensitivity range of 62 to 82 per cent. The highest resistance showed by MRSA isolates was 38.10 per cent against levofloxacin (Table 46.2).

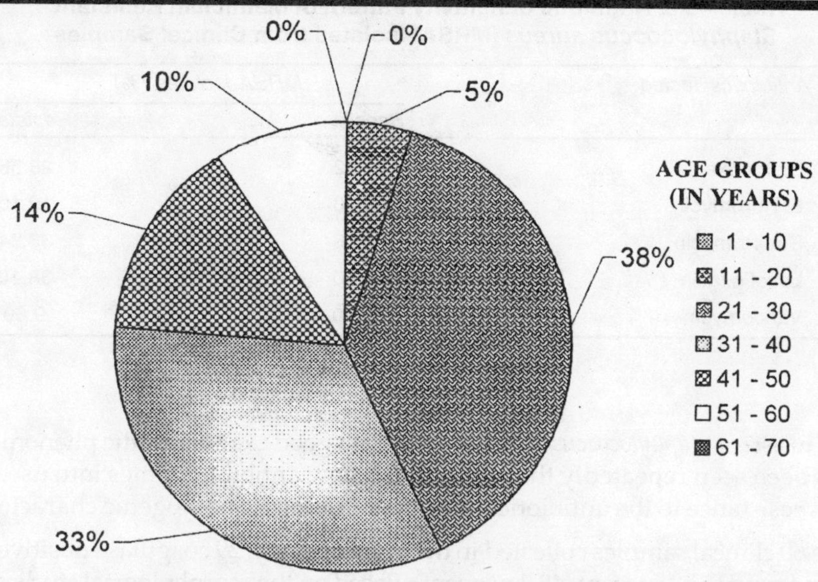

Figure 46.1: Prevalence of Methicillin Resistant *Staphylococcus aureus* (MRSA) among Different Age Groups of Sample Donors

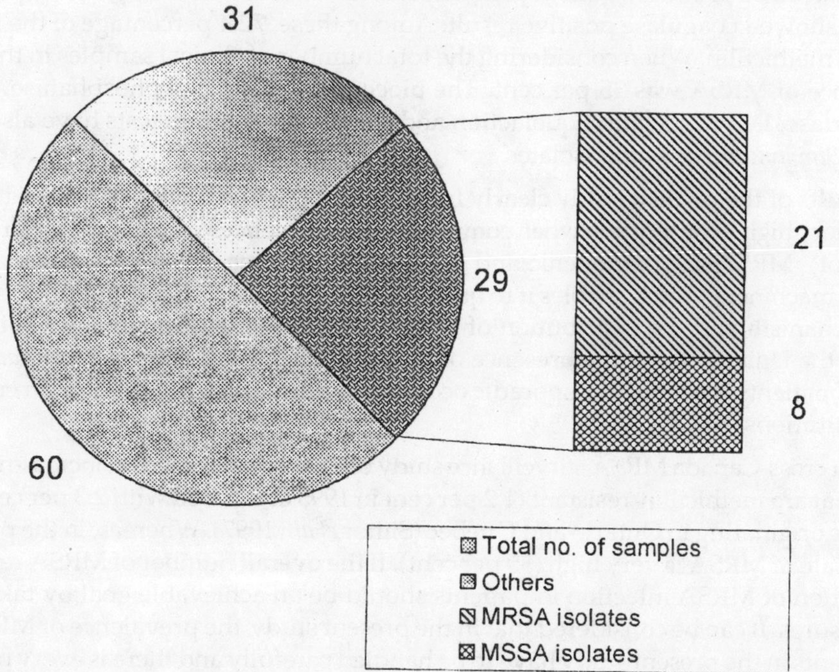

Figure 46.2: Number of *Staphylococcus aureus* Isolated from Total Number of Clinical Samples Characterized as Methicillin Resistant (MRSA) and Sensitive (MSSA)

Table 46.2: Antibiotic Sensitivity Pattern of Methicillin Resistant *Staphylococcus aureus* (MRSA) Isolates from Clinical Samples

Sl.No.	*Antibiotics Tested*	*MRSA Isolates (%)*	
		Sensitive	*Resistant*
1.	Ciprofloxacin	71.42	28.58
2.	Erythromycin	80.90	19.10
3.	Streptomycin	66.66	33.34
4.	Levofloxacin	61.90	38.10
5.	Vancomycin	90.40	9.60

Discussion

Methicillin resistant *Staphylococcus aureus* (MRSA) is a good example of the phenomenon of natural selection. It has been seen repeatedly that soon after a new antibiotic comes into use, *Staphylococcus aureus* develops resistance to the antibiotic while maintaining its pathogenic characteristics.

Among the 60 clinical samples collected in the present study, 29 coagulase positive *Staphylococcus aureus* have been found to be present (48.3 per cent). Based on the morphological studies, 58.33 per cent isolates were Gram positive cocci in clusters; 48.3 per cent of bacterial growth possessed β-hemolytic activity in blood agar; 58.3 of the bacterial colonies were yellow in colour on Mannitol Salt agar and the same isolates also produced golden yellow coloured colonies in nutrient agar. A percentage of 48.3 isolates also showed coagulase positive result. Among these, 72.4 percentage of the isolates showed resistance to methicillin. When considering the total number of clinical samples in the present study, the prevalence of MRSA was 35 per cent. The biochemical tests like phosphatase, DNase, urease, catalase, oxidase, IMViC, gelatin liquefaction and sugar fermentation tests have also confirmed the existence of *Staphylococcus aureus* isolates.

The results of the present study clearly indicate that the prevalence of MRSA in the study area seems to be very high in proportion when compared with the isolates which were sensitive to methicillin (13.3 per cent). MRSA are pathogenic and are common causes of *Staphylococcus aureus* due to its biochemical machinery which enables it to quickly colonize on host tissue and protect itself from host defense mechanism and the distribution of MRSA is worldwide (Zaman and Dibb, 1994). In both Europe and the United States, the presence of methicillin resistant *Staphylococcus aureus* (MRSA) in hospitalized patients is no longer a sporadic occurrence, and MRSA has become an endemic problem in many institutions (Boyce, 1990).

A recent cross-Canada MRSA surveillance study of 21 hospitals noted an increase in the proportion of isolates that are methicillin-resistant (1.2 per cent in 1995 compared with 2.3 per cent in 1996), with most isolates originating in Ontario and Quebec (Simor *et al.*, 1997). Whereas, in the present study, the prevalence rate of MRSA is very high (35 per cent). If the overall number of MRSA remains relatively low, prevention of MRSA infection in patients should be an achievable goal by taking appropriate control measures. It can be considered that in the present study, the prevalence of MRSA is very high, and the patients in the present study have to be handled carefully and there is every possibility for the spread of MRSA by the carriers. Hence it is suggested that the measures to control MRSA is an immediate need of this hour.

The MRSA isolates subjected to antibiotic sensitivity pattern in the present study reveals that the isolates have also developed resistance to other antibiotics tested. The MRSA of clinical isolates showed a maximum of 38.1 per cent resistance to levofloxacin which was followed by streptomycin (33.34 per cent) > ciprofloxacin (28.58 per cent) > erythromycin (19.1 per cent) > vancomycin (9.6 per cent). But the recent report by Deodhar and Vyas (2003) shows that out of 54.5 per cent MRSA isolates from pus samples, not even a single isolate showed resistance to vancomycin even though they showed 100 per cent penicillin resistance and 90.7 per cent tetracycline resistance. It is realizable that in the present study, 9.6 per cent of the isolates showed resistance to vancomycin.

Multiple drug resistance of *Staphylococcus aureus* is due to several drug resistant genes in a single plasmid, each with its own resistance markers. A bacterial cell may carry more than one plasmid with resistance markers. Genes in the chromosome which are variably expressed. The resistance development in *Staphylococcus aureus* dates back to 1940s. It has a long history of drug resistance can be explainable by the following data (IDIC, 1999).

Table 46.3

Antibiotic	*Year Introduced*	*Reports of Resistance*
Penicillin	1941	1940s
Streptomycin	1944	Mid 1940s
Erythromycin	1952	1950s
Gentamycin	1964	Mid 1970s
Ciprofloxacin	1988	Late 1980s
Vancomycin	1958	1997

Since the development of resistance to antibiotics by the pathogenic strains of *Staphylococcus aureus* is an ever increasing problem, a suitable and possible alternate chemotherapeutic compounds which are of plant origin *i.e.*, phytochemical compounds such as alkaloids, terpenoids, polyphenols, flavonoids and steroids may be tried for effective means of controlling drug resistant bacteria like MRSA.

Acknowledgement

The authors sincerely acknowledge the Directors, Principal and Head of the Department of Microbiology, Kanchi Shri Krishna College of Arts and Science for facilities.

References

Bauer, A.W., Kirby, W.M.N., Sherris, J.C. and M. Turck (1966). Antibiotic susceptibility testing by a standardized single disk method. *Am. J. Clin. Pathol.*, 45: 493–496.

Bhat, G.K., Nagesha, C.N., Joseph, K.M. and P.G. Shivanand (1990). Biochemical characteristics and drug resistance of pathogenic Staphylococci. *Indian. J. Microbiol.*, 30(1): 69–74.

Bhat, G.K., Joseph, K.M., Nagesha, C.N. and P.G. Shivanand (1991). Some properties of *Staphylococcus aureus* isolated from clinical sources. *Indian. J. Microbiol.*, 31(3): 313–315.

Boyce, J.M. (1990). Increasing prevalence of methicillin-resistant *Staphylococcus aureus* in the United States. Infect. Control Hosp. Epidemiol., 11: 639–642.

Coia, J.E., Carter, F.T., Baird, D. and D.J. Platt (1990). Characterization of Methicillin-Resistant *Staphylococcus aureus* by biotyping, immunoblotting and restriction enzyme fragmentation patterns. *J. Med. Microbiol.*, 31: 125–132.

Coimbra De Silva, M.V., Teixeira, L.A., Ramos, R.L.B., Predari, S.C., Castello, L., Familglieti, A., Vay, C., Klans, L. and A.M.S. Figueiredo (2000). Spread of the Brazilian Epidemic clone of a Multiresistant MRSA in two cities in Argentina. *J. Med. Microbiol.*, 49: 187–192.

Collee J.C., Dugaid, J.P., Fraser, A.G. and B.P. Marimio (1989). *Medical Microbiology, Vol 2: Practical Medical Microbiology*, 13th edn., Churchill Livingstone, Edinburgh.

Cookson, B.D. (1990). Mupirocin resistance in staphylococci. *J. Antimicrob. Chemother.*, 25: 497–501.

Gillespie, M.T., May, J.W. and R.A. Skurray (1985). Antibiotic susceptibilities.and plasmid profiles of nosocomial MRSA: A retrospective study. *J. Med. Microbiol.*, 17: 295–310.

Humphreys, H. (2002). Control of Methicillin–Resistant *Staphylococcus aureus* in hospital: An impossible dream? *J. Med. Microbiol.*, 51: 283–285.

IDIC (1999). Control of methicillin resistant *Staphylococcus aureus* in Canadian paediatric institutions is still a worthwhile goal. *Paediat. Child Health*, 4(5): 337–341.

Simor, A., Ofner-Agostini, M. and S. Paton (1997). The Canadian Nosocomial Infection Surveillance Programme: results of the first 18 months of surveillance for methicillin-resistant *Staphylococcus aureus* in Canadian hospitals. *Can. Commun. Dis. Rep.*, 23: 41–45.

Zaman, R. and W.L. Dibb (1994). Methicillin resistant *Staphylococcus aureus* (MRSA) isolated in Saudi Arabia: epidemiology and antimicrobial resistance patterns. *J. Hospit. Infect.*, 26: 297–300.

Chapter 47

Effect of Integrated Nutrient Management on Fertilizer Use Efficiency and Changes in Soil Fertility Status under Rice Based Cropping System

M. Chettri, S.S. Mondal and P. Bandhopadhaya

Department of Agronomy, Bidhan Chandra Krishi Viswavidyalaya, Mohanpur – 741 252, Nadia, West Bengal

ABSTRACT

Field experiment was carried out in entisol soils of neutral reaction having 1430 kg/ha total N, 18.10 kg/ha available P, 235 kg/ha available K and 18.92 kg/ha available S respectively, at University farm, Kalyani, West Bengal during the period of 2000–2001 and 2001–2002, to study the integrated effect of nutrient management with or without sulphur and FYM to increase the fertilizer use efficiency and soil fertility status with the inclusion of legume as fodder or green manure crop under rice based cropping system. Maximum grain yield of rice was recorded in the sequence where green gram or lathyrus was incorporated *in situ* before transplanting of rice crop in addition to 75 per cent of the recommended dose of N (urea), P (DAP), and K (MOP) + FYM @ 10 t/ha. Higher grain yield of rice was also observed in the same sequence with the application of 100 per cent of the recommended dose of N, P and K through ammonium sulphate, DAP and MOP, respectively. Higher agronomic efficiency (AE), physiological efficiency (PE) and recovery fraction (RF) of fertilizer use were recorded in the sequence where rice crop was grown after green gram or lathyrus as fodder crop with the application of 75 per cent of the recommended dose of N (urea), P (DAP) and K (MOP) in conjunction with FYM @ 10 t/ha. The soil nutrient status after harvest of 9th crop in the sequence was also improved where green manuring was done before transplanting of rice crop with same fertilizer management treatment (75 per cent NPK + FYM @ 10 t/ha).

Keywords: *Rice based cropping, Green manure, FYM, Grain yield, Soil nutrient status, Agronomic efficiency, Physiological efficiency recovery fraction, Green gram, Lathyrus, Oat, Maize.*

In India, food production has nearly doubled in the last two decades. This has been achieved through the adoption of high yielding crop varieties, intensive cropping, improved management practices and use of organic and inorganic fertilizers. Application of inorganic fertilizers though increased the yield substantially, but cannot sustains the soil fertility status (Bhardwaj and Omanwar, 1994). At present, much attention is given to the integrated use of organic or green manure and mineral nutrition for meeting the economic need of the farmers as well as for sustainability in terms of productivity and soil fertility. It has been realized that combined application of organic and inorganic fertilizers increased fertilizer use efficiency under intensive cropping system (Mondal and Mondal, 1996). In this context, an effort has been made to find out the best integrated system of fertilizer management with or without sulphur bearing fertilizers and FYM to increase the fertilizer use efficiency and fertility build up of soil with the inclusion of legume crop as fodder or green manure under rice based cropping system.

Materials and Methods

The field experiment was conducted during *kharif* and *boro* season of 2000–2001 and 2001–2002 at University farm, Kalyani, West Bengal. The soil of the experimental site was clay loam in texture having 1430 kg/ha total N, 18.10 kg/ha available P, 235 kg/ha available K, 18.92 kg/ha available sulphur and pH of 7.2. The experiment was laid out in split plot design with four different cropping sequence in main plots such as KR–O (F)–BR–G (Gm) [*Kharif* rice–oat (fodder)–*boro* rice–green gram (green manure)]; KR–L (Gm)–BR–M (F) [*Kharif* rice–lathyrus (green manure)–*boro* rice–maize (fodder)]; KR–L (F)–BR–G (F) [*Kharif* rice–lathyrus (fodder)–*boro* rice– green gram (fodder)] and KR–O (F)–BR–M (F) [*Kharif* rice–oat (fodder)–*boro* rice–maize (fodder)]. In the sub-plot, there were four different types of fertilizer management treatments such as F1 = 100 per cent of the recommended dose of N, P and K through non-sulphur bearing fertilizers (urea, DAP and MOP) applied to *kharif* (80 : 50 : 50, N : P : K) and *boro* rice (100 : 60 : 60, N : P : K), respectively, F2 = 75 per cent of the recommended dose of N, P and K through non-sulphur bearing fertilizers (urea, DAP and MOP) applied to *kharif* and *boro* rice, respectively, F3 = 75 per cent of the recommended dose of N, P and K through non-sulphur bearing fertilizer and (urea, DAP and MOP) along with FYM @ 10 t/ha applied to *kharif* and *boro* rice, respectively, F4 = 100 per cent of the recommended dose of N, P and K through sulphur bearing fertilizers (ammonium sulphate, DAP and MOP) applied to *kharif* and *boro* rice, respectively. The above treatments were replicated in three times randomly. According to treatment, FYM @ 10 t/ha (containing 0.44 per cent N, 0.22 per cent P, 0.38 per cent K and 0.13 per cent S) was applied in both *kharif* and *boro* rice during final land preparation. After harvesting of green gram (containing 0.80 per cent N, 0.25 per cent P, 0.76, K and 0.19 per cent S) and lathyrus (0.53 per cent N, 0.23 per cent P, 0.62 per cent K and 0.15 per cent S) as green manure crop were chopped in pieces and mixed with the soil by spading in presence of standing water. The legume and non-legume crop as fodder or green manure were grown on residual fertility of rice crop.

By considering only the grain yield of rice obtained in different sequence, the Agronomic efficiency (AE), Physiological efficiency (PE) and Recovery fraction (RF) of fertilizer use were determined as per the formulae suggested by Novoa and Loomis (1981).

$$\text{Agronomic Efficiency (AE)} = \frac{\text{Kg of Grain}}{\text{Kg of Nutrient Added}}$$

$$\text{Physiological Efficiency (PE)} = \frac{\text{Kg of Grain}}{\text{Kg of Nutrient Absorbed}}$$

$$\text{Recovery Fraction (RF)} = \frac{\text{Kg of Nutrient Absorbed}}{\text{Kg of Nutrient Added}}$$

The available nutrients in FYM and crop residues incorporated in the system have been calculated at 25 per cent of the total nutrient present in the materials incorporated as per suggestion of Daji and Iyenger (1971).

Results and Discussion

Grain Yield

The grain yield of rice increased significantly due to the combining effect of cropping sequence and fertilizer management (Table 47.1). Maximum grain yield of rice was recorded in the sequence where green gram and lathyrus were incorporated *in situ* preceding to rice crops and rice crops were fertilized with 75 per cent of the recommended dose of N, P and K through urea, DAP and MOP in conjunction with FYM @ 10 t/ha. It is clearly implied that FYM and green manuring has a marked effect on grain yield of rice. This results are in agreement with the findings of Mondal and Chettri (1998). Application of sulphur bearing fertilizers (ammonium sulphate, DAP and MOP) also gave higher grain yield of rice grown after legume crop as green manure or fodder. In the present investigation, it was also observed that any reduction of the recommended dose of fertilizer in absence of FYM, significantly reduced the grain yield of rice. Similarly, growing of non-legume crop (oat and maize) as fodder before rice crop drastically reduced the grain yield of rice.

Fertilizer Use Efficiency

From Table 47.1, it was observed that the agronomic efficiency, physiological efficiency and recovery fraction of fertilizer use were reduced with the application of increasing level of sulphur and non-sulphur bearing fertilizers to rice crops without FYM in all the cropping sequence. The maximum AE (29.28), PE (13.15) and RF (2.23) were recorded in rice (*kharij*)–lathyrus (fodder)–rice (*boro*)–green gram (fodder) sequence where both rice crop was fertilized with 75 per cent of the recommended dose of N (urea), P (DAP) and K (MOP) along with FYM @ 10 t/ha. Higher AE, PE and RF were also observed in the sequence where both rice crop was followed by legume crop (as green manure) with all fertilizer management treatments. Combined application of organic manure through FYM or green manure and inorganic fertilizers increased the availability of nutrients to the crop plants and thus, increase the fertilizer use efficiency. This view is supported by the findings of Mondal and Mondal (1996). Rice crop fertilized with 100 per cent of the recommended dose of sulphur containing fertilizer (ammonium sulphate, DAP and MOP in all sequence also gave higher AE, PE and RF value as compared to the same fertilizer management treatment without sulphur containing fertilizers (urea, DAP and MOP). Mondal and Chettri (1998) also found higher AE, PE and RF value when rice crop was fertilized with 100 per cent of the recommended dose of sulphur containing fertilizers.

Table 47.1: Agronomic Efficiency, Physiological Efficiency and Recovery Fraction of Fertilizer Use in Different Cropping Sequence at Different Levels of Fertilizer Management (Pooled Data of 2 Years)

Cropping Sequence	*Total Yield of Rice in Kg (X)*	*Total Nutrients (N, P, K and S Basis) in Kg/ha/annum*		*Agronomic Efficiency*	*Physiological Efficiency*	*Recovery Fraction*
		Added (Y)	*Absorbed (Z)*	*(X/Y)*	*(X/Z)*	*(Z/Y)*
At 100% of the recommended dose of N, P and K (Urea, DAP and MOP)						
KR–O (F)–BR–G (Gm)	8062	430	741.71	18.75	10.86	1.72
KR–L(Gm)–BR–M(F)	8205	430	751.79	19.08	10.91	1.75
KR–L (F)–BR–G (F)	7990	380	685.11	21.03	11.66	1.80
KR–O (F)–BR–M (F)	6714	380	644.34	17.67	10.41	1.69
At 75% of the recommended dose of N, P and K (Urea, DAP and MOP)						
KR–O(F)–BR–G (Gm)	7590	335	716.30	22.65	10.59	2.13
KR–L (Gm)–BR–M (F)	7620	335	729.50	22.75	10.44	2.17.
KR–L(F)–BR–G (F)	7345	285	632.14	25.77	11.62	2.22
KR–O (F)–BR–M (F)	6225	285	605.91	21.84	10.27	2.13
At 75% of the recommended dose of N, P and K (Urea, DAP and MOP) + FYM @ 10 t/ha						
KR–O(F)–BR–G (Gm)	9975	385	823.00	25.91	12.12	2.14
KR–L(Gm)–BR–M(F)	10820	385	850.94	28.10	12.71	2.21
KR–L (F)–BR–G (F)	9808	335	746.46	29.28	13.15	2.23
KR–O (F)–BR–M (F)	8008	335	704.95	23.90	11.05	2.10
At 100% of the recommended dose of N, P and K (Ammonium Sulphate, DAP and MOP)						
KR–O(F)–BR–G (Gm)	9730	454	789.21	20.64	11.87	1.73
KR–L(Gm)–BR–M (F)	9900	454	824.08	21.81	12.01	1.81
KR–L(F)–BR–G (F)	9165	404	746.12,	22.68	12.28	1.85
KR–O(F)–BR–M(F)	7505	404	712.63	18.58	10.53	1.76
CD (p = 0.05) for grain yield						
At the same level of C	0.219					
At the same level of F	0.292					

Nutrient Status after Harvest of 9th Crop in the Sequence

The N, P, K and S status of soil after harvest of 9th crop were improved in the sequence where at least one legume crop was grown as fodder or green manure crop before either *kharif* or *boro* rice with all fertilizer management except at the reduced level of nutrients (75 per cent N, P and K) without FYM (Figures 47.1–47.4). The extent of increase of N (+210), P (+24.3) and K (+41.5) were maximum with the combined application of organic (FYM) and inorganic (75 per cent N, P and K) fertilizers in KR–O (F)–BR–G (Gm) sequence where *kharif* rice was green manured with green gram *in situ*. However, the extent of increase of S (+9.58) was maximum with the application of 100 per cent of the recommended dose of sulphur bearing fertilizer (ammonium sulphate, DAP and MOP) in the same sequence. This might be due to the addition of 24 per cent sulphur through ammonium sulphate. This result confirms

T_1: 100% N (urea), P and K; T_2: 75% N (urea), P and K +FYM; T_3: 75% N (urea), P and K; T_4: 100% N (AS), P and K; Series 1: KR–O (F)–BR–G (Gm); Series 2: KR–L (Gm)–BR–M (F); Series 3:KR–L (F)–BR–G (F); Series 4: KR–O (F)–BR–M (F)

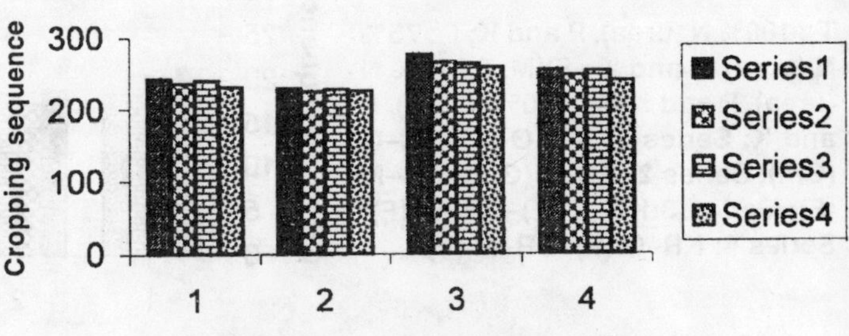

Figure 47.1: Effect of Cropping Sequence on Total Nitrogen Status in the Soil after Harvest of 9th Crop at Different Fertilizer Management Treatments

T_1: 100% N (urea), P and K; T_2: 75% N (urea), P and K +FYM; T_3: 75% N (urea), P and K; T_4: 100% N (AS), P and K; Series 1: KR–O (F)–BR–G (Gm); Series 2: KR–L (Gm)–BR–M (F); Series 3:KR–L (F)–BR–G (F); Series 4: KR–O (F)–BR–M (F)

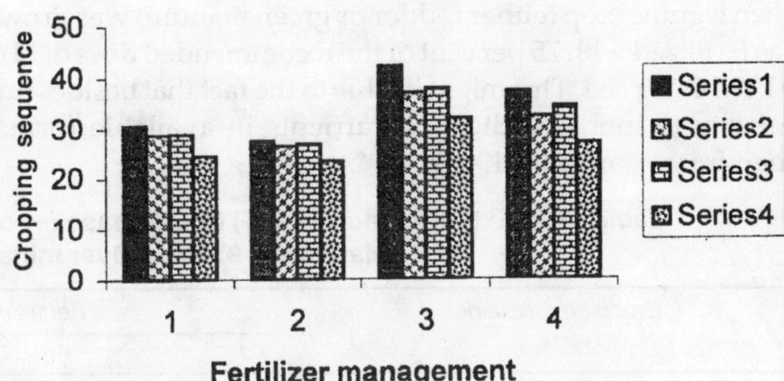

Figure 47.2: Effect of Cropping Sequence on Available Phosphorous Status in the Soil after Harvest of 9th Crop at Different Fertilizer Management Treatments

T_1: 100% N (urea), P and K; T_2: 75% N (urea), P and K +FYM; T_3: 75% N (urea), P and K; T_4: 100% N (AS), P and K; Series 1: KR–O (F)–BR–G (Gm); Series 2: KR–L (Gm)–BR–M (F); Series 3:KR–L (F)–BR–G (F); Series 4: KR–O (F)–BR–M (F)

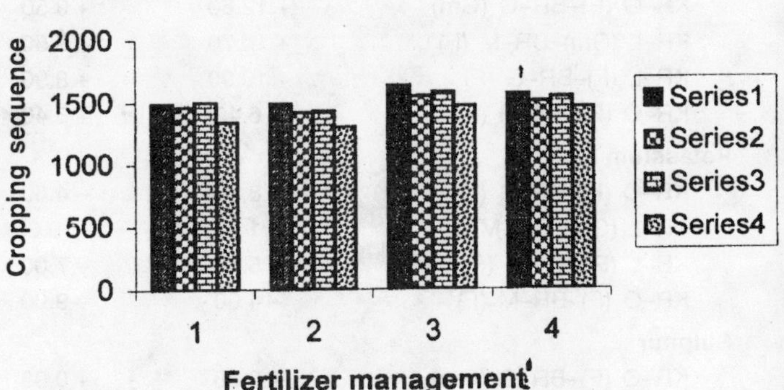

Figure 47.3: Effect of Cropping Sequence on Available Potassium Status in the Soil after Harvest of 9th Crop at Different Fertilizer Management Treatments

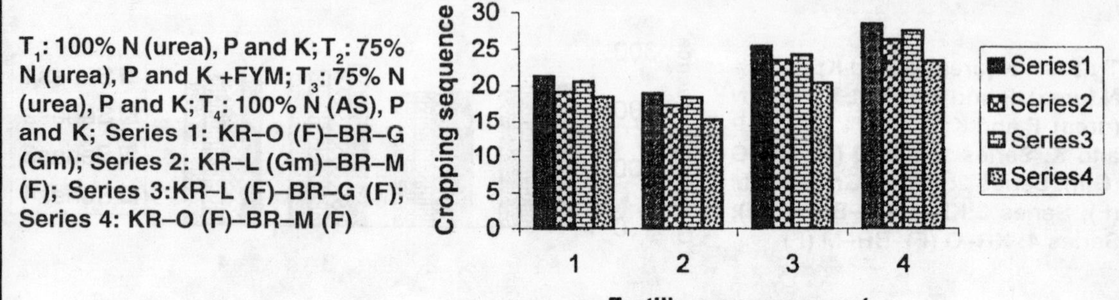

Figure 47.4: Effect of Cropping Sequence on Available Sulphur Status in the Soil after Harvest of 9th Crop at Different Fertilizer Management Treatments

the findings of Sarkar *et al.* (1994). It was also observed that soil fertility status improved appreciably when legume crop (either fodder or green manure) was grown before rice crop and both the rice crop was fertilized with 75 per cent of the recommended dose of N (urea), P (DAP) and K (MOP) in addition to FYM @ 10 t/ha. This might be due to the fact that besides nutrient content organic manures have the capacity to mobilize the soil nutrients in available form. Similar result was also observed by Chandrasoorian *et al.* (1994).

Table 47.2: Extent of Increase (+) or Decrease (–) of Nutrient Status of Soil after Harvest of 9th Crop Over Initial Value*

Cropping Sequence	*Fertilizer Management*			
	F1	*F2*	*F3*	*F4*
Nitrogen				
KR–O (F)–BR–G (Gm)	+ 60	– 20	+ 210	+ 140
KR–L (Gm)–BR–M (F)	+ 35	– 10	+ 120	+ 80
KR–L (F)–BR–G (F)	+ 70	0	+ 150	+ 115
KR–O (F)–BR–M (F)	– 90	– 130	+ 40	+ 30
Phosphorous				
KR–O (F)–BR–G (Gm)	+ 12.60	+ 9.50	+ 24.30	+ 9.40
KR–L (Gm)–BR–M (F)	+ 10.70	+ 8.50	+ 19.10	+ 14.40
KR–L (F)–BR–G (F)	+ 10.90	+ 8.90	+ 19.90	+ 16.50
KR–O (F)–BR–M (F)	+ 6.40	+ 5.40	+ 5.40	+ 9.10
Potassium				
KR–O (F)–BR–G (Gm)	+ 8.60	– 4.60	+ 41.50	+ 20.00
KR–L (Gm)–BR–M (F)	+ 1.50	– 10.00	+ 30.80	+ 13.00
KR–L (F)–BR–G (F)	+ 5.10	– 7.00	+ 27.60	+ 18.50
KR–O (F)–BR–M (F)	– 4.00	– 9.00	+ 23.40	+ 5.60
Sulphur				
KR–O (F)–BR–G (Gm)	+ 2.38	+ 0.08	+ 6.58	+ 9.58
KR–L (Gm)–BR–M (F)	+ 0.28	– 1.62	+ 4.58	+ 7.38
.KR–L (F)–BR–G (F)	+ 1.68	– 0.42	+ 5.38	+ 8.68
KR–O (F)–BR–M (F)	– 0.42	– 3.62	+ 1.48	+ 4.48

Initial soil nutrient status before transplanting of first crop during *kharif* season of 2000, where 1430 kg N, 18.10 kg P, 235 kg K and 18.92 kg S/ha.

References

Bhardwaj, V. and P.K. Omanwar (1994). Long-term effects of continuous rotational cropping and fertilization on crop yield and soil properties. *J. Indian Soc. Soil Sci.*, 42 (3): 387–392.

Chandrasoorian, S., Palaniappan, S.P. and G. James Martin (1994). Studies on soil fertility changes in a rice based cropping system. In: *Abstract of 13th National Symposium on Integrated Input Management for Efficient Crop Production*, organized by Indian Society of Agronomy and TNAU, 22–25th February, 1994, pp. 74.

Daji, J.A. and T. Iyenger (1971). Organic manures (others). In: *Handbook of Manures and Fertilizers*, Indian Council of Agricultural Research, New Delhi, pp. 81.

Mondal, S.S. and M. Chettri (1998). Integrated nutrient management for sustaining productivity and fertility building under rice based cropping system. *Indian. J. Agric. Sci.*, 68(7): 337–340.

Mondal, T.K. and S.S. Mondal (1996). Integrated nutrient management for rice (*Oryza sativa*) based cropping system. *Indian J. Agric. Sci.*, 66(12): 964–968.

Novoa, R. and R.S. Loomis (1981). Nitrogen and plant production. *Plant and Soil*, 58: 172--204.

Sarkar, S., Mondal, S.S., Maity, P.K. and B.N. Chatterjee (1994). Sulphur nutrition of crops with and without organic manure under intensive cropping. *Indian J. Agric. Sci.*, 64(2): 88–92.

Chapter 48

Study of the Combined Effect of Irrigation Scheduling and Plant Population Levels on Growth, Yield and Quality of Soybean

D.A. Sonawane, R.M. Gethe**, V.K. Thombre*** and D.K. Kambale*****

**Assistant Professor, **Senior Research Assistant, ***Junior Research Assistant of Agronomy*
*****Assistant Professor of ASDS*
Mahatma Phule Agricultural University, Rahuri – 413 722,
District Ahmednagar, M.S., India

ABSTRACT

A field experiment was conducted on soybean Cv. MACS–57 at Agricultural College Research Farm, Pune (M.S.) to identify appropriate scheduling of irrigation and plant population. The highest plant height, plant spread, No. of branches/plant, no. of leaves/plant, leaf area and dry matter was recorded in the treatment of 10 days irrigation interval. The highest seed yield (28.28 q/ha) and straw yield (49.77 q/ha) was also recorded in the treatment of 10 days irrigation interval while lowest in the treatment of 125 mm CPE. The plant population 3.33 lakh/ha recorded highest in all the growth contributing characters and seed yield (26.55 q/ha) but highest straw yield (48.12 q/ha) was recorded in the treatment of plant population of 6.66 lakh/ha. The oil content in seed was increased (19.36 per cent) due to irrigation scheduled at 75 mm CPE followed by 10 days irrigation interval (19.28 per cent), but protein content was declined with increase in levels of irrigation. The maximum consumptive use of water 637.35 mm was recorded due irrigation scheduling at 10 days interval and highest WUE (4.92 kg/ha/mm) was recorded at 125 mm CPE. The plant population 3.33 lakh/ha was recorded highest consumptive use of water 561.38 mm and WUE of 4.72 kg/ha/mm than other levels. The interaction effects for all the characters were found to be non-significant.

Introduction

Soybean [*Glycine max* (L.) Merrill] is endowed with high protein content as 36–42 per cent and plays three dimentional role as a pulse, oilseed and legume. It improves the soil fertility therefore it has got importance in crop rotations in irrigated as well as dry land area. Water is the most important input which increase the yield per hectare and there is relationship between available soil moisture and plant population per hectare. The optimum plant population with its proper plant geometry for achieving maximum production should be the basis for maximization of yield per hectare. In view of the above information there was a necessity to plan a research work on proper scheduling of irrigation and optimum plant population of soybean (*Glycine max*) in summer season for increasing productivity and hence, this study.

Materials and Methods

A field experiment was conducted on the Agricultural College Research Farm, Pune during summer season of 1999. A total of 15 treatment combinations were formed in which 5 levels of irrigation with 3 levels of plant population *viz.*, (*a*) Irrigation levels (CPE): 10 days interval, 75 mm CPE. 100 mm CPE, 125 mm CPE, 125–75–100 mm CPE and (*b*) Plant population 3.33 (30 × 10 cm), 4.44 (30 × 7.5 cm), 6.66 lakh/ha (30 × 5 cm) were evaluated in split plot design with three replications. The gross and net plot size was 5.40 × 3.60 m^2 and 4.20 × 2.40 m^2, respectively. Soybean variety of MACS-57 was selected for experimental study. The experimental field soil was medium black, low in available nitrogen, medium in available phosphorus and rich in available potash and it was slightly alkaline in reaction (pH 8.2). Total of 52 mm rainfall was distributed during the crop growth period. Soil moisture constant necessary for irrigation studies like, field capacity, permanent wilting point and bulk density were determined at 0-15 cm, 15–30 cm and 30–45 cm soil depth. Sowing was done on 23rd February, 1999 by dibbling method and harvesting was done on 7th June, 1999.

Results and Discussion

Growth Characters

Plant Height

The data presented in Table 48.1 showed that the maximum plant height (73.03 cm) was found with irrigation treatment of 10 days interval which was at par with 75 mm CPE and significantly superior over rest of all 100, 125 and 125–75–100 mm CPE treatments and also plant population treatments produced significantly maximum plant height over each other but the interaction between irrigation levels and plant population on the height of soybean plant was found to be non-significant at all the stages of crop growth. These results are in confirmative with the results of Restuccia *et al.* (1992).

Plant Spread

The maximum plant spread (53.90 cm), was observed with irrigation treatment of 10 days which was at par with 75 mm CPE treatment and significantly superior over 100, 125 and 125–75–100 mm CPE treatments and plant population of 3.33 lakh plants/ha and was also at par with 4.44 lakh plants/ha and both treatments were significantly superior over 6.66 lakh plants/ha. The interaction was found non-significant at all stages of observations. Similar results were obtained by Gomez *et al.* (1988) and Pritoni *et al.* (1990).

Table 48.1: Effect of Different Treatments on Growth, Yield Contributing Characters and Yield of Soybean

Treatments	*Plant Height (cm)*	*Plant Spread (cm)*	*Branches/ Plant*	*Leaves/ Plant*	*Leaf Area/ Plant (dm²)*	*Dry Matter/ Plant at Harvest (gm)*	*Seed Yield q/ha*	*Straw Yield q/ha*	*Protein Content %*	*Oil Content %*
Irrigation levels	73.03	53.90	11.74	36.12	43.00	60.19	28.28	49.77	36.62	19.28
10 days interval										
75mm CPE	72.97	53.89	11.25	35.94	42.00	59.95	28.08	49.42	37.72	19.36
100mm CPE	70.40	51.55	10.00	62.66	38.65	55.37	22.78	44.19	40.31	17.99
125mm CPE	61.74	50.28	8.99	31.60	34.24	45.70	20.71	49.01	41.78	17.37
125–75–100mm CPE	72.57	53.76	10.65	35.27	40.35	58.76	27.69	0.29	39.10	18.39
S.E. ±	0.080	0.40	0.46	0.18	0.65	0.78	0.80	0.29	0.048	0.29
C.D. at 5%	0.26	1.30	1.51	0.60	2.12	2.56	2.61	0.95	0.15	0.94
Plant Population Lakh/ha	68.38	53.88	11.73	35.28	42.64	57.77	26.55	45.29	40.43	17.69
3.33										
4.44	70.48	53.02	10.60	34.40	39.70	56.06	25.63	46.86	39.35	18.82
6.66	71.55	51.11	9.23	33.26	37.17	54.13	24.34	48.12	38.73	18.91
S.E. ±	0.14	0.39	0.19	0.18	0.15	0.50	0.50	0.39	0.094	0.26
C.D. at 5 %	0.41	1.17	0.58	0.53	0.45	1.49	1.48	1.17	0.27	0.78
S.E. ±	0.31	0.89	0.44	0.40	0.34	1.13	1.12	0.94	0.21	0.59
C.D. at 5%	–	–	–	–	1.02	–	–	–	0.62	–

Branches/Plant

The data on mean number of branches/plant at 70 days after sowing as periodically influenced by irrigation and plant population treatments are presented in Table 48.1. From the same table it could be seen that the maximum number of branches/plant (11.74) were obtained by irrigation treatment of 10 days interval and was at par with 75 and 125–75–100 mm CPE treatments and significantly superior over 100 and 124 mm CPE treatments. The irrigation treatment of 75 mm CPE was at par with 100 mm CPE 125–75–100 mm CPE however, it was significantly superior over 125 mm CPE irrigation treatment. The plant population level of 3.33 lakh plants/ha produced significantly more branches (11.73) than plant population of 4.44 and 6.66 lakh/ha and branches/plant of soybean remained unaffected due to interaction effects between irrigation and plant population levels at all the stages of crop growth. These results are in agreement with results of Gomez *et al.* (1988).

No. of Leaves and Leaf Area

Scheduling of irrigation at 10 days interval showed that number of leaves and leaf area/plant was at par with 75 CPE treatment and significantly superior over rest of the irrigation treatments. The differences due to irrigation treatment of 75, 100, 125–75–100 mm CPE were significant. Also maximum number of leaves (35.28) and leaf are per plant (42.64 dm²) was observed with the lowest plant population of 3.33 lakh/ha which was significantly superior over rest of the plant population treatment.

The medium plant population of 4.44 lakh/ha was significantly superior over higher plant population of 6.66 lakh/ha and influencing number of leaves and leaf area/plant. Under all the irrigation treatments the plant population of 3.33 lakh/ha produced significantly more number of leaves and leaf area per plant than 4.44 and 6.66 lakh plants/ha. These results are in confirmative with the results reported by Losavio *et al.* (1989).

Yield Characters

Dry Matter/Plant

The dry matter production in soybean at harvest was maximum from the irrigation treatment of 10 days fixed interval, 75 and 125–75–100 mm CPE which was significantly more than 100 and 125 mm CPE treatments. The plant population of 3.33 lakh plants/ha has recorded significantly maximum dry matter production/plant (57. 77 gm/plant) and lower dry matter (54.13 gm/plant) was recorded in 6.66 lakh plants/ha. The interaction effects between irrigation and plant population levels were found non-significant. These results are confirmative with the results of Purushothaman *et al.* (1992).

Seed and Straw Yield

The seed and straw yield obained due to irrigation treatments of 10 days fixed interval, 75 and 125–75–100 mm CPE were found at par with each other and significantly increased the seed and straw yield over 100 and 125 CPE. The highest seed yield of 28.28 and straw yield of 49.77 q/ha was obtained from the treatment of irrigation scheduling at 10 days interval. Similar results are inagreement with results obtained by Singh and Tripathi (1972). The plant population of 3.33 lakh/ha recorded higher seed yield (26.55 q/ha) than a plant population of 6.66 lakh/ha, but it was at par with 4.44 lakh plants/ha. The highest straw yield/ha (48.12 q/ha) was recorded due to plant population level of 6.66 lakh/ha which was significantly more than straw yield produced due to 3.33 and 4.44 lakh plants per hectare. The interaction effects between irrigation scheduling and plant population was found non-significant. Similar results were also reported by Chaudhari (1991) and Martin *et al.* (1994).

Quality

Protein Content

The protein content of soybean seed was significantly influenced due to various treatments (Table 48.2). The highest protein content of 41.78 per cent was recorded by irrigation scheduling at 125 mm CPE and the lowest protein content of 36.62 per cent was recorded by irrigation at 10 days interval. Plant population of 3.33 lakh plants per hectare was recorded highest protein content 40.43 per cent followed by 4.44 lakh plants per hectare (39.35 per cent) and 6.66 lakh plants per hectare (38.73 per cent). These results are inagreement with the results obtained by Arangino (1987).

Oil Per cent

The oil content in soybean seed was increased (19.36 per cent) due to irrigation scheduling at 75 mm CPE followed by irrigation at 10 day interval (19.28 per cent). The plant population level of 3.33 lakh plants per hectare was recorded lowest oil content (17.69 per cent) which was significantly lower than plant population of 4.44 and 6.66 lakh plants per hectare. These results are inagreement with the results obtained by Kolekar (1993) and Martin *et al.* (1994). The interaction effects between irrigation and plant population on oil content of soybean seed was found non-significant.

Irrigation Studies

Consumptive Use

The consumptive use of water was increased with increase in volume of water applied. The mean maximum consumptive use of water was estimated to be 637.35 mm and 561.38 mm through irrigation scheduled at 10 days interval and plant population of 3.33 lakh plants per hectare, respectively than other treatments of irrigation scheduling and plant population similar results were also obtained by Rajput *et al.* (1991).

Table 48.2: Mean Profile Water Depleted and Filled in Deficit, Effective Rainfall, Seasonal Consumptive Use, Consumptive Use per day, Yield in kg/ha, Water Use Efficiency, Total Pan Evaporation and ET/PE Ratio as Influenced by Different Treatments

Treatments	*Profile Water Depleted and Filled in Deficit (mm)*	*Effective Rainfall (mm)*	*Seasonal Consum-ptive Use (mm)*	*Consum-ptive Use per day (mm)*	*Yield kg/ha*	*Water Use Efficiency kg/ha/mm*	*Total Pan Evaporation (mm)*	*ET/PE Ratio*
Irrigation Levels	593.35	44	637.35	6.18	2828	4.43	869	0.73
10 days interval								
75mmm CPE	588.00	44	632.00	6.13	2808	4.44	869	0.72
100mm CPE	444.85	48	492.85	4.92	2278	4.62	869	0.56
125mm CPE	372.88	48	420.88	4.20	2071	4.92	869	0.48
125–75–100mm CPE	520.17	48	568.17	5.51	2769	4.87	869	0.65
Plant Population Lakh/ha	513.38	48	561.38	5.50	2655	4.72	869	0.64
3.33								
4.44	503.30	48	551.30	5.40	2563	4.64	869	0.63
6.66	494.88	48	542.88	5.32	2434	4.48	869	0.63
Mean	503.85	48	550.85	5.39	2550.66	4.64	869	0.62

Water Use Efficiency (WUE)

The average water use efficiency was observed to be 4.64 kg/ha/mm due to different treatments under study. The highest water use efficiency of 4.92 kg/ha/mm and 4.72 kg/ha/mm was recorded due to irrigation treatment of 125 mm CPE and plant population of 3.33 lakh plants per ha.

References

Arangino, R. (1987). The influence of irrigation regime on bio-agromatic characteristic of same soybean cultivars. *Informative Agrario*, 44(12): 159–165.

Chaudhari, M.A. (1991). Effect of scheduling of irrigation on yield performance of soybean. *M.Sc. Thesis,* Mahatma Phule Krishi Vidyapeeth, Rahuri, Dist. Ahmednagar (India).

Gomez, C.J.C., Pena, O.M.G. and B.B.R. Morquez (1988). Evaluation of two soybean cultivators under four population densities in villaflores chis mexico. *Revista Chapingo*, 12(58–59): 66–69.

Kolekar (1993). Effect of varieties and plant population on growth, yield and quality of soybean. *M.Sc. (Agri.) Thesis,* MPKV, Rahuri (M.S.).

Losavio, N., Mastroilli, M. and M.E. Venezian Scarascia (1989). Effect of seasonal irrigation level on growth and development of soybean in Southern Italy. *Irrigazionee Drenaggio*, 3693: 124–128.

Martin, D.E., Santa Olalla, F., Juan, Vale, J.A. and C. Fabeiro Cortes (1994). Growth and yield analysis of soybean (*Glycine max* (L.) Merill) under different irrigation schedules. *European J. of Agron.*, 3(3): 187–196.

Pritoni, G., Venturi, G. and M.T. Amaducci (1990). Effect of adequate or inadequate water supply at critical growth phases of soybean. *Irrigazionee Drenaggio*, 37(4): 29–34.

Purushothaman, S., Jeyaraman, S. and M. Muthian (1992). Effect of irrigation regimes on growth and yield of soybean. *Madras Agril. J.*, 79(8): 473–476.

Rajput, R.L., Kaushik, J.P. and O.P. Verma (1991). Consumptive use, water use efficiency and moisture extraction pattern of soybean as influenced by irrigation and row spacing. *Haryana J. Agron.*, 7(1): 1–5.

Restuccia, G., Mauromicale, G. and A. Irna (1992). Effect of irrigation regime on the agronomic behaviour of soybean grown in the Mediterranean. *Rivista Di agronomia*, 26: 777–784.

Singh, A. and N.C. Tripathi (1972). Effect of moisture stress on soybean [*Glycine max* (L.) Merill]. *Ind. J. Agric. Sci.*, 42(7): 582–585.

Chapter 49
Effect of Pre-sowing Seed Treatments on Germination and Chloroplast Pigments in Early Seedlings of *Glycine max* L. cvs. KHSB–2 and Hardee

G. Panduranga Murthy, M.S. Sudarshana and Prakasha

Tissue Culture and Crop Physiology Laboratory, P.G. Department of Studies and Research in Botany, University of Mysore, Manasagangotri, Mysore – 570 006, Karnataka, India

ABSTRACT

Laboratory experimentation on the effect of pre-sowing seed treatments with Gaucho (an insecticide) and Krilaxyl (a fungicide) on two cultivars of *Glycine max*, L. cvs KHSB–2 and Hardee, were undertaken using various concentrations and combinations. The response of the crop plant was evidenced by reduction on germination, early seedling growth and chloroplast pigments (chl 'a', 'b' total and carotenoids) at higher and in-combination concentrations of both the chemicals. Among two varieties the relative susceptibility in Hardee have been noticed than KHSB–2 is concerned.

Keywords: *Pre-sowing seed treatments, Germination, Early seedling growth, Chloroplast pigments, Glycine max (Soybean).*

Introduction

Several reports indicate that the adverse effects of pesticidal application on various agricultural crop plants (Pradhan and Basu, 1980; Prasad and Mathur, 1983; Somashekar and Sreenath, 1987).

Besides, controlling pests and diseases, the toxicity and persistence of pesticides reacts with cellular constituents producing variations on physiological, biochemical and cytological level (Zutshi and Kaul, 1975; Shrisat and Kale, 1979; Krishna Murthy and Rao, 1980; Patil and Shirashyad, 1989; Panduranga Murthy *et al.*, 2002 and 2003). In addition to other methods of application of pesticides, seed treatment and seed dressing have been proved to be potential carcinogenic and mutagenic agents which, affects the seed health and quality. Pesticides poisoning in human beings and high levels of their residues in food items are also of great concern (Sreenivasa Rao and Ramamohana Rao, 2000). Hence, the present study was instigated to determine whether, the individual and in-combination seed treatment with Gaucho–70 ws (an insecticide) and Krilaxyl–35 ws (a fungicide) would have an additive or synergistic effects on germination, seedling growth and chloroplast pigments in Soybean cvs. KHSB–2 and Hardee.

Materials and Methods

Initially, the seeds of Soybean cvs. KHSB–2 and Hardee were obtained from National Seed Project, G.K.V.K. University of Agricultural Sciences, Bangalore – 65. Insecticide, Gaucho and Fungicide, Krilaxyl were purchased from authorized dealers, Mysore.

Healthy and uniform sized seeds of same weight (25 gms) divided into eight batches and were surface sterilized with Sodium hypo-chloride for 5–10 minutes. Then rinsed repeatedly with sterile distilled water and thoroughly mixed with prepared slurry of both the pesticides of various concentrations *viz.*, recommended dosage (Gaucho 10 gms/kg and Krilaxyl 7 gms/kg of seeds), below recommended (25 per cent less), above recommended (25 per cent more) and in combination of both the chemicals (50 per cent among recommended). One batch of seeds was kept as control, by giving treatment with distilled water. After treatment all the batches were kept for 24 hours incubation. Then, they were allowed to germinate in germination towels at room temperature.

The germinating seeds of 8th day old seedlings (ISTA, 1995) were taken up for determining the per cent germination, seedling growth (root and shoot length). Chlorophyll (Arnon, 1949) and Carotenoids (Kirk and Allen, 1965) were estimated in leaves. Vigor index (Abdul Baki and Anderson, 1973), tolerance index (Turner and Marshal, 1972) and per cent phytotoxicity (Chiou and Muller, 1972) were calculated and the data were subjected for analysis of arithmetic mean and standard deviation followed by Newman (1939) and Keul (1957).

Results and Discussion

Germination percentage of KHSB–2 and Hardee in accordance to both the pesticides show drastic reduction at higher and in-combination dosages.

The inhibitory and injurious action might have disturb the osmotic relationship of the seed and water, thus reducing the amount of absorbed water and thereby retarding seed germination (Reddy and Vidyavathi, 1984; Chakravarthi, 1986; Kamble and Sabale, 1999 and Panduranga Murthy *et al.*, 2002 and 2003).

Almost the above same results were found with respect to root length and shoot length of crop treated with higher and in-combination dosages. This growth inhibition of radicle and plumule was due to their concentrations and toxic effects on cell division of shoot and root meristems as well as elongation (Somashekar and Sreenath, 1986).

The ratio of shoot and root showed non-significant differences at higher and in-combination treatment of both the pesticides suggesting, an inhibition of shoot development (Benjamini, 1986). Thereafter, a harmful phytotoxic effect was noticed.

Table 49.1: Effect of Presowing Seed Treatments with Gaucho and Krilaxyl on Germination, Early Growth, Phytotoxicity, Chlorophyll Content and Dryweight in Soybean cvs. KHSB–2 and HARDEE

Crop Treatments		CVS	Germ (%) AM±SD	MR.L. (cms) AM±SD	MS.L (cms) AM±SD	R/S AM	VI AM	TI AM	P.P (%) AM	CHL. 'a' (mg.g–1) AM	CHL. 'b' (mg.g–1) AM	Total CHL. (mg.g–1) AM	CAR (mg.g–1) AM	SFW (gms) AM±SD	SDW (gms) AM±SD
		Soybean (Glycine max, L.)													
Control		KHSB–2	100.0±0.02	8.6160±0.26	10.199±0.94	0.8447	1881.50	–	–	0.0880	0.0745	0.1627	0.0660	9.21±0.21	0.94±0.02
		Hardee	96.22±1.18	7.2440±0.27	9.8118±0.36	0.7826	2068.40	–	–	0.0760	0.0755	0.1517	0.0509	9.00±0.16	0.88±0.01
	B R	KHSB–2	89.70±1.69	7.6430±0.04	9.1106±0.21	0.8389	1502.79	88.70	11.30	0.0771	0.0586	0.1356	0.0601	8.34±0.35	0.44±0.01
		Hardee	72.66±1.60	6.3410±0.10	8.2900±0.26	0.7648	1063.08	68.79	31.41	0.0602	0.0476	0.1080	0.0703	7.47±0.18	0.39±0.03
Gaucho	R D	KHSB–2	71.33±1.85	6.7001±0.60	8.0806±0.50	0.8291	1054.30	77.76	22.24	0.0719	0.0440	0.1159	0.0516	7.11±0.03	0.29±0.02
Dosages		Hardee	58.66±1.50	6.0293±0.06	6.2706±0.37	0.9615	0721.63	65.22	34.78	0.0564	0.0317	0.0880	0.0560	5.86±0.20	0.17±0.04
	A R	KHSB–2	65.00±0.76	5.5564±0.20	7.3809±0.27	0.7528	0840.92	64.48	35.52	0.0524	0.0360	0.0885	0.0440	6.24±0.14	0.14±0.02
		Hardee	51.60±0.88	4.0170±0.80	5.4138±0.02	0.7419	0486.62	43.45	56.55	0.0407	0.0289	0.0696	0.0496	4.96±0.35	0.10±0.01
	B R	KHSB–2	74.00±1.00	6.7267±0.62	8.1231±0.09	0.8280	1098.88	78.07	21.93	0.0747	0.0517	0.1264	0.0465	7.86±0.41	0.26±0.02
		Hardee	59.33±1.50	5.9602±0.06	6.9149±0.14	0.8619	0763.87	64.47	35.53	0.0602	0.0414	0.1018	0.0504	6.46±0.10	0.15±0.02
Krilaxyl	R D	KHSB–2	60.33±0.97	5.6636±0.49	7.1111±0.11	0.7964	0770.69	65.73	34.27	0.0617	0.0484	0.1102	0.0419	4.45±0.12	0.13±0.00
Dosages		Hardee	51.90±0.80	3.8672±0.16	5.4700±0.10	0.7069	0484.60	41.83	58.17	0.0427	0.0396	0.0822	0.0414	3.86±0.18	0.08±0.00
	A R	KHSB–2	56.33±1.24	4.2626±0.27	6.6656±0.40	0.6394	0615.58	49.47	50.53	0.0492	0.0404	0.0899	0.0335	3.90±0.20	0.11±0.02
		Hardee	45.99±1.09	2.8118±0.32	4.3696±0.12	0.6440	0330.10	30.41	69.59	0.0360	0.0297	0.0659	0.0374	2.82±0.14	0.07±0.00
Gaucho +	I+F	KHSB–2	47.00±1.14	3.7710±0.05	5.0010±0.47	0.7540	0412.28	43.76	56.24	0.0325	0.0327	0.0653	0.0400	3.12±0.09	0.08±0.00
Krilaxyl		Hardee	39.33±1.47	2.2424±0.25	3.6700±0.09	0.6110	0232.53	24.25	75.75	0.0202	0.0401	0.0602	0.0433	1.84±0.09	0.03±0.01

Germ: Germination %; RL: Root Length; SL: Shoot Length; R/S: Root Shoot Ratio; VI: Vigour; TI: Tolerance Index; PP: Percent Phytotoxicity; CHL: Chlorophyll; CAR: Carotenoid; SFW: Seedling Fresh Weight; SDW: Seedling Dry Weight; CV: Cultivar Varieties; BR: Below Recommended Dosage; RD: Recommended Dosage; AR: Above Recommended Dosage; I+F: Insecticide + Fungicide. AM: Arithmetic Mean; SD: Standard Deviation.

Values are Arithmetic mean of ± SD based on three determinations.

Both the varieties of Soybean showed maximum vigour index and tolerance index at control and very least at higher and combined treatment of both the chemicals (Dhakshinamoorthy and Sivaprakasam, 1989; Nijenstein and Ester, 1990).

Carotenoid and chlorophyll pigments (chl 'a', 'b' and total) reveals gradual decline as the pesticidal concentrations increases. It is understandable that, the higher dosages of above pesticides are injurious with regard to the pigment presence (Ramulu and Rao, 1987).

The fresh weight and dry weight were also reduced at greater extent along with increased concentrations over control (Ramadoss and Sivaprakasam, 1994).

The higher and in-combination treatments of the both the pesticides affect severely on germination and other above said parameters. Although, further elaborate research in this field is requisite to confirm it.

Acknowledgement

We are grateful to University of Mysore, for providing necessary facilities and also owe thanks to Sri G. Parashurama Murthy, *MA. P.G.D.L.* for financial assistance to complete this research paper.

References

Abdul Baki, A.A.V. and J.D. Anderson (1973). Vigour determination of Soybean seed by multiple criteria. *Crop Sci.*, 3: 630–633.

Anonymous (1995). International rules for seed testing. *Seed Sci. and Tech.*, 13: 421–463.

Arnon, D.I. (1949). Copper enzymes in isolated chloroplast, Polyphenoloxidase in *Beta vulgaris. Plant Physiol.*, 24: 1–15.

Benjamini, L. (1986). Effect of Carbofuran on the germination rate and initial development of seedling. *Phytoparasitica.*

Casida, J.E. and L. Lykken (1969). Mechanism of organic pesticide chemicals in higher plants *Ann. Rev. Pl. Physiol.*, 20: 607–636.

Chakravarthi, S.K. (1986). Effect of Bavistin 25 DS on germination and seedling vigour of Wheat. *Pesticides*, 3: 23–26.

Chiou, C.H. and C.H. Muller (1972). Allelopathic Mechanism of *Archtostaphylous glandulosa* variety. Zazaerisis, *Am. Mid Nat.*, 88: 324–347.

Dakshinamoorthy, T. and K. Sivaprakasam (1989). Effect of seed treatment with fungicides and insecticides on the seedling vigour (*pearl millet*). *Madras. Agric. J.*, 76: 166–169.

Gupta, R.C., Beg, M.U. and P.S. Chandel (1983). Effect of Endosulfan on seed germination and seedling growth of *Vigna radiata*, Linn. *Pestology*, 7: 25–28.

Kamble, A.B. and A.B. Sabale (1999). Influence of Bavistin and Monocrotophos on seed germination and seedling growth of *Trigonella foenum-graecum. Poll. Res.*, 18(1): 61–65.

Keul, M. (1952). The use of studentized range in connection with an analysis of variance. *Euphytica*, 1: 112.

Kirk, J.T.O. and R.I. Allen (1965). Dependence of Chloroplast pigment synthesis on protein synthesis, effect of actidione. *Biochem. & Biophy. Res. Comm.*, 21(6): 523–530.

Krarup, H.A and S.C. Kivera (1982). Effect of fungicides and insecticides on the germination and emergence of peas (*Pisum sativum*, L.). *Agro. Sur.*, 10: 75–78.

Krishnamurthy. P. and D. Rao (1980). Effect of two fungicides on germination and growth in *Brassica nigra Koch. Ibid*, 7: 160–161.

Moti, U.N. (1978). Effect of systemic insecticides on the germination and subsequent growth of Pea (*Pisum sativum*, L.) seed. *Seed Res.*, 6(1): 62–66.

Newman, D. (1939). The distribution of range of samples from a normal population expressed in terms of an independent estimate of Standard Deviation. *Biometrica*, 31: 20.

Nijenstein, J.H. and A. Ester (1990). Method of Evaluation as a factor in the Determination of Insecticide Phytotoxicity in field beans (*Vicia faba*, L.). *Seed Sci. & Tech.*, 18: 597–607.

Panduranga Murthy, G. and S. Leelavathi (2002). Effect of insecticidal and fungicidal seed treatments on seedling vigour and chlorophyll content in red gram and sunflower during seed germination. *Asian. J. Microbiol. Biotech. & Env. Sci.*, 4(2): 271–275.

Panduranga Murthy, G., Vijay Kumar, D.R. and M.S. Sudarshana (2003). Effects of different exposure periods of Gaucho and Apron on seedling morphology, Chlorophyll a/b and phytotoxicity in Sorghum var. BS–7707 during seed germination. *Poll. Res.*, 22(1): 51–53.

Patil, T.M. and V.S. Shirashyad (1989). Effect of Methylparathion and Phosphamidon on seed germination and on the activity of peroxidase and α-amlylase in some vegetable seeds. *Geobios*, 16(2–3): 57–60.

Pradhan, J.P and P.K. Basu (1980). Effect of Rutin on seed germination and growth of *Tephrosia vogeli. Geobios*, 7: 232–233.

Prasad, B.N. and S.N. Mathur (1983). Effect of Metasystox and Cuman-L. on seed germination, reducing sugar content and amylase activity in *Vigna mungo* (L.) Hepper. *Indian J. Pl. Physiol.*, 26: 209–213.

Ramadoss, S. and K. Sivapraskasam (1994). Effect of Cowpea seed treatment with fungicides and insecticides on the seedling vigour. *Madras Agri. J.*, 81: 297–299.

Ramulu, C.A. and D. Rao (1987). Effect of Monocrotophos on seed germination, growth and chlorophyll content of Custered bean. *Comp. Physiol Ecol.*, 12: 102–105.

Reddy, J.M.K. and P. Vidyavathi (1984). Effect of fungicide on the growth and seedling metabolism of *Dolichos biflorus. Geobios*, 10: 174–178.

Somashekar, R.K. and K.P. Sreenath (1987). Effect of fungicide Benlate on germination and seedling growth of plants. *Geobios*, 14: 93–96.

Shrisat, A.M. and V.V. Kale (1979). Effect of fungicide treatment on germination and seedling vigour of Chick pea (*Cicer arietinum*). *Tropical Grain Legume Bull.*, 16: 29–32.

Sreenivasa Rao, A. and P. Ramamohana Rao (2000). Study on Pesticide residues in Vegetables. *Poll Res.*, (19(4):661–664.

Turner, L.G. and C. Marshal (1972). Accumulation of zinc by subcellular fraction of root of *Agrostis tennis sibth* in relation to zinc tolerance. *New Phytol.*, 71: 671–676.

Zutshi, U. and S.L. Kaul (1975). Cytogenetic activity of some common a fungicides in higher plants. *Cytobios*, 12: 61–67.

Chapter 50

Mass Propagation of Bamboo (*Dendrocalamus hamiltonii* Nees and Ex Munro) in Response to Plant Growth Regulators and Fertilization

S.K. Kaushal and Usha Rana

College of Basic Sciences, CSKHPKV, Palampur – 176 062, India

ABSTRACT

Dendrocalamus hamiltonii Nees and Ex Munro like other bamboo is a multipurpose, fast growing and high yielding variety with large sized and thick walled culms having excellent fodder quality. It has been propagated exclusively by vegetative means but the produced plants are of the same physiological age. Hence, they are bound to flower terminally leading to large scale death. It is imperative to raise new planting material from seeds well in time. Therefore, studies were conducted to improve seedling vigour by applying hormones and fertilizers for the mass multiplication of elite seedlings. Different concentrations of growth regulator GA_3 and single dose of kinetin were sprayed on one year old seedling growing in plots. Foliar application of growth regulators (50 ppm GA_3 and 2×10^{-6} g/l kinetin in seedlings significantly increased the number of tillers, internode length and tiller heights to almost double. A significant reduction in sprouting, rooting rhizogenesis days and an increased percentage of rooting and sprouting was observed with farm yard manure (FYM 1.5 kg/cutting) in combination with different doses of nitrogen (10 g/cutting) and phosphorus 10 g/ cutting. Plants raised from seedlings usually requires a period of ten years to attain normal size. Therefore, to reduce this long period, cutting segments from juvenile seedlings were taken with single, bi and trinodal segments and grown horizontally. The culms raised from these segments were very healthy and single node segments showed best result in culm emergence.

Keywords: *Vegetative propagation, Juvenile seedlings, Dendrocalamus hamiltonii.*

Introduction

Dendrocalamus hamiltonii (Nees and ex Munro) a native of tropical Eastern Himalaya and Nepal, is cultivated in Himachal Pradesh and has been put to so many uses. It has entered the culture of hilly peoples who call it "Maggar" or "Phargalu" (Sharma and Kaushal, 1985). It is planted from 350 to 1500 m altitude and gives its best performance in Kangra valley of H.P. marked by 2000–2800 mm annual rainfall mostly concentrated during rainy season. Once established, it is tolerant to moisture stress and has adaptability to wide range of edaphic conditions, except water logging. It is a good source to improve the economy of the people, conserve environment, prevent soil erosion, proper utilization of waste and marginal lands to reduce pressure over valuable timber and generate local employment as a cottage industry.

Bamboo is monocarpic and bears sporadically flowering culms after 40–50 years. The existing clones in Himachal Pradesh are physiologically very old. Plant growth regulators are used in improving relative growth rate (Guttridge and Thompson, 1964; Dijkstra *et al.*, 1990). Vegetative propagation of bamboo is performed by seedling division, seedling multiplication, culm cuttings and nodal segment cuttings (Bennett and Gaur, 1990; Singh 1999). Bamboo improvement has been initiated in Arunachal Pradesh (Banwal and Singh, 1988) but no work has been taken up on the selection and clonal multiplication of elite individual of seminal origin. Naturally growing seedlings never survive due to overcrowding with grassy weeds, post-monsoon dry spell and severe winter. However, such seedlings were transplanted in the nursery and for their healthy growth present investigation was designed to find out the efficacy of different treatments (plant growth regulators and fertilizers) to improve relative growth rate of seedling to attain normal culm size in short duration and to use appropriate plantation technology to multiply elite seedlings on a large scale.

Material and Methods

Viable seeds produced during April-June in sporadically flowering culms germinate during June–July were taken out from the field and were transplanted in nursery where they perform well under care. Different doses of growth regulators of GA_3 (10, 25, 100, 200, 300, 400 and 500 × 10^{-6} g/l) and single dose of kinetin (2 × 10^{-6}g/l) were sprayed on one year old seedlings growing in plots. Fifteen seedlings of uniform height and tillers were selected for each treatment and replicated thrice. For growth hormones were sprayed on seedlings in the month of March and observations on growth were taken in April and July.

To study the effect of different fertilizers on the growth, one year old culm cuttings were given following treatments.

T_1 (Control), T_2 (FYM alone), T_3 (FYM+N_1P_1), T_4 (FYM+N_1P_2), T_5 (FYM+N_1P_3), T_6 (FYM+N_2P_1), T_7 (FYM+N_2P_2), T_8 (FYM+N_2P_3), T_9 (FYM+N_3P_1), T_{10} (FYM+N_3P_2), T_{11} (FYM+N_3P_3), T_{12} (FYM+N_4P_1), T_{13} (FYM+N_4P_2) and T_{14} (FYM+N_4P_3). Levels of nitrogen (urea g/cutting) applied were N_1 =2.5, N_2=5, N_3=10, N_4 = 15 and P (SSP g/cutting), P_1 =5, P_2 = 10, P_3 = 15. The fertilizer trials were conducted in RBD with plot size 1 m × 1 m with 10 cuttings per treatment and three replications. The nitrogen (urea) and phosphorus (SSP) were applied in two equal split doses, first half at the time of rooting and another half at rhizogenesis stage. The cuttings were planted in March and observations were taken till July.

For the vegetative propagation, seedlings raised from seeds were bisected in such a way that each part comprises root, rhizomes and tillers. Seedlings were placed in poly bags and kept in shade/ polyhouse to avoid direct sunlight on young seedlings. The seedlings were kept hydrated till their division performed another seedlings division and so on and thus one seedling produced 3–4 plants in a year. Seedlings were transplanted after two years and allowed to proliferate.

For the mass multiplication and to obtain normal size of bamboo culm from seedling in short duration, single, bi and trinodal segment cuttings of juvenile elite seedlings were performed. The nodal segments were grown horizontally in FYM @ 1.5 kg alongwith different doses of nitrogen (3, 6, 9, 12g/cutting) and phosphorus (5 g, 10 g/cutting). Nitrogen was applied as half dose during planting and another half at rhizogenesis stage. Phosphorus was applied as single dose at the time of planting.

To evaluate the performance of bamboo culms from juvenile seedlings, various growth parameters like, number of tillers, internode length, tiller height, number of new culms, wall thickness, culm girth and number of nodes of new sprouts of bamboo raised from horizontally planted seedlings (15 cm depth) were recorded.

Results and Discussion

Response of Seedlings to Foliar Application of Growth Regulators

Number of Tillers

In April, one month after spray, non-significant changes in the number of tillers were observed (Table 50.1). However, in July a significant increase in tillers was found with all treatments as compared to control. Maximum tillers were produced by 50 × 10^{-6}g/l GA_3 that were significantly higher than control and other treatments. Lower concentration were more effective than higher treatments as compared to control. On the other hand, earlier workers (Nanda *et al.*, 1973) reported non-significant differences in the number of tillers after foliar applications of regulators. These differences in the results may be attributed to different timings of spraying.

Table 50.1: Response of Different Foliar Applications of GA_3 at Constant Concentration of Kinetin (2 × 10^{-6} g/l) on the Growth of Seedlings

GA_3 (10^{-6} g/l)	*Number of Tillers*			*Internode Length (cm)*			*Average Tiller Height (cm)*		
	March	*April*	*July*	*March*	*April*	*July*	*March*	*April*	*July*
10	3.2	3.2	4.9	3.6	4.1	8.7	21.5	22.6	41.7
25	3.2	3.3	5.1	3.6	4.8	8.5	20.2	23.0	47.5
50	3.5	3.5	6.0	3.5	4.8	10.6	20.1	28.4	71.9
100	3.5	3.5	6.1	3.7	4.2	7.2	21.2	27.5	46.4
200	3.5	3.7	6.0	3.8	4.7	7.7	21.4	22.8	46.3
300	3.3	3.3	6.0	3.7	4.8	6.4	20.8	21.6	37.2
400	3.4	3.4	5.8	3.6	4.2	6.2	21.6	24.3	37.2
500	3.2	3.2	5.7	3.5	4.6	6.2	20.4	24.2	38.1
Control	3.5	3.5	4.5	3.6	4.4	5.1	21.3	22.1	35.4
CD at 5%									
Months		0.29			0.32			2.16	
Concentrations		0.62			0.68			4.56	
Interactions		0.88			0.97			6.46	

Internode Length

Average length of internode increased significantly in April over March by the application of all concentrations of GA_3. However, there were non-significant differences among different doses of hormone. On the other hand, in July the internode length increased significantly over April. All the concentrations exhibited significant increase in the internode length over control, whereas GA_3 (50×10^{-6} g/l) showed the highest increase.

Average Height of Tillers

All the concentrations showed significant increase in the average tiller heights in the month of April over March and maximum height of tiller was observed with the application of GA_3 (50×10^{-6}g/l). In the month of July, tiller height increased significantly over April in all the treatments. Again, GA_3 (50×10^{-6}g/l) was most promotory as compared to other treatments and control. The higher concentration (more than 200×10^{-6}g/l) were not effective in stimulating the tiller height. Nandi *et al.* (1996) have also observed increase in average height with foliar application of GA_3.

Response of Culm Cuttings to Fertilizer Application

Days to Sprouting

While the untreated cutting took 16.2 days to sprout, a significant reduction in sprouting days was observed with all the treatments (Table 50.2). T_{10} caused maximum decrease over the control and other treatments. The effect of T_2 to T_8 was significantly lower than T_{10} to T_{14}.

Table 50.2: Effect of Fertilizer Application on the Growth of One-year Old Culm Cuttings

Treatment	*Days to Sprouting*	*Days to Rooting*	*Days to Rhizogenesis*	*Rooting (%)*	*Sprouting (%)*
T_1	16.2	47.4	74.8	51.2	82.6
T_2	13.4	44.2	75.1	54.2	84.4
T_3	13.6	45.2	74.1	53.1	85.1
T_4	12.9	45.3	68.3	60.2	84.9
T_5	12.2	43.1	68.1	60.6	85.7
T_6	12.8	38.7	65.8	63.4	85.1
T_7	11.3	38.4	65.9	68.6	88.2
T_8	10.2	35.1	63.2	68.6	87.6
T_9	10.5	35.4	60.2	72.1	85.4
T_{10}	7.2	30.1	58.2	86.4	90.2
T_{11}	7.8	30.8	58.6	85.2	91.2
T_{12}	9.4	32.2	61.2	85.8	86.2
T_{13}	7.6	34.2	60.2	82.1	90.6
T_{14}	7.8	33.5	60.4	86.6	88.4
CD at 5%	2.1	4.3	4.7	4.1	4.3

Days to Rooting

Treatments, T_{10} and T_{11} took significantly lesser days for rooting as compared to control and T_2 to T_9 treatments. T_{12}, T_{13} and T_{14} also caused a significant reduction over the control and their effect were at par with T_{10} and T_{11}. T_2 to T_5 treatments resulted in non-significant decrease in rooting days over the control.

Days to Rhizogenesis

A significant reduction in days to rhizome initiation was observed over control with T_4 to T_{14} treatments. T_{10} and T_{12} were most effective and had the same impact for inducing early rhizome formation. The effects of T_9 and T_{12} to T_{14} were at par with T_{10} and did not differ significantly among themselves.

Rooting

Higher doses of nitrogen and phosphorus (above T_3) significantly increased the rooting in the cutting. Maximum rooting (86.6 per cent) was observed with T_{14}, which was non-significantly different from T_{10} and T_{12} treatments but was significantly higher than control (51.2) and T_2 to T_9 treatments.

Mass Multiplication of Elite Seedlings

The elite seedlings selected over the period of last ten years has been multiplicated using vegetative propagation technology. Bisected division of the seedling were repeated to increase the number of elite seedlings for mass multiplication. Juvenile seedlings were also used for mass multiplication because they are thinner and easier to cut. Single, bi, and trinodal segmental cuttings were used to evaluate root proliferation by growing them horizontally. It has been observed that single node cutting gave best result as flow of nutrient was more concentrated on one potential node than the segment having many buds and node. Plants raised from such selection are genetically improved, live for many years and provide propagating material of young physiological age.

References

Banwal, B.S. and N.B. Singh (1998). Role of promoting substances in Bamboo cultivation. *Indian Forester*, 114(9): 549–559.

Bennett, B.S. and R.C. Gaur (1990). *Twenty Six Bamboo* spp. *Cultivated in India*. Forest Research Institute, Dehra Dun, India.

Dijkstra, P.H., Ter Reegen and P.J.C. Kuiper (1990). Relation between relative growth rate, endogenous gibberellins, and the response to applied giberellic acid for *Plantago major. Physiol. Plant*, 79: 629–634.

Guttridge, C.G. and P.A. Thompson (1964). The effect of gibberelins on growth and flowering of *Fragaria* and *Duchenea*. *J. Exp. Bot.*, 15: 631–46.

Nanda, K.K., P. Kumar and V. Kocchar (1973). Role of auxins, antioxins and phenols in the production and differentiation of callus on stem cuttings of *Populus robusta*. *NZ. J. For. Sci.*, 4: 338–346.

Nandi, S.K., L.M.S. Palani and H.C. Rikhari (1996). Chemical induction of adventitious root formation in *Taxus baccata*. *Plant Growth Regulation*, 19: 117–122.

Sharma, O.P. and S.K. Kaushal (1985). Exploratory propagation of *Dendrocalamus hamiltonii* ex Munro by one node culm cuttings. *Indian Forester*, 111(3): 135–139.

Singh, Balbir (1999). Studies on macro propagation of bamboo with relation to plant growth regulators. *M.Sc. Thesis*. H.P. Krishi Vishvavidyalaya, Palmpur, H.P., India.

Chapter 51

Combined Effect of Organic and Inorganic Fertilizer on Growth and Yield of Sugarcane

D.A. Sonawane, R.N. Sabale**, R.M. Gethe*** and S.B. Kharbade*****

**Assistant Professor (Agronomy), **Professor of Agronomy,
***Senior Research Assistant, ****Assistant Professor of Entomology,
Mahatma Phule Agricultural University, Rahuri – 413 722,
District Ahmednagar, M.S., India*

ABSTRACT

Among the treatment of application of nitrogen through different sources in combinations, the treatment of application of 250 kg N/ha, 50 kg through pressmud and 200 kg through urea showed a significant increase in growth characters and was significantly superior over all the treatments in increasing the yield contributing characters and yield of sugarcane. The maximum sugarcane yield of (101.54 t/ha) was obtained due to the same treatment, besides obtaining the maximum sugarcane trash yield of 7.89 t/ha. The yield obtained due to this treatment was 12.27 per cent more than the treatment of urea alone. The second best treatment in order of merit was the treatment of application of 250 kg N/ha @ 50 kg through vermin-compost and 200 kg through urea.

Keywords: Sugarcane, Nitrogen sources.

Introduction

Sugarcane is important cash crop of the world and Cuba is the largest producer of white sugar followed by India, Brazil and Australia. Island produces 60 per cent of the total cane sugar of the world.

Presently the nitrogen is mainly supplied through chemical fertilizers. The continuous use of chemical fertilizers have made hazards to soil health, mainly the physical, chemical, and biological properties of the soil and there by disturb the sustainability. The concept of organic farming is gaining importance. Some of the farmers from USA, Canada and Europe are changing their approach of supplying nutrients only through chemical fertilizers and now converting their farms based on only organic farming.

India is a tropical country, the total conversion from chemical to organic may not be possible because of fast degradation of organic matter due to oxidation process of organic matter in India. Therefore, partly supply of nutrients through organic and partly through chemical wild find way to achieve sustainability in the production of sugarcane (Suinde *et al.*, 1990). With this view the present experiment was planned on the supply of nutrients through different sources to test their efficacy on the production of sugarcane.

Materials and Methods

The experiment was laid out in Randomized Block design with three replications and ten different treatment, sources of nitrogen in Suru season. The treatments under study were supply of 250 kg nitrogen through different sources *viz.*

1. 250 kg Nitrogen through urea (NU)
2. 250 kg N/ha, 50 kg through FYM and 200 kg through Urea (NFU)
3. 250 kg N/ha, 50 kg through compost and 200 kg through urea (NCU)
4. 250 kg N/ha, 50 kg through vermicompost and 200 kg through urea (NVU)
5. 250 kg N/ha, 50 kg through Sugarcane trash and 200 kg through Urea (NSU)
6. 250 kg N/ha, 50 kg through poultry manure and 200 kg through urea (NPoU)
7. 250 kg N/ha, 50 kg through pressmud and 200 kg Urea (NPrU)
8. 250 kg N/ha, 50 kg through green leaves and 200 kg through Urea (NG_1U)
9. 250 kg N/ha, 100 kg through green leaves and 150 kg through Urea (NG_2U) and
10. 250 kg N/ha, 150 kg through green leaves and 100 kg through Urea (NG_3U).

Uniform basal dose of 115 kg P_2O_5 and K_2O/ha was applied through the source of single super phosphate and Murriate of Potash, respectively. While Nitrogen was applied in four installments.

Nitrogen content of different sources was determined before application and were applied on the basis of actual content of nitrogen. The nitrogen content of different sources was Urea (46.4), FYM (0.51), Compost (0.40), Vermicompost (0.62), Sugarcane trash (0.30), Poultry manure (1.4), Pressmud 1 and green leaves 0.53 per cent. The experimental soil was medium black with pH 7.7. The gross and net plot size was 9×6 m^2 and 7×4 m^2, respectively. Normal planting was done in ridges and furrows. Total 18 irrigations were given as and when required considering the water received through rainfall. Planting was done on 10th January, 1997 and harvesting on 24 February, 1998. Observations during different growth stages were recorded on four randomly selected plants. The cane yield data were recorded on net plot basis.

Results and Discussion

Plant Height

The data presented in Table 51.1 indicates that the plant height was significantly influenced by different treatments. All the treatments with the application of nitrogen through different sources in combination were produced significantly more total and milliable plant height throughout the growth of crop over supply of nitrogen through urea alone. The maximum total plant height of (311.82 cm) and milliable plant height of (269.98 cm) was recorded (Table 51.1) with the treatment of application of 250 kg N/ha, 50 kg through pressmud and 200 kg through urea at harvest. It was significantly more than other treatments and was 9.16 and 5.81 per cent, more respectively than application of nitrogen through urea alone. These results are in conformity with the results of Ingole *et al.* (1989) and Tanilselven and Jaybal (1993). The second and third best treatment observed was the application of 250 kg N/ha, 50 kg through vermicompost or sugarcane trash and 200 kg through urea.

Table 51.1: Combined Effect of Different Nitrogen Sources on Growth and Yield of Sugarcane

Treatment	Total Plant Height (cm) at Harvest	Milliable Plant Height (cm) at Harvest	No. of Leaves/ Plant at 270 days	Leaf Area/ Plant (dm^2) at 240 days	Dry Matter/ Plant at Harvest (g)	Milliable Cane Population/ ha at Harvest	Yield (t/ha)	
							Cane	Trash
NU	283.25	255.15	14.47	87.14	303.25	116349	90.44	6.72
NFU	297.90	265.81	14.95	94.70	320.08	122308	93.12	7.49
NCU	293.87	264.01	14.78	91.73	315.12	119744	92.48	7.21
NVU	305.91	268.44	15.86	100.48	325.41	126200	97.86	7.81
NSU	300.48	267.16	15.80	94.94	322.37	124895	96.34	7.63
NpoU	295.29	264.37	14.89	94.05	317.89	121095	92.89	7.34
NPrU	311.82	269.98	16.62	101.77	331.29	128897	101.54	7.89
NG_1U	291.69	263.11	14.69	90.46	312.23	118902	92.23	7.13
NG_2U	289.03	261.15	14.64	89.07	310.34	118154	92.04	7.04
NG_3U	287.86	259.21	14.51	88.17	307.10	117487	91.64	6.97
S.E.±	1.771	1.968	0.236	0.948	1.938	468567	1.064	0.342
C.D. at 5%	5.262	5.846	0.701	2.816	5.770	1392.02	3.160	N.S.

Number of Leaves and Leaf Area

The maximum number of leaves (16.62) and leaf area (101.77 dm^2 per plant) was recorded at crop growth stage of 270 and 240 days, respectively due to treatment of application of 250 kg N/ha, 50 kg through pressmud and 200 kg through urea. An increase in leaf number per plant is obtained from the increase in the plant height and increase of leaf area per plant was obtained from the increase in the plant height, number of leaves, leaf size and crop age upto 270 days. The application of 250 kg N/ha, through urea alone was recorded lowest number of leaves and leaf area than other. The results are confirmative with the results obtained by Patil *et al.* (1978).

Dry Matter/Plant at Harvest

The increase in plant vigour in terms of plant height, leaf number, leaf size and crop age were found useful in accumulation of more dry matter. These results are in agreement with the finding of Patil *et al.* (1978). The maximum dry matter accumulation per plant (331.29 g/plant) at harvest was highest by application of 250 kg N/ha, 50 kg through pressmud and 200 kg through urea which was significantly superior over other treatments followed by application of 250 kg N/ha, 50 kg through vermicompost or sugarcane trash and 200 kg through urea, while the application of nitrogen through urea alone was found to be lowest.

Milliable Canes/ha.

From table it is observed that the maximum number of milliable canes (1,28,897) per hectare were recorded due to application of 250 kg N/ha, 50 kg through pressmud and 200 kg through urea than other treatments. The application of nitrogen 50 kg through vermicompost or sugarcane trash with 200 kg urea was found second and third best treatments, respectively.

Cane and Trash Yield

The partly application of nitrogen through organic and inorganic fertilizers significantly influenced the growth attributes which subsequently increase the yield contributing characters and yield of sugarcane. The highest cane (101.54 t/ha) and trash (7.89 t/ha) yield was obtained due to application of 250 kg N/ha, 50 kg through pressmud and 200 kg through urea than other treatments. The alternate application of pressmud and urea significantly influenced the growth and yield contributing characters and helped in increasing physical, chemical and biological properties of soil in addition to the increment in yield. These results are confirmative with results obtained by Duraiswany *et al.* (1987) and Hapse (1993). The second and third best treatment was obtained with application of 250 kg N/ha, 50 kg through vermicompost or sugarcane trash and 200 kg through urea and lowest cane and trash yield was recorded by application of nitrogen through urea alone.

References

Duraiswamy, P., Perumal Rani, Baskaran, S., Mani, S. and S. Chellanuthu (1987). Effect of NPK and compost on yield and quality of sugarcane. *Madras Agric. J.*, 74(3): 522–527.

Hapse, D.G. (1993). Organic farming in the light reduction in use of chemical fertilizers. *DSTA 43rd Annu. Conf.*, Part I–8881: 37–51.

Ingole, P.G., Lakhdive, B.A. and D.M. Mahulikar (1989). Effect of application of P.M.C. on phosphorus uptake by sugarcane and soil properties. *Anonymous*, pp. 33.

Patil, S.P., Bawaskar, V.S., Randive, S.J. and G.K. Zende (1978). Response of sugarcane pressmud cake. *India Sugarcane*, 27: 711–714.

Shinde, D.B., Nayale, A.M., Vaidya, B.R. and S.B. Jadhav (1990). Effect of incorporation of sugarcane trash into field soil on cane yield and soil fertility. *India Sugar*, 12(3): 905–911.

Tamilselvan, N. and V. Jaybal (1993). Maximization of cane yield of sugarcane (*Saccharum officinarum*) under the planted condition. *Indian J. Agron.*, 38(1): 78–81.

Chapter 52

Use of Biopesticides and Biocontrol Agents for the Management of Collar Rot Pathogen *Phytophthora cactorum*

Bhupesh Gupta, L.N. Bhardwaj, Anil Handa and Usha Sharma

Department of Mycology and Plant Pathology,
Dr Y.S. Parmar University of Horticulture and Forestry, Nauni, Solan – 173 230, H.P.

ABSTRACT

Collar rot of apple caused by *Phytophthora cactorum*, a major soil borne disease in Himachal Pradesh is presently managed by application of copper oxychloride or metalaxyl or mancozeb. Non-judicious use of fungicides has many negative effects. Therefore, biopesticides and biocontrol agents were evaluated *in vitro* for their effectiveness against the pathogen. Among different biopesticides clove oil resulted in maximum inhibition (97.59 per cent) followed by Ovis (93.93 per cent) though statistically at par with each other. Among biocontrol agents tested *Trichoderma longibrachiatum* resulted in maximum inhibition (70.00 per cent) of mycelial growth followed by *T. viride* (68.33 per cent) and *T. harzianum* (68.17 per cent). All these treatments were statistically at par with each other.

Keywords: *Apple, Collar rot, Biopesticides.*

Introduction

Apple (*Malus domestica* Borkh.) is one of the important temperate fruit crops grown throughout the world wherever agro-climatic conditions are met with for its cultivation. In India, it is commercially grown in the state of Himachal Pradesh, Jammu and Kashmir, Kumaon hills of Uttranchal and parts of Arunachal Pradesh and Sikkim on an area of 231000 ha with an annual production of 1380000 MT (FAO, 2001), however, area under apple cultivation in Himachal Pradesh is about 88673 ha with an annual production of 49129 MT (FAO, 2001). Apple suffers heavily due to variety of diseases which cause serious damage to crop under favourable conditions, thus affecting fruit quality. Of various diseases attacking apples, collar rot is the most important soil-borne diseases following white root rot, prevalent throughout the apple growing areas. Presently, collar rot caused by *Phytophthora cactoum* is managed by application of copper oxychloride or metalaxyl or mancozeb and use of resistant rootstocks like M2, M4, M9 etc., however, non-judicious use of chemical fungicides has many negative effects. Therefore, search for an alternative means of disease control is of utmost importance. Biopesticides and biocontrol agents offer best environment friendly answer to this problem.

Materials and Methods

Evaluation of some biopesticides *viz.* Ovis, Nimbicidine, clove oil, ginger oil, Neem Jeevan Spray oil, mentha oil and eucalyptus oil were evaluated for their inhibitory effect on mycelial growth of *P. cactorum* by poisoned food technique (Falck, 1907) under *in vitro* conditions. Double strength oat meal agar medium prepared in distilled water was sterilized at 15 p.s.i. for 20 minutes. Simultaneously, double concentrations of biopesticides were also prepared in sterilized water. The suspensions were added separately into equal quantities of double strength agar medium aseptically before pouring into petriplates. These were then inoculated with mycelial discs of 7 mm diameter of vigorously growing fungal culture. Each treatment was replicated thrice. Data on diametric growth of fungus was recorded ten days after incubation at 25±1°C when control plates were fully covered with fungal mycelium. Per cent inhibition is calculated as Vincent (1947)

$$I = \frac{C - T}{C} \times 100$$

where,

I = Per cent growth inhibition
C = Growth of fungus in control
T = Growth of fungus in treatment.

Five fungal antagonists namely, *Trichoderma viride, T. harzianum, T. hamatum, T. longibrachiatum* and *T. virens* were evaluated for their inhibitory effect against *P. cactorum* under *in vitro* conditions by dual culture method. Mycelial discs (7 mm diameter) of all the antagonists and test pathogen were taken from the margins of actively growing cultures (8 days old) and transferred to petriplates containing oat meal agar. A check having only mycelial discs of test pathogen was also kept for comparison. All the treatments were replicated thrice. Observations on diametric growth were taken after ten days of incubation at 25 ± 1°C. Per cent inhibition was also calculated as per Vincent (1947). Four bacterial antagonists namely, *Pseudomonas fluorescens* IISR–8, *P. fluorescens* $T_1R_2K_4$, *P. fluorescens*–81 and *Bacillus subtilis* were screened for their inhibitory effect against *P. cactorum* causing collar rot of apple. A loopful of each bacterial culture (48 hours old) was streaked in the centre of the petriplates containing oat meal agar medium. µcelial discs (7mm) of test fungus cut from 8 days old culture was placed simultaneously on either sides of the petriplates where streaking was done with antagonists at the centre. Each treatment was replicated thrice and kept at 25 ± 1°C for ten days. Per cent inhibition was calculated as described earlier in the text.

Result and Discussion

Biopesticides are the safest means of plant disease control and are supposed to have inhibitory effect on vegetative growth of plant pathogens. Thus the inhibitory effect of some biopesticides and biocontrol agents were studied separately under present investigations. Data presented in Table 52.1 revealed that all the biopesticides tested at different concentrations inhibited the mycelial growth of *P. cactorum*, however, clove oil resulted in maximum inhibition of the test fungus. This was followed by Ovis (a formulated product from *Lantana camera*) which gave 93.93 per cent inhibition, though statistically at par with clove oil. Mentha oil followed by Neem Jeevan Spray oil were the next best giving 58.10 and 52.59 per cent inhibition, respectively. However, minimum inhibition (7.08 per cent) was recorded in eucalyptus oil. The concentrations of different fungicides tested show that higher concentrations of fungicides were more effective than lower concentrations, however, maximum mycelial inhibition was achieved with 1000 ppm (63.49 per cent) followed by 750 ppm (55.08 per cent) and 500 ppm (50.08 per cent), while the least inhibition was recorded with 250 ppm (37.94 per cent) concentration. The interaction studies showed that the increase in concentration of fungicides from 250 to 1000 ppm, the percentage of mycelial inhibition also increased. Ovis and clove oil gave cent per cent mycelial inhibition at 500 ppm, however, the same amount of inhibition was achieved with mentha oil at 1000 ppm. Clove oil at 250 ppm was found significantly superior to all other biopesticides tested at this concentration, this was followed by Ovis which gave 75.74 per cent inhibition, however, least inhibition was recorded with eucalyptus oil. These findings corroborate the observations of Sharma (1998) and Gupta (2001) who also reported Ovis, clove oil and mentha oil against *P. cactorum* causing leather rot of strawberry.

Table 52.1: *In vitro* Evaluation of Biopesticides and Plant Oils Against *Phytophthora cactorum*

Treatment	Mycelial Inhibition (%)				Overall Mean
	250 ppm	500 ppm	750 ppm	1000 ppm	
Clove Oil	90.37	100.00	100.00	100.00	97.59
	(72.00)	(90.00)	(90.00)	(90.00)	(85.50)
Neem Jeevan Spray Oil	18.52	20.93	21.30	31.67	23.10
	(25.39)	(27.05)	(27.26)	(34.22)	(28.48)
Ginger Oil	44.63	52.96	53.33	59.44	52.59
	(41.92)	(46.71)	(46.93)	(50.46)	(46.50)
Nimbicidine	19.07	25.37	34.45	43.70	30.65
	(25.74)	(30.08)	(35.93)	(41.38)	(33.28)
Ovis	75.74	100.00	100.00	100.00	93.93
	(60.67)	(90.00)	(90.00)	(90.00)	(82.67)
Mentha Oil	13.33	44.26	74.81	100.00	58.10
	(19.22)	(40.14)	(60.08)	(90.00)	(52.53)
Eucalyptus Oil	3.89	7.04	7.78	9.66	7.08
	(9.25)	(15.38)	(16.11)	(18.06)	(14.70)
Overall Mean	37.94	50.08	55.96	63.49	
	(36.41)	(49.47)	(52.33)	(59.15)	

Figures in parentheses are arcsin transformed values $CD_{0.05}$

Treatment = 4.50; Concentration = 3.40; Treatment × concentration = 6.37.

Five fungal and four bacterial antagonists were evaluated against collar rot pathogen *Phytophthora cactorum* under *in vitro* conditions by dual culture method. The per cent mycelial inhibition is presented in Table 52.2. A perusal of the data in Table 52.2 depicts that all the fungal as well as bacterial antagonists, in general, inhibited the vegetative growth of the test fungus, *P. cactorum*, however, the percentage of inhibition varied with the antagonists. *Trichoderma longibrachiatum* resulted in maximum growth inhibition (70.00 per cent) followed by *T. viride* (68.33 per cent) and *T. harzianum* (68.17 per cent), though satistically at par with each other. Amongst the bacterial antagonists *Pseudomonas fluorescens* IISR–8 was significantly superior in inhibiting the vegetative growth of *P. cactorum* followed by *Bacillus subtils* and *P. fluorescnes* $T_1R_2K_4$, though statistically at par with each other *P. fluorescens* 81 was the least effective antagonist in inhibiting the mycelial growth of the test fungus. Inhibitory effect of *Trichoderma* sp. was also reported by Chambers and Scott (1995) against *Phytophthora cinnamomi*. Recently, Rajan *et al.* (2002) reported the management of foot rot disease (*P. cacpcisi*) of black pepper with *Trichoderma* spp. Several workers have also reported *Psuedomonas* species to be inhibitory to vegetative growth of different species of *Phytophthora* including *P. cactorum* (Utkhede and Smith, 1991; Sauer and Zeller, 1992; Grosch and Grote, 1998; Gulati *et al.*, 1999). *Bacillus subtilis* and other species of *Bacillus* have been found effective biocontrol agents against *P. cactorum*, collar rot pathogen of apple (Gupta and Utkhede, 1986; Utkhede and Smith, 2000).

Table 52.2: *In vitro* Evaluation of Antagonists Against *Phytophthora cactorum*

Antagonist	*Growth Inhibition (%)*
Trichoderma virens	63.33
Trichoderma viride	68.33
Trichoderma harzianum	68.17
Trichoderma hamatum	65.83
Trichoderma longibrachiatum	70.00
Bacillus subtilis	59.83
Pseudomonas fluorescens 81	55.00
Pseudomonas fluorescens IISR–8	64.67
Pseudomonas fluorescens $T_1R_2T_4$	58.50
$CD_{0.05}$	3.05

References

Chambers, S.M. and E.S. Scott (1995). *In vitro* antagonism of *Phytophthora cinnamomi* and *P. citricola* by isolates of *Trichoderma* spp. and *Gliocladium virens*. *J. Phytopath.*, 143: 471–477.

Falck, R. (1907). Wachstomgesetze, Wachstm-taktoren und temperature Wertder holzesterenden Mycelien, 1: 43–154.

FAO (2001). FAO Website, www.fao.org.

Grosch, R. and D. Grote (1998). Suppression of *Phytophthora nicotianae* by application of *Bacillus subtilis* in closed soil less culture of tomato plants. *Gartenbauwissenschaft*, 63: 103–109.

Gulati, M.K., Koch, E., Zeller, W. and H.D. Sisler (1999). Isolation and identification of antifungal metabolites produced by fluorescent Pseudomonas, antagonist of red core disease of strawberry. In: 12th *International Reinhardsbrunn Symposium*, Germany. 24th–29th May, 1998, pp. 437–444.

Gupta, M. (2001). Studies on *Phytophthora* diseases of strawberry. *Ph.D. Thesis*, Dr. Y.S. Parmar UHF, Nauni, Solan, H.P., pp. 83.

Gupta, V.K. and R.S. Utkhede (1986). Factors affecting the production of antifungal compounds by *Enterobacter aerogenes* and *Bacillus subtilis*, antagonists of *Phytophthora cactorum*. *J. Phytopathol.*, 117: 9–16.

Rajan, P.P. and Y.R. Sarma (1997). Compatibility of potassium phosphonate (Akomin–40) with different species of *Trichoderma* and *Gliocladium virens*. In: *Proceedings of the National Seminar on Biotechnology of Spices and Aromatic Plants*, Calicut, India, 24th–25th April, 1996, p. 150–155.

Sauer, H. and W. Zeller (1992). Biological control of collar rot on apple (*Phytophthora cactorum* (Leb. et Cohn) Schroet.) by bacteria of soil and bark. *Acta Phytopathologica er-Entomologica Hungarica*, 27: 1–14.

Sharma, A. (1998). Studies on *Phytophthora* rot of strawberry. *M.Sc. Thesis*. Dr Y.S. Parmar University of Horticulture and Forestry, Nauni, Solan, H.P., pp. 53.

Utkhede, R.S. and E.M. Smith (1991). Biological and chemical treatments for control of *Phytophthora* crown and root rot caused by *Phytophthora cactorum* in a high-density apple orchard. *Can. J. Plant Path.*, 13: 267–270.

Utkhede, R.S and Smith, E.M. (2000). Impact of chemical, biological and cultural treatments on growth and yield of apple in replant-disease soil. *Aus. Plant Path.*, 29: 129–136.

Vincent, J.M. (1947). Distortion of fungal hyphae in the presence of certain inhibitors. *Nature*, 150: 850.

Chapter 53

Efficacy of Cashewnut Shell Liquid as Seed Protectant of Cowpea, *Vigna unguiculata* (Lin.) Against its Pest *Callosobrunchus maculatus* (Fab.)

Binu N. Nair and V.R. Prakasam

Department of Environmental Sciences, University of Kerala, Kariavattom, Thiruvananthapuram – 695 581

ABSTRACT

In this investigation on the effect of Cashew Nut Shell Liquid (CNSL) on oviposition of the pest insect *Callosobrunchus maculatus* on cowpea seeds, it was found that there was reduction in egg laying by 66.02 per cent at a dose rate of 1.25 ml CNSL/kg; by 93.10 per cent at a dose rate of 2.5 ml CNSL/kg; and 100 per cent at a dose rate of 5 ml CNSL/kg. Similarly adult emergence from the eggs deposited on cowpea was reduced by 82.07 per cent at the lowest experimented dose of 1.25 ml CNSL/kg while it was nil at higher doses.

The percentage germination of CNSL treated cowpea seeds was reduced by 4.4 per cent at a dose rate of 1.25 ml CNSL/kg; 5.2 per cent at dose rate of 2.5 ml/kg and 8.9 per cent at a dose rate of 5 ml/kg. The study revealed the possibility of using CNSL as an ecofriendly substitute for harmful chemical pesticides in seed protection.

***Keywords**: Callosobrunchus maculatus, Cowpea, Seed protectant, Germination.*

Introduction

Cowpea weevil, *Callosobrunchus maculatus* is a serious pest of stored pulses in tropical and sub-tropical regions. Both the larvae as well as the adults are infective (Raja and Ignacimuthu, 2001). Various chemical pesticides like malathion have been used as seed protectant against pulse beetle. Uncontrolled use of these chemicals can be detrimental to man and non-target species. Moreover phytotoxic effects are also produced by these chemical pesticides (Casida and Lykken, 1969; Chakravarthi, 1986). Therefore application of natural pesticides for seed protection is becoming more and more popular. Seed protection using Neem oil, Castor oil, Mustard oil, Sesame oil have been studied (Raghavani and Kapadia, 2003).

Surface application of CNSL, obtained through perforation of fresh cashewnuts, was found superior to ginger and neem against infestation and damage by *C. maculatus* in cowpeas without mentioning exact doses used (Echendu, 1991). The present study is an attempt to find out the efficacy of Cashewnut Shell Liquid (CNSL) extracted by roasting of cashewnuts (heat extracted CNSL) as a seed protectant of cowpea *Vigna unguiculata* against a pest weevil, *Callosobrunchus maculates.* The chemical composition of heat extracted CNSL is different from natural CNSL. During heat extraction the anacardic acid component of CNSL is decarboxylated to form cardanol and, therefore, the effect of such CNSL is likely to be different from natural CNSL.

Materials and Methods

Cowpea *Vigna unguiculata,* seeds of variety Sharika, purchased from a local market in Thiruvananthapuram, were used from the experiment. Initial stocking of *Callosobrunchus maculatus* was made from infested cowpea seeds obtained from the same market. Subsequent stocks were raised in the laboratory. Technical grade CNSL (heat extracted) obtained from Vijayalakshmi cashew industries, Kollam was used for the study.

Effect of CNSL on Oviposition and Adult Emergence

The cowpea seeds were surface sterilized with 0.02 per cent $HgCl_2$ for 5 minutes and washed thoroughly with distilled water. One kg of seeds was put in a glass jar of 1 litre capacity. Required amounts of CNSL (1.25, 2.5 and 5 ml) were added separately and shaken manually until a uniform smear of oil was produced on seed surfaces. These treated seeds were stored in the laboratory. A control consisting of untreated seeds was also maintained.

Three samples of 25 g seeds were then drawn from each treated stock into suitable plastic jars (6.00 × 4.00 cm) with perforated lid. Into them two pairs of adults *C. maculatus* were released. The number of eggs laid on treated and control seeds were recorded after 15th day. The experimental setup was kept undisturbed for 15 days in the laboratory allowing the development of the grub into the adults (Raja and Ignacimuthu, 2001). The number of F_1 adults emerged were then recorded on 30th day and till the day of last emergence (Approximately 40 days).

The data obtained were analysed using 2 way ANOVA and significant differences were noted using least significance difference test (LSD).

Germination of CNSL Treated Cowpea Seeds

A sample of 70 seeds was randomly drawn from each set of treated and control seeds six months after storage. Ten seeds each were sown in a petridish containing two layered filter paper and kept in the laboratory at 30 ± 1°C. It was moistured daily. Percentage germination of seeds was calculated on 4th day.

Results and Discussion

The results of the effect of CNSL on oviposition and adult emergence of *C. maculatus* are given in Table 53.1. It showed that at doses of 1.25 and 2.5 ml CNSL/kg seeds, the mean number of eggs laid were 21.2 ± 1.94 and 4.3 ± 1.74 respectively, whereas the control value was 62.4 ± 3.73. It indicated that CNSL had affected the normal deposition of eggs. Further, at a higher dose (5.0 ml) the weevil failed to lay eggs on treated seeds.

Table 53.1: Showing Number of Eggs Laid by *Callosobrunchus maculatus* on Cowpea Seeds Treated with CNSL and Control

Dose of CNSL ml/kg	*Eggs Laid (Mean ± S.D on Seed Surfaces)*
1.25	21.2 ± 1.94
2.5	4.3 ± 1.74
5	Nil
Control	62.4 ± 3.73

($p < 0.05$) by LSD.

It can be noted from Table 53.2 that at a dose of 1.25 ml/kg, the number of adults emerged was 6.4 ± 2.37, whereas it was 37.6 ± 1.67 in control. But at higher doses of 2.5 and 5 ml/kg there was no emergence of adults. It is likely that the cardol fraction of CNSL might have produced toxicity leading to death of embryos. Cardol is a common constituent of both natural CNSL and heat extracted CNSL.

Table 53.2: Effect of CNSL on Adult Emergence of *Callosobrunchus maculatus*

Dose of CNSL ml/kg	*No. of Adults Emerged (Mean ± S.D)*
1.25	6.4 ± 2.37
2.5	nil
5	nil
Control	37.6 ± 1.67

($p < 0.05$) by LSD.

The result of germination test is given in Table 53.3. It is seen that at a dose of 1.25 ml/kg, the percentage germination was 85.3 ± 3.6; at a dose of 2.5 ml/kg percentage germination was 84.6 ± 41 and at the highest dose of 5 ml/kg the percentage germination was 81.2 ± 3.8. It is evident that CNSL has not affected the germination of treated seeds when compared to control. In this context it may be recollected that many of the synthetic seed protectants have interfered with germination of seeds through their phytotoxicity (Casida and Lykken, 1969; Patil and Shirashyad, 1989). From the reduced deposition of eggs, poor emergence of young ones from eggs and normal germination of CNSL treated seeds, it may be derived that technical CNSL could be a safe, economical and ecofriendly substitute for chemical pesticides to be used for seed protection from insect pests.

Table 53.3: Mean Percentage Germination of Seeds (CNSL Treated and Control)

Dose of CNSL ml/kg	*Mean ± S.D % Germination*
1.25	85.3 ± 3.6
2.5	84.6 ± 4.1
5	81.2 ± 3.8
Control	89.3 ± 4.6

($p < 0.05$) by LSD.

Acknowledgement

Sincere thanks are due to Mr. Thomas George and Mr. Abilash R., Research Scholars, Department of Environmental Sciences for their co-operation and help in carrying out this work.

References

Casida, J.E. and L. Lykken (1969). Mechanism of organic pesticide chemicals in higher plants. *Ann. Rev. Pl. Physiol.*, 20: 607–636.

Chakravarthi, S.K. (1986). Effect of Bavistin 25 DS on germination and seedling vigour of wheat. *Pesticides*, 3: 23–26.

Echendu, T.N.C. (1991). Ginger, cashew and neem as surface protestants of cowpea against infestation and damage by *C. maculatus. Trop. Sci.*, 31: 209–211.

Patil, T.M. and V.S. Shirashyad (1989). Effect of Methyl parathion and phosphamidon on seed germination and on the activity of peroxidase and α-amylase in some vegetable seeds. *Geobios*, 16(2–3): 57–60.

Raghavani, B.R. and M.N. Kapadia (2003). Efficacy of different vegetable oils as seed protectants of pigeon pea against *C. maculatus. Ind. J. Plant Prot.*, 31(1): 115–118.

Raja, N. and S. Ignacimuthu (2001). Use of madhuca longifolia seed oil in controlling pulse beetle *Callosobrunchus maculatus*, L. *Entomon.*, 26(384): 279–283.

Chapter 54
Effects of Brewery Effluent on Photosynthetic Pigments, Starch, Nitrate Reductase Activity and Protein Content of *Vigna mungo*

A. Pragasam, R. Praveena and J. Prasena

Department of Botany, Kanchi Mamunivar Centre for Post Graduate Studies, Pondicherry – 605 008, India

ABSTRACT

The photosynthetic pigments, starch, nitrate reductase activity and protein content of leaves of *Vigna mungo,* treated with different, concentrations (10, 25, 50, 75 and 100 per cent) of brewery effluent and mere water, were estimated. All the parameters studied were found to be more than control in the plants treated with 10, 25, 50 and 75 per cent concentrations of the effluent and less than control in those treated with 100 per cent concentration. Maximum response was recorded in the plants treated with 50 per cent concentration. The present work suggests that the diluted effluent can be used for better growth of the test crop.

Keywords: Brewery effluent, Vigna mungo, Photosynthetic pigments, Starch, NRA, Protein.

Introduction

A number of industries such as dairies, distilleries, breweries, tanneries, refineries, pharmaceutical factories, textile mills, sugar mills, paper mills, card board industries, automobile industries and sewage release an enormous quantity of effluents into the ecosystem. Though the effluents of many industries are known for the presence of toxic substances those of some industries contain the essential elements required for better growth of plants. Because of the presence of such elements, the effluents offer a possibility to make use of them for irrigation after proper treatment and dilution. A number of studies revealed the growth promoting effects of the effluents (Ajmal and Khan, 1984: brewery effluent; Gautam and Bishnoi, 1990: dairy effluent; Misra and Behera, 1991: paper factory effluent; Medhappan, 1993: tannery effluent; Vijayarengan and Lakshmanachary, 1993: textile mill effluent; Kumar and Bhargava, 1998: sugar mill effluent; Pragasam and Kannabiran, 2001: distillery effluent).

Perusal of literature revealed that there are several works with reference to the effluents of various industries on the growth of plants, but work on the effects of brewery effluent on plants is meagre. So the present work has been taken up to analyse the effects of brewery effluent on pigments, starch, nitrate reductase activity and protein contents of *Vigna mungo* (L.) Hepper cv. Vamban.

Materials and Methods

The South India Corporation Agencies (SICA) breweries limited, Pondicherry is the first in India to be awarded the ISO 9002 and 14001 certification. It has the production capacity of 2.5 million cases of beer per annum. For every litre of beer produced about 10 litres of effluent is released. An average of 650 kilolitres of effluent is released on every working day.

The effluent was collected from the main outlet of the brewery and analysed for its physico-chemical properties using standard methods as recommended by American Public, Health Association (APHA, 1975). Different concentrations of the effluent (10, 25, 50, 75 and 100 per cent) were prepared by mixing the raw effluent with water (V/V). Certified seeds of *Vigna mungo* were obtained from Perunthalaivar Kamaraj Krishi Vigyan Kendra, Pondicherry. Twenty seeds in each pot were sown and treated with 500 ml of each concentration of the effluent daily. A control set treated with water was also maintained. Three sets were maintained for each concentration. The leaves of 15 days old seedlings were used for analysis. The parameters such as photosynthetic pigments, starch, nitrate reductase activity and protein content were estimated. They were extracted and estimated following the standard procedures proposed by Hiscox and Israeltam, 1979 (pigment extraction); Harborne, 1984 (Pigments estimation); Mc Cready *et al.*, 1950 (Starch); Jaworski, 1971 (Nitrate Reductase Activity) and Lowry *et al.*, 1951 (Protein).

Results and Discussion

The effluent was found to be very light brown in colour and had no objectionable smell. The temperature of the effluent was 30°C, pH 8.77, total suspended solids 24 mg/l, biochemical oxygen demand 20 mg/l, chemical oxygen demand 44 mg/l, oil and grease 3 mg/l, nitrogen 12 mg/l, phosphorus 1.8 mg/l, Chloride 12 mg/l and sulphate 20 mg/l. All the parameters observed were found to be within tolerance limits. The pigment contents were more in the plants treated with 10, 25, 50 and 75 per cent concentrations of the effluent. The contents were less than control in the plants treated with 100 per cent concentration and maximum in those treated with 50 per cent concentration of the effluent (Figure 54.1). Starch, nitrate reductase activity (NRA) and protein contents also showed similar results (Figures 54.2 and 54.3).

An increase of 27.273 per cent total chlorophyll, 36.165 per cent carotenoid, 36.446 per cent starch, 56.976 per cent NRA and 46.391 per cent protein and a decrease of 17.531 per cent total chlorophyll, 4.576 per cent carotenoid, 13.323 per cent starch, 9.431 per cent NRA and 7.293 per cent protein were observed in the plants treated with 50 and 100 per cent concentrations of the effluent respectively over control.

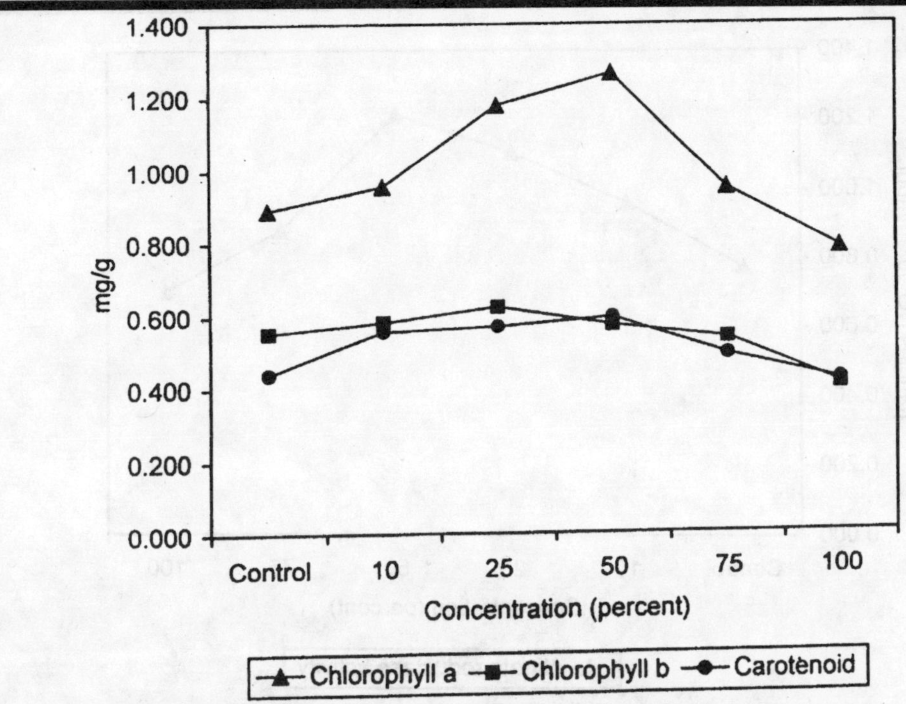

Figure 54.1: Effects of SICA Brewery Effluent on the Photosynthetic Pigments of Leaves of *Vigna mungo* (mg/g fresh weight)

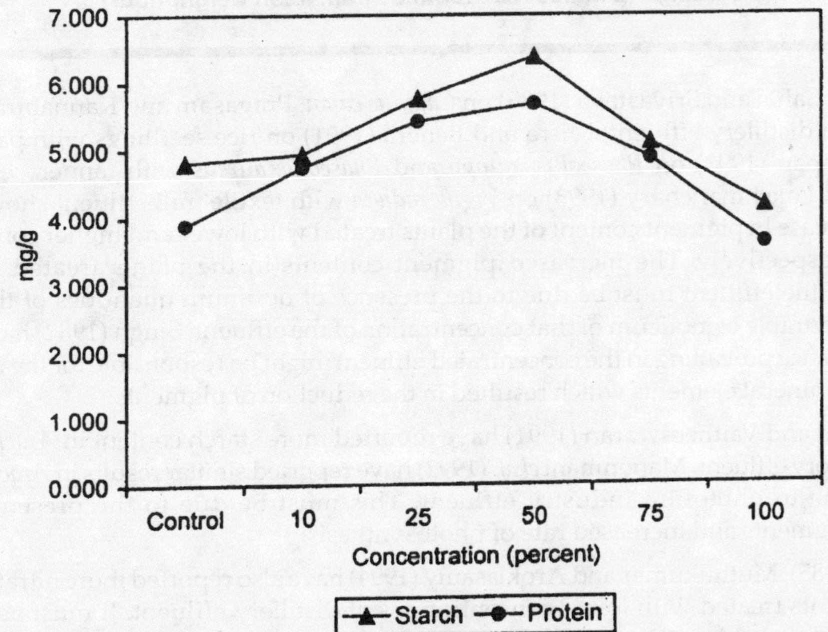

Figure 54.2: Effects of SICA Brewery Effluent on the Starch and Protein Contents of Leaves of *Vigna mungo* (mg/g fresh weight)

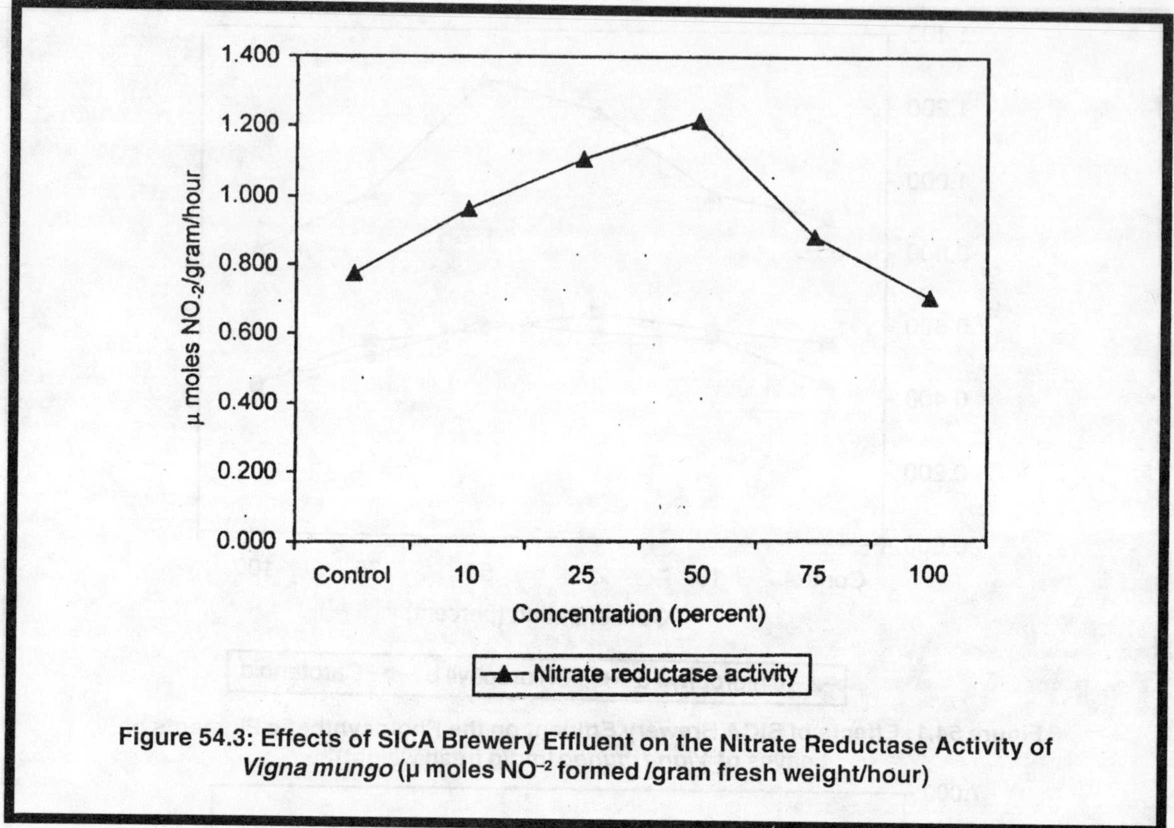

Figure 54.3: Effects of SICA Brewery Effluent on the Nitrate Reductase Activity of *Vigna mungo* (μ moles NO^{-2} formed /gram fresh weight/hour)

The works of Sahai and Srivastava (1986) on *Cajanus cajan,* Pragasam and Kannabiran (2001) on *Vigna mungo* with distillery effluent; Misra and Behera (1991) on rice seedlings with paper factory effluent; Madhappan (1993) on *Phaseolus mungo* and *Phaseolus aureus* with tannery effluent and Vijayarengan and Lakshmanchary (1993) on *Vigna radiata* with textile mill effluent showed similar increase and decrease in pigment content of the plants treated with lower and higher concentrations of the effluent respectively. The increased pigment contents in the plants treated with lower concentrations of the effluent must be due to the presence of optimum quantities of the nutritive elements and favourable osmoticum of that concentration of the effluent. Singh (1982) has suggested the osmotic imbalance prevailing in the concentrated effluent might be responsible for the reduction in the uptake of the mineral elements which resulted in the reduction of pigments.

Swaminathan and Vaitheeswaran (1991) have reported more starch content in *Arachis hypogaea* due to dyeing factory effluent. Manonmani *el al.* (1992) have reported similar results in *Arachis hypogaea* and *Zea mays* due to photofilm industry effluent. This must be due to the presence of more photosynthetic pigments and increased rate of photosynthesis.

Sahai *et al.* (1985), Muthukumar and Arokiasamy (1994) have also reported more nitrate reductase activity in the plants treated with lower concentrations of distillery effluent. It must be due to the presence of more absorbable nitrogen. Sahulka (1978) has reported reduced rate of NRA in the plants treated with higher concentrations of chloride. According to Pandey and Srivastava (1993) it was due to the restricted transport of nitrate to the metabolic pool.

Experiments of Kadiaglu and Algur (1990) with distillery effluent on *Helianthus annuus* and *Pisum sativum* have revealed that nitrate nitrogen present in optimum quantities in the diluted effluent was responsible for increased protein content in the plants. In the plants treated with cent per cent concentration of the effluent, the absorption and metabolism of nitrogen might have been reduced leading to the reduction in total physiological activities and resulted in the reduction of protein content.

Thus the present work suggests the use of diluted brewery effluent up to 75 per cent, especially 50 per cent, concentration for better growth of the test crop.

References

Ajmal, M. and A.U. Khan (1984). Effect of brewery effluent on soil and crop plants. *Environ. Pollut.*, 33: 341–352.

APHA, AWWA and WPCF (1975). *Standard Methods for the Examination of Water and Wastewater*, 14th ed. American Public Health Association, Inc., New York.

Gautam, D.D. and S. Bishnoi (1990). Studies on the effects of Urmul dairy effluent on soil characteristics and growth of wheat plant. *Ad. Plant Sci.*, 3: 236–244.

Harborne, J.B. (1984). *Phytochemical Methods: A Guide to Modern Techniques of Plant Analysis*, 2nd ed. Chapman and Hole, London, p. 215–217.

Hiscox, J.B. and G.M. Israelstam (1979). A method for the extraction of chlorophyll from leaf tissue without maceration. *Can. J. Bot.*, 57: 1332–1334.

Jaworski, E.G. (1971). Nitrate reductase in intact plant tissue. *Biochem. Biophys. Res. Commun.*, 43: 1274–1279.

Kadioglu, A. and O.F. Algur (1990). The effect of vinasse on the growth of *Helianthus annuus* and *Pisum sativum*. Part I: The effects on some enzymes and chlorophyll and protein contents. *Environ. Pollut.* 67: 223–232.

Kumar, Rajesh and A.K. Bhargava (1998). Effects of Sugar mill effluent on the vegetative growth and yield of *Triticum aestivum* cv. U.P. 2003. *Ad. Plant Sci.*, 11(2): 221–227.

Lowry, O.H., Rosenbrough, N.J., Farr, A.L. and R.J. Randall (1951). Protein measurement with the folin-phenol reagent. *J. Biol. Chem.*, 193: 265–275.

Madhappan, K. (1993). Impact of tannery effluent on seed germination, morphological characters and pigment concentration of *Phaseolus mungo* L. and *Phaseolus aureus* L. *Poll. Res.*, 12: 159–163.

Manonmani, K., Murugeswaran, R. and K. Swaminathan (1992). Effect of photofilm factory effluent on seed germination and seedling development of some crop plants. *J. Ecobiol.*, 4: 99–105.

Mc Cready, R.M., Guggole, J., Silviera, V. and H.S. Owners (1950). Determination of starch and amylase in vegetables: Application to peas. *Anal. Chem.*, 29: 1156–1158.

Misra, R.N. and B.K. Behera (1991). The effect of paper factory effluent on growth, pigments, carbohydrates and protein of rice seedlings. *Environ. Pollut.*, 72: 159–167.

Muthukumar, B. and D.I. Arokiasamy (1994). Effect of distillery waste on the nitrate reduction system in *Vigna mungo* (Black gram). *J. Ecotoxicol. Environ. Moni.*, 4: 27–33.

Pandey, M. and H.S. Srivastava (1993). Inhibition of nitrate reductase activity and nitrate accumulation by mercury in maize leaf segments. *Indian J. Environ. Hlth.*, 35: 110–114.

Pragasam, A. and B. Kannabiran (2001). Effect of distillery effluent on the morphology, biochemistry and yield of *Vigna mungo. Ad. Plant Sci.*, 14(11): 547–552.

Sahai, R., Jabeen, S. and P.K. Saxena (1985). Nitrate reductase activity of *Vigna mungo* (L.) Hepper as affected by distillery waste. *J. Environ. Biol.*, 6: 205–209.

Sahai, R. and N. Srivastava (1986). Effect of distillery waste on the seed germination, seedling growth and pigment content of *Cajanus cajan. J. Indian Bot. Soc.*, 65: 208–211.

Sahulka, J. (1978). The effect of chloride on nitrate reductase level as anaerobic nitrate reduction and on nitrate content in excised *Pisum sativum* L. roots. *Biol. Plant* (PRAHA), 20: 201–209.

Singh, D.K. (1982). Effect of distillery effluent on some germination parameters of certain legumes. *Sci. and Environ.*, 4: 61–67.

Swaminathan, K. and P. Vaidheeswaran (1991). Effect of dyeing factory effluent on seed germination and seedling development of groundnut (*Arachis hypogaea*). *J. Environ. Biol.*, 12: 353–358.

Vijayarengan, P. and A.S. Lakshmanachary (1993). Effect of textile mill effluent on growth and development of green gram seedlings. *Ad. Plant Sci.*, 6: 359–365.

Chapter 55

Bioconversion of *Parthenium hysterophorus* as an Organic Manure for Chilli (*Capsicum annuum* L.)

B. Vijayakumari and R. Hiranmai Yadav***

**Reader in Botany, ** Ph.D. Scholar*

Department of Botany, Avinashilingam Deemed University, Coimbatore – 641 043

ABSTRACT

A pot culture experiment was conducted at Avinashilingam Deemed University, Coimbatore to assess the impact of fresh, composted and vermicomposted parthenium, on chilli. The germination per cent of chilli was observed on 14, 21 and 28 DAS (Days After Sowing). The biometric attributes like root length, shoot length, fresh weight, dry weight and vigour index were observed on 30, 60 and 90 DAS. The germination per cent was more in vermicomposted parthenium treatment on 21st and 28th DAS compared to control. The increased root length of chilli on 30, 60, 90 DAS and shoot length on 30 and 60 DAS was due to composted parthenium application and on 90 DAS, fresh parthenium application registered longer shoot. The fresh and dry weight on 30 and 90 DAS were increased in composted parthenium and on 60 DAS in fresh parthenium. The vigour index was more in composted parthenium applied pots on 30 and 60 DAS and in fresh parthenium on 90 DAS.

Keywords: Compost, Vermicompost, Parthenium, Chilli.

Introduction

Chilli is an indispensable condiment of every home. Besides, imparting pungency and red colour to the dishes, it is a rich source of vitamin–A, C, E and P and has certain medicinal properties. Chilli stimulates saliva and gastric juices and helps digestion. It is used as a counter irritant in prickly heat powders, cosmetics, skin ointments and pain balms (Anand Singh, 2001).

Carrot grass more popularly known as 'Parthenium' (*Parthenium hysterophorus*) is one of the most noxious weed found increasingly in agricultural and plantation fields. Eco-friendly use of parthenium should be considered to reduce this menace. Considering the positive effect and applying profitable alternate use can solve the problem of spread (Sukanya, 2002).

In the present study an attempt was made to compost and vermicompost the parthenium plants and applying it to chilli as an organic manure to find its impact on the crop.

Materials and Methods

An experiment was conducted at Avinashilingam Deemed University, Coimbatore to study the effect of fresh, composted and vermicomposted parthenium on biometric attributes of chilli.

The treatment details:

T_0: Control–Red soil
T_1: Fresh parthenium @ 35.0 g/pot
T_2: Composted parthenium @ 26.25 g/pot
T_3: Composted parthenium @ 35.0 g/pot
T_4: Composted parthenium @ 43.75 g/pot
T_5: Vermicomposted parthenium @ 26.25 g/pot
T_6: Vermicomposted parthenium @ 35.0 g/pot
T_7: Vermicomposted parthenium @ 43.75 g/pot
T_8: NPK @ 60 : 30 : 30.

The pots of 7 kg capacity were used and the design adopted for the study was completely randomised design. The pots were filled with red soil and sand in 2 : 1 ratio. The manures were mixed into the soil and chilli seeds were sown after moisturing the soil.

The germination per cent was calculated 14, 21 and 28 DAS using the formula of Abdul Baki and Anderson (1973).

Biometric characters like root length, shoot length, fresh weight, dry weight and vigour index were observed 30, 60 and 90 DAS.

Statistical Analysis

The data obtained from the various biometrical observations were subjected to the statistical analysis (Panse and Sukhatme, 1978).

Results and Discussion

Germination Per cent

The germination per cent was observed on 14, 21 and 28 DAS (Table 55.1).

The highest germination per cent of 78.33 per cent was in T_5 compared to lower germination per cent of 40.00 per cent (21 DAS) and 50.00 per cent (28 DAS) in control.

Table 55.1: Impact of Fresh, Composted and Vermicomposted Parthenium on the Biometric Characters of Chilli (*Capsicum annuum* L.) on 30, 60 and 90 DAS

	Germination (%)			Root Length (cm)			Shoot Length (cm)			Fresh Weight (g)			Dry Weight (g)			Number of Leaves			Vigour Index		
	14 DAS	21 DAS	28 DAs	30 DAS	60 DAS	90 DAS	30 DAS	60 DAS	90 DAS	30 DAS	60 DAS	90 DAS	30 DAS	60 DAS	90 DAS	30 DAS	60 DAS	90 DAS	30 DAS	60 DAS	90 DAS
T_0	31.67	40.00	50.00	5.90	9.97	24.20	5.17	20.40	40.20	0.26	1.93	12.33	0.03	0.39	2.47	5.67	19.67	82.67	545.50	1495.60	3185.50
T_1	35.00	73.33	73.33	6.93	10.60	26.80	6.57	24.80	63.53	0.47	4.19	24.67	0.04	1.15	4.93	7.67	27.00	108.00	1085.00	2640.83	6664.83
T_2	46.67	56.67	63.33	7.10	6.97	29.67	6.87	25.53	57.00	0.38	2.26	22.33	0.04	0.45	4.47	7.00	21.00	87.00	886.33	2054.00	5480.00
T_3	53.33	58.33	63.33	5.77	14.30	26.07	6.03	26.33	63.00	0.28	3.24	29.33	0.04	0.65	5.80	5.67	21.33	126.66	754.67	2589.50	5647.67
T_4	63.33	75.00	75.00	7.23	12.50	27.93	7.70	27.77	56.33	0.58	3.66	23.67	0.05	0.73	4.75	7.33	22.67	79.67	1118.50	3006.50	6336.67
T_5	55.00	78.33	78.33	6.40	9.03	25.90	5.70	20.00	50.83	0.25	2.40	16.00	0.03	0.48	3.88	7.00	18.66	72.00	891.67	2131.33	5619.00
T_6	61.67	70.00	76.67	5.17	11.37	21.30	5.87	22.33	38.73	0.28	2.13	14.33	0.04	0.43	4.08	6.33	20.67	80.00	845.17	2586.17	4606.10
T_7	31.67	51.67	51.67	5.83	11.00	18.43	6.23	17.80	42.17	0.25	2.38	13.33	0.04	0.48	2.95	6.66	17.00	58.00	502.50	1400.50	3033.17
T_8	41.67	55.00	63.33	4.37	7.02	20.17	7.73	18.20	48.20	0.40	2.85	16.33	0.04	0.19	5.24	9.00	13.00	50.67	766.83	1625.00	4314.83
S.E.d	NS	9.40	8.39	0.40	2.11	3.07	0.49	3.05	3.52	0.07	0.37	2.88	0.01	0.08	0.80	0.67	3.41	10.41	121.64	445.01	738.27
C.D. (0.05)	NS	19.74	17.62	0.83	4.44	6.46	1.03	6.40	7.39	0.14	0.78	6.04	0.03	0.16	1.68	1.40	7.16	21.87	255.56	934.95	1551.07

DAS: Days After Sowing.

Biometric Parameters

The biometric parameters observed were significant (Table 55.1).

Root Length

The root length was more in T_4 (7.23 cm) on 30 DAS, T_3 (14.30 cm) on 60 DAS and T_2 (29.67 cm) on 90 DAS. The least values were observed in T_8 (4.37 cm) on 30 DAS, T_2 (6.97 cm) on 60 DAS and T_7 (18.43 cm) on 90 DAS.

Shoot Length

The shoot length was more in T_4 (7.70 and 27.77 cm) on 30 and 60 DAS. On 90 DAS, it was in T_1 (63.53 cm) immediately followed by T_3 (63.00 cm). The shoot length was minimum in control on 30 DAS (5.17 cm) in T_7 on 60 DAS (17.80 cm) and T_6 on 90 DAS (38.73 cm).

Fresh and Dry Weight

The fresh and dry weights were more in T_4 (0.58 and 0.05 g) on 30 DAS, T_1 (4.19 and 1.15 g) on 60 DAS and in T_3 (29.33 and 5.80 g) on 90 DAS. On 30 DAS, the lesser fresh weight of 0.25 g was in T_7 and dry weight of 0.03 g in T_0 and T_5. On 60 and 90 DAS, least values for fresh weight (1.93 and 12.33 g) was in control. The minimum dry weight was in T_0 (0.19 g) on 60 DAS and in control (2.47 g) on 90 DAS.

Vigour Index

The vigour index was more in T_4 on 30 (1118.50) and 60 (3006.50) DAS. On 90 DAS, T_1 (6664.83) registered higher vigour index. The vigour index was least in control on 30 DAS (545.50). On 60 (1400.50) and 90 (3033.17) DAS it was least in T_7.

Govindan *et al.* (1995) has also reported that application of 100 per cent vermicompost resulted in maximum vegetative growth of bhindi.

Mayalagu (1983) observed increased growth of groundnut due to the composted coir pith application.

Any organic waste application aids in moisture conservation which is utilized for better root penetration and crop growth.

References

Abdul Baki, A.A. and J.D. Anderson (1973). Vigour determination in soybean seed by multiple criteria. *Crop Sci.*, 13: 360–363.

Govindan, M., Muralidharan, P. and S. Sasikumaran (1995). Influence of vermicompost in the field performance of bhindi (*Abelmoschus esculentus* L.) in a laterite soil. *J. Tropic. Agric.*, 33: 173–174.

Mayalagu, K. (1983). Influence of different soil amendments on the physical properties of a heavy black soil and yield of groundnut, TMV7 in the Periyar-Vaigai command area. *Madras Agric. J.* 70: 304–308.

Panse, V.G. and P.V. Sukhatme (1978). *Statistical Methods for Agricultural Workers*. ICAR, New Delhi, pp. 327.

Singh, Anand (2001). Chillies: An important cash crop. *Kisan World*, 28: 47–48.

Sukanya, T.S. (2002). Parthenium: The most noxious weed. *Kisan World*, 29: 47–49.

Chapter 56

Correlation and Path Analysis for Yield and Other Economic Traits in White X Colour Linted Crosses of American Cotton (*G. hirsutum* L.)

B. Subba Reddy and N. Nadarajan***

**Research Associate, NATP RCPS, 8 Regional Agricultural Research Station, Lam, Guntur – 522 0034*

***Professor, Department of Genetics and Plant Breeding, AC & RI, Madurai*

ABSTRACT

For character association and path analysis for yield and economic traits, in 8 × 8 diallel mating design involving colour linted and high seed oil content genotypes of cotton were undertaken. Seed cotton yield per plant exhibited significant positive association with number of bolls per plant, plant height, number of sympodia per plant, boll weight and lint index. Number of bolls per plant, plant height and boll weight exhibited positive high direct effects on seed cotton yield whereas, lint index had moderate positive direct effect on seed cotton yield. Hence number of bolls per plant is the major determinant for improving seed cotton yield.

Keywords: *Gossypium hirsutum, Correlation, Path coefficient analysis, Yield attributes.*

Introduction

Cotton, a crop of prosperity having a profound influence on men and matter, is an industrial commodity of worldwide importance. Cotton cultivation in India encompasses economic, political and social affairs of the world. Besides being a money spinner, it is also an employment generator.

Higher yield is the main breeding objective in all the crops and yield is regarded as a complex character or super character which is influenced by many component traits.

Selection for seed cotton yield will be effective only when it lies along with the yield components rather than relying on yield alone (Grafius, 1960). Hence, the knowledge about relationship between yield and its contributing characters is needed for an efficient breeding programme. Correlation provides information on the nature and extent of relationship among characters. Therefore, this technique aids in the isolation of superior yielding genotypes from genetically variable population. The direction and degree of association of each component trait with seed cotton yield is ascertained by correlation coefficients between these traits. However, path coefficient analysis (Wright, 1921) provides an effective means of finding direct and indirect causes of association.

Materials and Methods

The experimental material consisted of eight genotypes of which included two colour linted genotypes *viz.*, Arkansas green, Algerian brown, and six white linted genotypes *viz.*, SVPR 2, MCU 11, MCU 5, MCU 12, MCU 7, Sahana. Among these MCU 7, Sahana had high seed oil content. These eight parents were crossed in half diallel fashion (excluding reciprocals) during summer 2001–02 at AC and RI Madurai. Twenty eight resultant crosses (F_1) were evaluated along with their parents in randomized block design with two replications: Recommended package of practices and plant protection measures were followed to raise good crop. The biometrical observations were recorded on five randomly selected competitive plants for the characters *viz.*, days to 50 per cent of flowering, plant height (cm), number of sympodia per plant, number of bolls per plant, boll weight (g), seed cotton yield, ginning outturn, lint index, seed index and oil content. Genetic correlation coefficients were estimated as per the procedures of Johnson and Path coefficient analysis was made following the procedure suggested by Dewey and Lu (1959).

Results and Discussion

The analysis of variance exhibited that, high significant differences was observed among the genotypes for all the ten traits studied. Genotypic correlation coefficients between yield and nine other component traits and also inter-correlation among the component traits are furnished in Table 56.1. In the present investigation, the association between seed cotton yield and its component traits *viz.*, plant height, number of sympodia per plant, number of bolls per plant, boll weight and lint index were positive and significant. The positive association between seed cotton yield and plant height was already reported by Sumathi and Nadarajan (1995) and Sambamurthy (1999). Positive association of number of sympodia per plant, number of bolls per plant and boll weight with seed cotton yield was earlier confirmed by many researchers (Shanti and Selvaraj, 1993; Basha, 1997 and Mandoli *et al.*, 1998). Positive contribution of lint index to seed cotton yield was in accordance with the results obtained by Shanti and Selvaraj (1993) and Sambamurthy (1999). This indicated that high yielding genotypes possessed more boll number per plant, number of sympodia per plant and plant height were rather tall in habit.

Knowledge on inter-correlation between yield attributes revealed the strength and direction of association with each other. This could facilitate effective selection for simultaneous improvement of one or more yield contributing characters. The inter-correlation among the yield contributing traits *viz.*, plant height, number of sympodia per plant and number of bolls per plant were positive and significant. This was in accordance with the results obtained by Shanti and Selvaraj (1993) and Gnana Arul Samuel Rajan (1997). Thus, selection of anyone of these traits might be useful to improve the seed cotton yield substantially. Besides this, the inter-relationship of lint index with seed index was significantly positive. Hence simultaneous selection for lint index or seed index will improve the seed cotton yield substantially. These results are in conformity with the results obtained by Tyagi (1994), Sumathi and Nadarajan (1995) and Basha (1997).

Table 56.1: Genotypic Correlation Coefficients for Yield and Yield Components

Character	*Plant Height*	*Number of Sympodia per Plant*	*Number of Bolls per Plant*	*Boll Weight*	*Ginning Outturn*	*Lint Index*	*Seed Index*	*Oil Content*	*Seed Cotton Yield per Plant*
Days to 50% flowering	– 0.37*	– 0.017	– 0.045	– 0.057	– 0.396*	– 0.452*	0.017	0.081	0.022
Plant height		0.508*	0.448*	0.090	– 0.091	0.326*	0.074	– 0.090	0.460*
Number of sympodia per plant			0.369*	0.085	– 0.061	0.262	0.026	– 0.074	0.433*
Number of bolls per plant				– 0.362*	0.148	0.229	0.024	0.790	0.670*
Boll weight					0.063	0.134	0.140	– 0.196	0.409*
Ginning outturn						0.223	– 0.081	0.330*	– 0.185
Lint index							0.721*	0.246	0.314*
Seed index								0.074	0.160
Oil content									0.042

* Significant at 5 per cent level.

The genotypic path coefficient analysis of different yield contributing characters traits are furnished in Table 56.2. In the present investigation number of bolls per plant had high direct effect on seed cotton yield. Similar results were reported by Sumathi and Nadarajan (1995) and Rao *et al.* (2001). The direct effect of boll weight on seed cotton yield was high which was in accordance with the results obtained by Basha (1997) and Sangeetha (1998). Lint index showed positively moderate direct effect on seed cotton yield, which was in accordance with the results obtained by Sumathi and Nadarajan (1995). None of the quality traits including lint colour showed direct effect on seed cotton yield.

The indirect effect of plant height through number of bolls per plant was positive and high. Similar findings were quoted by Sumathi and Nadarajan (1995). Number of sympodia per plant showed positive indirect effect through number of bolls per plant. Similar findings were also reported by Sangeetha (1998). Indirect effect of lint index through number of bolls per plant was positive and moderate. Similar findings were quoted by Rao and Mary (1996). Hence, number of bolls per plant is the important character on which selection pressure is to be applied for improving the seed cotton yield.

Correlation studies inferred that intensive selection for plant height, number of sympodia per plant and number of bolls per plant will enhance the seed cotton yield substantially. Path analysis indicated that number of bolls per plant has a prime role in determining seed cotton yield per plant involving colur linted genotypes of cotton which might be considered as a selection criteria including *per se* in a selection of a selection of a breeding programme. Some of the parents identified having superior performance for the important component characters for increasing yield were Algerian brown, SVPR 2, MCU 7 and Sahana.

Table 56.2: Genotypic Path Coefficients for Yield and Yield Components

Character	Days to 50% Flowering	Plant Height	Number of Sympodis per Plant	Number of Bolls per Plant	Boll Weight	Ginning Outturn	Lint Index	Seed Index	Oil Content	Correlation with Seed Cotton Yield per Plant
Days to 50% flowering	0.216	0.003	0.000	– 0.041	– 0.040	– 0.030	– 0.034	– 0.002	– 0.050	0.022
Plant height	– 0.008	– 0.086	– 0.004	0.435	0.062	– 0.007	0.083	– 0.010	– 0.005	0.460*
Number of sympodia per plant	– 0.004	– 0.044	– 0.007	0.388	– 0.059	– 0.005	0.097	– 0.003	0.070	0.433*
Number of bolls per plant	– 0.010	– 0.039	– 0.003	0.917	– 0.250	0.051	0.018	– 0.003	– 0.011	0.670*
Boll weight	– 0.012	– 0.020	– 0.001	– 0.332	0.691	0.055	0.034	– 0.018	0.012	0.409*
Ginning outturn	– 0.095	0.051	– 0.006	0.136	– 0.043	– 0.077	0.052	– 0.011	– 0.090	– 0.185
Lint colour	– 0.098	– 0.028	– 0.002	0.210	0.093	0.017	0.232	– 0.095	– 0.015	0.314*
Seed index	0.004	– 0.006	0.000	0.074	0.047	– 0.006	0.183	– 0.132	– 0.004	0.160
Oil content	0.018	0.008	0.001	0.132	– 0.135	0.025	0.062	– 0.010	– 0.059	0.042

Residual effect: 0.04

Diagonal values are direct effects.

References

Dewey, D.R. and H.K. Lu (1959). A correlation and path coefficient analysis of components of crested wheat grass seed production. *Agron. J.*, 51: 515–516.

Basha, M. (1997). Genetic analysis of yield and yield components of intraspecific hybrids of cotton (*Gossypium hirsutum* L.). *Ph.D. (Ag.) Thesis.* ANGR Agric. Univ., Rajendranagar, Hyderabad.

Grafius, J.E. (1960). Does over dominance exist for yield in corn. *Agron. J.*, 52: 361.

Gnana Arul Samuel Rajan (1997). Studies on heterosis and combining ability in cotton (*Gossypium hirsutum* L.). *M.Sc. (Ag.) Thesis*, Tamil Nadu Agric. Univ., Coimbatore.

Mandoli, K.C., G.K. Koutu, U.S. Mishra, R. Julka, S. Holkar and S.K. Pandey (1998). Character association in coloured cotton. *J. Indian Soc. Cotton Improv.*, 23(2): 172– 175.

Rao, K.V.K. and T.N. Mary (1996). Variability, correlation and path analysis of yield and fibre traits in upland cotton. *J. Res. Andhra Pradesh Agric. Univ.*, 24(3– 4): 66–70.

Sambamurthy, J.S.V. (1999). Character association and component analysis in upland cotton. *Madras Agric. J.*, 86(1–3): 39– 42.

Sangeetha, K. (1998). Character association of quality traits on yield and yield components in cotton (*Gossypium hirsutum* L.). *M.Sc. (Ag.) Thesis*, ANGR Agric. Univ., Rajendranagar, Hyderabad.

Shanti, N. and U. Selvaraj (1993). Association of yield related characters in cotton (*G. hirsutum* L.). *J. Indian Soc. Cotton Improv.*, 18(2): 152–154.

Sumathi, P. and N. Nadarajan. (1995). Character association and component analysis in upland cotton. *Madras Agric. J.*, 82(4): 225–228.

Tyagi, A.P. (1994). Correlation coefficients and selection indices in upland cotton (*Gossypium hirsutum* L.). *Indian J. Agric. Res.*, 28(3): 189–196.

Wright (1921). Correlation and causation. *J. Agric. Res.*, 20: 557–587.

Chapter 57
Environmental Pollution and Sustainable Development: An Economic Analysis

Kamal Ray

Reader in Economics, Katwa College, Katwa, Burdwan – 713 130, West Bengal

ABSTRACT

Economists till 1950 did not take into account environmental factors to formulate economic models, and environment, as it is variable, was not introduced in the classical, neo-classical and even Keynesion model (1932). But the incidents of London smog in 1952, deadsea in Europe, nuclear explosion in Japan during the second world war, minamata disease in Japan, Creronobyl disaster in the USSR in 1986, rapid industrialisation and population growth in general all were causing severe damage on environment, which affected the ecosystem badly, and hence climate change, loss of agricultural productivity, water and air pollution, deforestation, desertification etc all are aftermath. The ecosystem may allow industrialisation and urbanisation up to threshold extent, beyond that a conflict between economists and ecologists may arise. Economists expect profit maximising output in capitalism but ecologists expect socially optimum output to maintain inter-generational equity or sustainable development. To get the benefit of reduced pollution firms should produce less than the profit maximising output which signifies the reduction of supply of goods leading to a market price rise and resultant loss in producers surplus. The protection of environment at the cost of loss of producers surplus may be the solution to the economist ecologist collusion. To keep the economy pollution free and ecosystem intact the conflict between economist and ecologist should be reduced to sustainability of human interest.

Environmental pollution arises when there is a divergence between private cost and social cost. Private cost is the cost of production/consumption borne by the producer/consumer privately and the social cost may be defined as the sum of private cost and environmental cost [damage/benefit], social cost is larger than the private cost as damage/environmental cost is positive, and private cost is larger than the social cost as damage is negative signifying benefit, for example, good looking building on roadside, site selection of flower garden, planting of trees in the common place with the benevolence of private a man. Thus pollution arises when there is a stress on environment due to pressure of economic activities such as production, consumption etc. A resource crisis arises if the rate of regeneration of resources falls short of the required flow of resources from the ecosystem signifying unsustainablity. If the rate of regeneration of pollutants/wastes exceeds the absorption rate of wastes by nature or nature's assimilative capacity, the remaining waste would be deposited in the ecosystem as pollutants. The stock of pollutants would accumulate in the ecosystem and it would adversely affect the productivity of natural system, human health etc. If the waste products are not biodegradable, they will disintegrate very slowly, the man-made polymer, hazardous chemicals, radioactive wastes may require thousands of years for degradation. Thus we are utilising the resources of nature and throwing back wastes to nature that are not biodegradable and hinder sustainable growth. The resource crisis and environmental crisis become a severe threat to sustainability of development. There is an inverse relationship between economic development and quality of environment up to a certain point of per capita GDP as referred to Kuznets, and after that threshold value, both move in the same direction, a part of additional per capita GDP is used up for the abatement of pollution/emission. The biodiversity is important in the context of measuring pollution of economy. The biodiversity index, one of its them called Simpson index, may be taken as a measure of the level of pollution due to the impact of economic activities on the ecosystem. Sewage and industrial waste and overdose of insecticides in agriculture and forest affect species composition of the ecosystem. Many species in the biotic components (living components) in the ecosystem are destructive and causes biodiversity loss and in turn aggravate pollution. In the industrial sector, every firm wants to produce profit maximising output but that profit maximising output may not be socially optimum as far as pollution is concerned. So to get the benefit of reduced pollution in terms of reduction of firm level output it is desirable for a firm to reduce the volume of output which is not market determined rather derivative of resilience of ecosystem. The Organisation of Economic Cooperation and Development has adopted polluter pay principle–those who use society's scarce environmental resources should compensate the public for their use. Economists always see the market solution but ecosystem may not allow market solution from the standpoint of sustainability. Agricultural farmer often use pesticides and high doses of chemical fertilisers in order to enhance their production, but sustainability does not allow it as it pollutes both biotic and abiotic (non-living component) materials of ecosystem. The cost of reduced output is to be borne by the whole society to keep the natural environment intact. Human beings like all other biotic species of ecosystem depend on nature for life support, that is oxygen, water and some other natural resources. Unlike other species, homosapiens transform materials and energy through production process into consumable goods and services. The size and growth rate of economy would cause strain and pressure of nature. The poverty affects the economy in many ways, forest burning, forcible occupation of open access property and its conversion into cropland, lead to ecological imbalance in land use. The deforestation and farming in hill areas causes soil erosion and flood. The over use of land for cultivation is one of the indicators of unsustainable agricultural practice. It is true that environmental quality is determined by the concentration of pollutants, and the demand for environmental quality is income elastic. The rich have a strong preference for cleaner environment. The rich society has access to the latest developed technology and better infrastructural arrangement for waste disposal and treatment. The poor are not able, and no question of willing, to spare any amount of income to improve the environmental quality, and hence they have no preference for quality of environment. But everybody irrespective of their income level, has to face almost the same natural environment, breathe the same air, use the same road, the same drinking water etc. We cannot make distinction between the environment of the poor and the environment of the rich. In global context, pollutant generated in the thermal power station in the USA flows to the air of Canadian territories. The damage to the natural air caused by the emission of CO_2 from the power plants is the example of international negative externalities, negative externalities even affect the people across the national borders. The developed nation induces developing countries to produce pollution intensive or dirty goods in their territories, because it has a higher social cost degrading natural environment of developing countries, and finally import that dirty goods at a price which are lower than its own cost, for the sake of benefit in terms of environment; the developing countries make benefit as they have comparative advantages, but this is only at the cost of environment. The pollution of dirty goods generated from production, concentrated in the developing region, but consumption, which has the smallest pollution, is in the developed region. Hence the developing countries are unable to sustain their existing level of production, export and income. In this situation strict regulations are called for in developing countries to ensure long term sustainability of its ecosystem. During the WTO meeting (World Trade Organisation) in December 1999, the non-government organisations from all over the world put up demonstration and fought with the police over the issue on the streets of Seattle, USA, where the meeting was held. But the stand taken by WTO on this issue was puzzling.

The different types pollution in Indian context are to be noted as follows: The emission of CO_2 from the developed countries is appreciably larger than that of developing countries. The emission of per capita CO_2 in India in 1995 is 0.81 tonnes as compared to 19.53 tonnes in the USA. In the UK, it is 10.00 tonnes, in Australia 15.10 tonnes, in Japan 8.79 tonnes. The value of estimated damage of environment in India was found to be 9.7 billion dollar in 1992, which is approximately 4.5 per cent of GDP in 1992 values. This is quite high as compared to other countries, 2.6 per cent in China, 3.3 per cent in Mexico. During 1989–90 the indicators of air pollution in India are to be examined, total physical emissions amounted to 6012 thousand tones of SO_2, 2578 thousand tons of NO_x, 2186 thousand tons CO_2, 8517 thousand tons of SPM. Signs of appreciable control were absent till very recently. In case of water pollution, report suggests that the surface water and ground water pollution are significantly active in India. The sources are domestic waste water, industrial waste water and agricultural run-off. All these have adverse impact on human health. Biodiversity loss is an important issue, 21 plants and 3 vertebrates are already bloted out, and many other species are fast disappearing, which inevitably adversely affect our ecosystem. Deforestation, pollution, climate change, desertification, loss of wetlands, ocean warming have all been destroying biodiversity. Over population, over harvesting, over grazing have been a threat to the ecosystem. According to a recent estimate, two to eight per cent of earth's species will be extinct if the economic activities are going on in such a way.

Development is concerned with the enhancement of human welfare. Sustainable development is one of the processes of development which generally looks after the future generation, the development not at the cost of trouble of future generation. According to the Brandtland Committee Report, 1987 sustainable development responds to the present day needs without compromising the prospect of future generation meeting their own needs. It requires that the natural capital remain intact over generations. Actually ruin is the destination towards which all men rush (Hardin, 1968) as we are not thinking about sustainable development. Environmental goods like air, water are free but they are gradually becoming scarce as industries and automobiles pollute water, air, fisherman over harvests lake, deforestation, conversion of wetlands, global warming due to emission of green house gases etc. Freedom of commons brings ruin to all (Hardin, 1968). The development, as determined by the market, does not provide the best solution to mankind, rather present development must keep some space for future generation to accommodate comfortably. The sustainable development for human welfare required production and distribution of good and services. They in turn depend on the availability of four major factors: human capital, man-made capital, renewable resources and non-renewable resources. Sustainable development differs from traditional classical models of development, it includes both economic and ecological aspects of resources. The polluting of air and its provision for cleaning, creation of waste and the need of waste management, conservation of natural capital and biomass, extraction of exhaustable resources, welfare of future generation, all are important factors which make development eco-friendly. Proper valuation of resourses are required in order to understand the relative importance of consumption. Sustainable development is a process of economic activities which leaves the environmental quality level intact and maximisation of net benefit of economic development of present and future generation subject to the maintenance of natural environment. There must be some combination of economic efficiency and other attributes for sustainability which makes the definition comprehensive. The present generation does not have the right to exhaust existing resource base since it does not own it. It is very important that what you are leaving for the future generation by maintaining present generation's well being, the maximisation of conservation of natural environment corresponding to optimum present development may lead to sustainable development. The principal indicators can be considered so far as sustainability is concerned. Air pollution may be taken into account in terms of emission of CO_2, SO_2, NO_x, CFC, CO etc. The DO, COD, BOD in water, indicators of soil degradation, deforestation rates, mineral depletion, desertification are ecological attributes. The import or export of emissions may be explained by the direction of wind, we cannot clamp a curtain between the nations to stop transfer of clean/polluted air.

Sustainable development may be defined by consideration of inter-generational equity in resource use. Emphasis has been on the use of natural resources that would enable the future generations to enjoy at least the present generation's level of utility. The stock of natural resources at the terminal year would be determined by the relative strength of the rate of extraction and the rate of regeneration for renewable resources during the relevant time horizon. For example, the stock of groundwater in any region would decline if the rate of extraction of groundwater exceeds the recharging rate of groundwater in that region. For exhaustible resources like coal and oil stock, sustainability may be explained by the discovery of new sources of resource or that of alternative resource. To maintain the inter-generational equity the development process does not end up the with decline of human well being of the society after some period of growth. The need of research and new technology is very much essential in case of the exhaustible resources or when the alternative resources are used with the decline of reserve of exhaustible resources. There is a conflict between economist and ecologist in the context of sustainability, the economist estimates the sustainability in terms of well being but according to the ecologist the stocks of natural resources would be non-declining over time. Sustainable development must not endanger the natural system that supports life on earth. In the conference of Reo, 1992, sustainable development emerged as a common theme, linking convention to reduce green house gases and to preserve biodiversity, each generation to care about the welfare of next the generation.

References

Bhattacharya, R.N. (Edited) (2001). *Environmental Economics: An Indian Perspective.*

Bowers, J. (1997). *Sustainability and Environmental Economics: An Alternative Text.* Longmen.

Chopra, K. and G.K. Kadekodi (1999). *Operationalising Sustainable Development: Economic and Ecological Modelling for Developing Countries.* Sage Publication.

Conrad, J.M. (1999). *Resource Economics.* Cambridge University.

Kolstad, C.D. (2000). *Environmental Economics*. Oxford University Press.

Pearce, D.W. and R.K. Turner (1990). *Economics of Natural Resources and Environment*.

Pearce, D.W. and D. Moran (1994). *The Economic Value of Biodiversity*. Earthscan Publication.

Sankar, U. (Edited) (2001). *Environmental Economics*. Oxford University Press.

Sengupta, R. (2001). *Ecology and Economics*. Oxford University Press.

Stavins, R.N. (Edited) (2000). *Economics of Environment*. W.W. Norton and Company.

Index

C